"Earth's wetlands are among the most precious of ecosystems, for they are both beautiful and an essential part of the architecture that is Nature. Marothia's moving love letter to wetlands masquerades as a scientific document, but it is all the better for it, because it hides the painstaking empirical work the author has undertaken to show what the wetlands offer the human economy. The book is a landmark."

Sir Partha Dasgupta, *Frank Ramsey Professor Emeritus of Economics, Fellow of St John's College, University of Cambridge, Cambridge, UK*

"Professor Marothia has written an exhaustive and impressive account of the manifold threats to the coherent management of fragile wetland habitats. Here we find careful attention to the institutional arrangements suited to the protection of these essential natural resources. No longer are such wet areas regarded as 'swamps' or economically useless. Now, their role in integrated environmental management is well understood. Marothia's exceptional treatise promises to be of enduring value in rescuing these habitats from mismanagement or continued destruction."

Daniel W. Bromley, *Anderson-Bascom Professor of Applied Economics (Emeritus), University of Wisconsin-Madison, USA*

"Governing wetlands sustainably and well requires the appropriate institutions that recognize their characteristics as common assets and their huge value in producing ecosystem services. Dinesh K. Marothia's summary of these challenges and choices is required reading for anyone interested in governing our valuable wetlands."

Robert Costanza, *Professor, Institute for Global Prosperity, University College London, London, UK*

"For millennia, wetlands and wetland habitats were universally regarded as the foundation of human civilisation. Yet, throughout the imperial age and the industrial revolution, vast tracts of wetlands were ignored, devalued, drained, filled, and otherwise repurposed because of their perceived inability to generate short-term economic gain. Over the last five decades, the consequences of these oversights have grown increasingly apparent. Professor Marothia convincingly presents the fact that wetlands and wetland habitats are socially, ecologically, environmentally, and economically important and require thoughtful and informed management. Maratha's comprehensive and detailed treatment of these complex issues provides us with an indispensable roadmap for wetland researchers, managers, and policy makers."

William E. Phillips, *Professor (Emeritus), Department of Resource Economics and Environmental Sociology, University of Alberta, Edmonton, Canada*

"This is one of the most comprehensive, fascinating excavations of the complexities and interdependencies related to multitude wetlands and their users and non-users. Read this study. You will LOVE it."

Nighisty Ghezae, *Director*, *International Foundation for Science (IFS)*, *Stockholm*, *Sweden*

Multiuse Wetlands Governance

The Ramsar Convention was established in 1971 to ensure the conservation and wise use of wetlands across the world. India joined the Convention in 1982, however, in the past 50 years, despite Ramsar's incredible achievements, the threat to wetlands across the globe, including India, has not diminished. This book studies the governance of multiuse wetlands in India.

The volume provides an exhaustive analysis of rural, peri-urban, and urban human-made wetlands to establish the relevance of institutional design and the effective role of authority in governing multiuse wetland ecosystem services. The author argues that the most challenging task in governing wetlands is to frame institutional choices that users and non-users can comprehend and agree to pursue under alternative property rights regimes.

Drawing on extensive fieldwork, the book provides a broad look into the causes and consequences of wetland ecosystem degradation and offers insights into improved sustainable management systems for different types of multiuse wetlands. It will be indispensable for students and researchers of environmental studies, sustainable development, biodiversity, conservation, and agricultural, natural, and environmental resource economics.

Dinesh K. Marothia is President, Indian Society of Agricultural Economics (Mumbai, India), Member Technical Advisory Board of Global EverGreening Alliance-an international collaborative platform to Restore South Asia and Coordinator, Centre for Integrated Natural Resources Management, National Institute of Ecology, Raipur/Delhi/Jaipur.

Multiuse Wetlands Governance

Challenges and Institutional Choices

Dinesh K. Marothia

Routledge
Taylor & Francis Group
LONDON AND NEW YORK

First published 2024
by Routledge
4 Park Square, Milton Park, Abingdon, Oxon OX14 4RN

and by Routledge
605 Third Avenue, New York, NY 10158

Routledge is an imprint of the Taylor & Francis Group, an informa business

British Library Cataloguing-in-Publication Data
A catalogue record for this book is available from the British Library

ISBN: 978-1-032-59514-6 (hbk)
ISBN: 978-1-032-59896-3 (pbk)
ISBN: 978-1-003-45674-2 (ebk)

DOI: 10.4324/9781003456742

Typeset in Sabon
by Apex CoVantage, LLC

Dedicated to my
Parents
Dariyav Devi and Poonam Chand
&
Grandparents
Gayatri Devi and Baldev Chand

Contents

Tables

Illustrations

Maps

Figures

Photos

Preface

For nearly four decades now, I have been concerned with the role of institutions in the sustainable management of renewable common-pool resources (CPRs), in general, and multiple uses of wetlands, in particular. In this book, I have put together my thoughts and perspectives, which have evolved over years of research, and the lessons learned from recently conducted case studies. Multi-use wetlands (MWs) are undeniably important environmental resources with high preservation, conservation, and utilization values. However, the existing evidence suggests that wetlands in many countries, including India, are still not being managed sustainably. A few attempts have been made in India to understand the causes and consequences of wetland ecosystem degradation, but there is little explicit research available to resolve the issues of governance and for framing appropriate institutional choices. Such an understanding is extremely important as wetlands ecosystems in India are de-facto CPRs. Governing MWs is an intricate issue as it can be influenced by disproportionate spatial and temporal externalities because of the typical characteristics of technical, socio-economic-cultural-political, and environmental attributes in an interdependent situation. The most challenging task in wetlands governance is to frame institutional choices that users and non-users can comprehend and agree to pursue under alternative property rights regimes. In spite of Ramsar's International Convention on wetlands, international development agencies, research organizations, and the national and state governments of various countries are confronted with the following questions:

(i) what policy mechanisms are needed to reverse the trend of wetlands degradation or loss and sustain restoration efforts or gains?
(ii) what policy actions at different governance hierarchy levels are required to ensure the 'wise use' of wetlands without alteration in ecological characteristics?
(iii) what policy actions need to be initiated to identify wetlands for getting tag/s of nationally and internationally important wetlands?
(iv) what types of research input are needed to improve understanding of wetland researchers, policy makers, and wetland managers to conserve and sustain MWs in the changing social-ecological-political environment?

This book provides answers to most of these questions and seeks to establish the relevance of institutional design and the effective role of governance systems in managing MWs and ecosystem services through the exhaustive analysis of the rural, peri-urban, and urban, human-made wetlands of Chhattisgarh, a state in eastern India. Each MW case study reported in this book is spelt out, including the kind of governance and institutional design that worked or did not work. Comparative case studies of wetlands from other regions were also drawn upon to the maximum possible extent. The case studies provide research and policy inputs to address the challenges and firm-up institutional choices for sustaining wetlands for inter-generational use. A synthesis of institutional analysis framework (an interdisciplinary theoretical framework that accommodates plural methodology) is used to analyse multiuse and multifunctional wetland ecosystems and other categories of multiple-use CPRs. The institutional analysis framework, coupled with real-world insights, recent policy decisions of the Ramsar Convention and the Government of India, is expected to generate interest in wetland researchers, policy makers, and wetland managers, who are concerned with current affairs in the wise use of wetlands and sustaining ecosystem services.

I completed my PhD dissertation at the University of Alberta, Edmonton, Canada, which dealt with the institutional analysis of soil-erosion problems in the Peace River region of Alberta (Riverine ecosystem – an important category of Ramsar Convention classification). I wish to express my gratitude to William E. Phillips, who has been my mentor since the 1970s. I am grateful to Daniel W. Bromley, University of Wisconsin-Madison, USA, for shaping my thoughts about the institutional economics of natural and environmental ecosystems. I also had an opportunity to discuss various aspects of urban and peri-urban wetlands governance with Partha Dasgupta, University of Cambridge, and Karl Goran Mäler, Beijer Institute – The International Institute of Ecological Economics, Stockholm, way back in 1995. I am grateful to both of them. I have benefited greatly over the years with the joint work I did with a dear colleague and friend, Brij Gopal, an internationally known wetland ecologist. With his passing away I lost a personal friend, but his thoughts helped me in designing the structure of this book.

I am also extremely grateful to Sir Partha Dasgupta, Daniel W. Bromley, Robert Costanza, William E. Phillips, and Nighisty Ghezae for writing blurbs for this book.

During the writing phase of this book, I have greatly benefited from the series of interactions with my friends. Evelyn Phillips and David N. McElhaney have made critical suggestions on several portions of the manuscript. Ramakrishnaiah Duvvuri, Debashish Sircar, Malavika Chauhan, and Manoj Misra have taken time out to read all the chapters of the book and to discuss them with me. P.K. Katiha, P.C. Mishra, B.P. Nonhare, H.S. Shylendra, Manohar S. Rathore, and S.K. Tiwari read and commented on an earlier draft of one or two chapters. Ramakrishnaiah Duvvuri, Sukhpal Singh, and V.P. Singh provided unwavering support throughout the project.

I sincerely thank all of them. I am grateful to late Deodutta Shukla for his unconstrained support and appreciation for my efforts to save the historical wetlands through research. I would like to acknowledge the encouraging feedback from referees who reviewed the manuscript for Routledge.

V.K. Shukla, Former Director and N.S. Nag, present Director Fisheries, SDF, GoCG, provided remote sensing data, information related to fish and aquatic crops, and permitted the production of maps of the Ratanpur wetland complex. Without the generous support of Sukhdev Prasad Kashyap during field work, it would not have been possible to cover a large number of issues facing the Ratanpur wetlands. G.L. Sharma provided timely and unconstrained support during the preparation of this text. I am grateful to all of them.

Chapter 5 of the book is a synthesis, but is a revised version of earlier published papers – an article published in *The Challenge of the Balance: Environmental Economics in India (Proceedings of a National Environment and Economics Meeting 1997* – a publication of Centre for Science and Environment CSE), *Indian Journal of Agricultural Economics (2001)*, and *International Journal of Ecology and Environmental Sciences (2004)*. Chapter 8 is a modified and updated version of a paper that appeared in the *International Journal of Ecology and Environmental Sciences (2012)*. Portions of the following two articles are also used in Chapters 1, 2, 3 and 9: an invited article presented at the *12th Agricultural Science Congress (Proceedings published in 2015)* and the Presidential Address delivered at the 29th Annual Conference (27 October 2021) of the Agricultural Economics Research Association (AERA), also published in *the Agricultural Economics Research Review*, 35(1). I am grateful to journals and publishers for permission to reproduce parts of these papers.

Satya Prakash Sahu typed the manuscript. His patience and skills helped me to make writing this book a pleasure. Special thanks to him for an excellent job done.

My most sincere thanks go to my family members. In all my endeavours, my wife, Sneh, stands out for her abundant and selfless support. My daughter Juhi offered sustained support during the manuscript preparation and ensured that I remained stress-free. My other two daughters and their husbands, Shivli-Rajesh and Molshree-Dheeraj, deserve great credit for keeping my focus on the topic of the book. My Nani, Radha Ji, and Nana, Deva Ji, nurtured a spirit in me of loving mother earth. This book is dedicated to my parents and grandparents who taught me the invisible values of wetlands in sustaining the very survival of our planet. I hope my grandchildren – Ahaana, Artham, Ayaana, and Anaisha – representing the next generation, will be able to visit a few historical wetlands, the crucial elements of our planet, struggling for existence.

Dinesh K. Marothia

Abbreviations

ACRP	Agro-Climatic Regional Planning Approach
ACZs	Agro-Climatic Zones
AMRUT	Atal Mission for Rejuvenation and Urban Transformation
ASCO	Assistant Soil Conservation Officer
BHEL	Bharat Heavy Electricals Limited
BKIT	Boriya Khurd Irrigation Tank
BMC	Bilaspur Municipal Corporation
BOD	Biological Oxygen Demand
CA	Comprehensive Assessment
CACP	Commission for Agricultural Costs & Prices
CAGR	Compound Annual Growth Rate
CCOST	Chhattisgarh Council of Science & Technology
CG	Chhattisgarh
CG FIMS Act	Chhattisgarh Farmer's Participation in Management of Irrigation System Act
CGPCB	Chhattisgarh Pollution Control Board
CGSPC	Chhattisgarh State Planning Commission
CGTB	Chhattisgarh Tourism Board
COD	Chemical Oxygen Demand
COPs	Conference of Parties
CPRs	Common-Pool Resources
CREDA	Chhattisgarh State Renewable Energy Development Agency
CSFCF	Chhattisgarh State Fish Cooperative Federation
CSR	Corporate Social Responsibility
CVM	Contingent Valuation Method
DCs	Distributaries Committees
DDA	Deputy Director Agriculture
DO	Dissolved Oxygen
DOF	Directorate of Fisheries
DPSIR	Drivers, Pressures, State, Impacts, Responses
DS	Dissolved Solids
DWA	District Wetland Authority

DWR	Department of Water Resources
ECCs	Environment Conscious Citizens
FARs	Freshwater Aquaculture Resources
FCSs	Fisheries Cooperative Societies
FFDA	Fish Farmers Development Agency
FGDs	Focus Group Discussions
FGs	Fishers Groups
FHs	Fishermen Households
FMIS	Forest Management Information System
FOs	Farmers' Organizations
FPOs	Farmers Producer Organizations
FRA	Forest Rights Act
FTL	Full Tank Level
GAPs	Good Agricultural Practices
GDP	Gross Domestic Product
GIS	Geographic Information System
GoCG	Government of Chhattisgarh
GoI	Government of India
GP	Gram Panchayat
GSDP	Gross State Domestic Product
ha	hectare
HRIDAY	National Heritage City Development and Augmentation Yojana
HWSA	Hectare Water-Spread Area
IBAs	Important Bird Areas
IFS	Integrated Farming System
IFs	Individual Fishers
IGKV	Indira Gandhi Krishi Vishwavidyalaya
IGP	Indo Gangetic Plain
INTACH	Indian National Trust for Art and Cultural Heritage
ISRO	Indian Space Research Organisation
ITCM	Individual Travel Cost Method
IUCN	International Union for Conservation of Nature
IWMI	International Water Management Institute
KVKs	Krishi Vigyan Kendra
LSL	Lowest Still Level
LV	Lake Vivekananda
MEA	Millennium Ecosystem Assessment
MGNREGA	Mahatma Gandhi National Rural Employment Guarantee Act
MIWSTs	Minor Irrigation Water-Storage Tanks
MMITs	Micro-Minor Irrigation Tanks (*Dabreis*)
MMPI	Modern Maintenance Products India Pvt. Ltd
MOEFCC	Ministry of Environment, Forest and Climate Change

MP	Madhya Pradesh
MPN	Most Probable Number
MPPCB	Madhya Pradesh Pollution Control Board
MSL	Mean See Level
MSP	Minimum Support Price
MTL	Maximum Tank Level
MTT	Mahamaya Temple Trust
MWAIs	Multiuse Wetlands-Agriculture Interactions
MWs	Multiuse Wetlands
NCSCM	National Centre for Sustainable Coastal Management
NEERI	National Environmental Engineering Research Institute
NGOs	Non-Governmental Organizations
NGT	National Green Tribunal
NMDC	National Mining Development Corporation
NPCA	National Plan for Conservation of Aquatic Ecosystems
NTFPs	Non-Timber Forest Products
NTPC	National Thermal Power Corporation Limited
OLS	Ordinary Least Square
PA	Protected Areas
PCs	Project Committees
PIM	Participatory Irrigation Management
PLIPLV	Population Living on the Periphery of Lake Vivekananda
PMKSY	Pradhan Mantri Krishi Sinchayee Yojana
POP	Privately Owned Ponds
PPM	Parts Per Million
PPP	Public-Private Partnership
PRIs	Panchayat Raj Institutions
PTs	Perennial Tanks
RF	Rainfed
RMC	Raipur Municipal Corporation
RNP	Ratanpur Nagar Palika
RRR	Repair, Renovation & Restoration
RSCL	Raipur Smart City Limited
RTI	Right To Information Act
SAC	Space Applications Centre
SACA	Space Application Centre Ahmedabad
SC	Scheduled Caste
SCOs	Soil Conservation Officers
SDA	State Department of Agriculture
SDF	State Departments of Fisheries
SDGs	Sustainable Development Goals
SHGs	Self-Help Groups
SID	Simpson Index for Crop Diversification
SIT	Serikhedi Irrigation Tank

SPIS	Solar Power Irrigation System
SRI	System of Rice Intensification
SS	Suspended Solids
ST	Scheduled Tribe
STs	Seasonal Tanks
SWA	State level Wetland Authority
SWC	Soil and Water Conservation
TBL	Top Bank Level
TCM	Travel Cost Method
TCs	Territorial Constituencies
TDS	Total Dissolved Solids
TPR	Transplanting
TS	Total Solids
UNCCD	UN Convention to Combat Desertification
UNEP	United Nations Environmental Program
VWA	Village Wetland Authority
WAIs	Wetlands-Agriculture Interactions
WHS	Water-Harvesting Structures
WSA	Water-Spread Area
WTP	Willingness to Pay
WUAs	Water Users Associations
WWF	World Wide Fund
WWT	Wildfowl & Wetlands Trust
ZTCM	Zonal Travel Cost Method
ZWA	Zonal Wetland Authority

1 Introduction

1.1 Wetlands and the Ramsar Convention

Wetlands are places where water accumulates for sufficiently long periods to allow the occurrence of plants and animals specially adapted to the water-dominated environment. Water bodies, called *sar, sarovar, taal, jheel* (lakes and ponds) as well as marshes and swamps (Sanskrit: *anoop*), which support a large diversity of plants (like rice, lotus, lilies, reeds) and animals (like fish, turtles, frogs, birds, molluscs), are now called wetlands. The foundation of human civilisations across the world was laid in wetlands along the rivers. Wetlands occur in all climatic zones and in practically all parts of the world. They occur along rivers, streams, lakes, reservoirs, and sea coasts, in arid deserts (oases and saline lakes), from below the sea's surface (coral reefs) to snow-clad mountains (glacial lakes), and even below the ground (caves and Karst systems). However, during British rule in India, these habitats, making no contribution to imperial revenue, were declared as 'wastelands' and were gradually ignored, reclaimed, drained, filled, and converted to other land uses (cited from Marothia, 2019, 2021; Gopal, 2015; Gopal and Marothia, 2017; see also Chatterjee et al., 2008; Finlayson et al., 2005).[1]

Wetlands drew attention in Western countries during the 1950s and 1960s for their huge migratory water bird populations. To ensure the conservation of these habitats worldwide, representatives of a few international organizations and national governments signed, on 2 February 1971 (in Ramsar, Iran), an agreement, called the Ramsar Convention on Wetlands of International Importance, especially as habitats for waterfowl. The Ramsar Convention has adopted a rather broad definition of "wetlands" to cover almost the whole range of inland and coastal aquatic ecosystems. Under the text of the Ramsar Convention (Article 1.1), wetlands are defined as:

> Areas of marsh, fen, peat land or water, whether natural or artificial, permanent or temporary, with water that is static or flowing, fresh, brackish or salt, including areas of marine water the depth of which at low tide does not exceed six metres. . . . [and] may incorporate riparian and

DOI: 10.4324/9781003456742-1

coastal zones adjacent to the wetlands, and islands or bodies of marine water deeper than six metres at low tide lying within the wetlands.

(*Ramsar Convention Secretariat, 2013*)

For practically the same reason, the Ramsar Convention considers lakes and rivers also as "wetlands in their entirety, regardless of their depth".

(*Ramsar Convention Secretariat, 2013*)

The Ramsar Convention (2013) recognized five major wetland types: marine (coastal wetlands including coastal lagoons, rocky shores, seagrass beds and coral reefs); estuarine (including deltas, tidal marshes and mud-flats, and mangrove swamps); lacustrine (wetlands associated with lakes); riverine (wetlands along rivers and streams); palustrine (meaning "marshy" – marshes, swamps, and bogs).

In addition, there are human-made wetlands such as fish and shrimp ponds, farm ponds, irrigated agricultural land including rice paddies, salt pans, dams, reservoirs, gravel pits, wastewater treatment ponds, and canals. The Convention has adopted a Ramsar Classification of Wetland Type, which includes 42 types, grouped into three categories: Marine and Coastal Wetlands, Inland Wetlands, and Human-made Wetlands (Ramsar Convention, 2016).

The term "wetlands" has been in use for more than six decades but there remains a lot of confusion about what is or is not a wetland. The Ramsar Convention, which has promoted "wise use" and "conservation", including the "restoration" of wetlands, considered all inland and coastal aquatic ecosystems as wetlands. The Ramsar Convention brings together diverse environments, spatially and temporally, but also in terms of physical size, ecology, hydrology and geomorphology, into a single definition, grouping together a wide variety of landscapes whose ecosystems share the fundamental wetland characteristic of being strongly influenced by water (Davis, 1994).[2]

Some 171 countries have adopted a common definition by signing the Ramsar Convention on Wetlands of International Importance. The Convention therefore adopts an exceptionally broad approach in determining the "wetlands" which come under its aegis (Ramsar Convention Secretariat, 2018).

1.2 Wetlands Definition and Classification in India: Contradiction and Confusion

In India, wetlands or multiple-use water bodies form an integral part of the socio-cultural ethos. In fact, the construction of water bodies to meet the water needs of people and animals alike during prolonged dry periods of the year was considered to be *"Raj Dharma"*. Wetlands in India have been home to the world's most important wetland plant – rice. Another wetland plant native to India – the lotus (*Nelumbo nucifera*) – is embedded in mythology and held sacred in Hinduism and Buddhism, symbolizing purity and "spontaneous" generation (Gopal, 2015; Gopal and Marothia, 2017).

India has been a Contracting Party to the Ramsar Convention since 1982 and has until now added 75 designated Ramsar wetland sites spread over 13,26,677 hectares (see Chapter 9 for India and the Ramsar Convention details). Despite several wetland-related programmes and activities at the central government and state levels, there has been increasing concern about the degradation and loss of wetlands in different parts of the country. Loss and unsustainable use of wetlands may arguably be attributed to the fact that India has no wetland-specific policy or law. The ecosystem services of wetlands are recognized in India's National Environment Policy, 2006, which also mentions the need for maintaining their ecological characters. The Government of India (GoI, 2010) adopted in 2010 the Wetland (Conservation and Management) Rules under the Environment Protection Act, 1986, to regulate various activities affecting wetlands adversely. These rules were replaced by the Wetland (Conservation and Management) Rules, 2017 (GoI, 2017). The Wetland Rules, which refer to the ecosystem services of wetlands and the value of conservation, do not prohibit or restrict hydrological alteration despite identifying various threats to the wetlands. Most of the provisions in the Rules do not support in any manner wetland conservation. The scope of the Rules is itself restricted to only a few wetlands in the country. The Rules define wetlands along similar lines as the Ramsar Convention's definition, but clearly exclude "river channels, paddy fields, human-made water bodies/tanks specifically constructed for drinking water purpose and structures specifically constructed for aquaculture, salt production, recreation and irrigation purpose". Effectively, it means that all human-made wetlands are afforded no legal protection. All small, medium, or large reservoirs are constructed for irrigation or drinking water supplies. The recent National Wetlands Inventory and Assessment has identified 122,370 tanks and 14,894 reservoirs among the 145,641 human-made wetlands in India (both inland and coastal). The Rules may be interpreted to exclude all of them from the purview of regulations. Notwithstanding the designation of the Wular lake (the River Jhelum flowing through it) and Upper Ganga River (the stretch from Brij Ghat to Narora), the Government does not consider rivers to be wetlands. The fact that the Bhoj wetland – a Ramsar site – was created centuries ago specifically for drinking water and irrigation, may also bring its status to dispute (cited from Gopal, 2018, critical observations on Rules 2010 and Rules 2017 are made in Chapter 9).

For the purpose of this book, I have included all inland wetlands categories defined under the Ramsar Convention, caves, and similar subterranean wetlands. In addition, a few indigenous wetlands, wetland complexes (a chain of wetlands or interconnected wetlands/water bodies), temples, and sacred wetlands have also been included for analysis. All the wetland types included in this book are multiuse and multifunctional (see Chapter 3), and fit well within the broad classification of the Ramsar Convention (2013, 2018).

1.3 Wetlands: Extent, Threats, Benefits and Sustainable Development Goals (SDGs)

1.3.1 Extent of Wetlands

Estimates of wetlands coverage of the Earth's surface is not known precisely. However, refinement in methodology has recently improved the accuracy of global wetland area data. According to a recent report of the Ramsar Convention (2018), global inland and coastal wetlands cover over 12.1 million km^2, an area almost as large as Greenland, with 54 percent of it permanently inundated and 46 percent seasonally inundated. However, natural wetlands are in long-term decline around the world; between 1970 and 2015, inland and marine/coastal wetlands both declined by approximately 35 percent, where data are available, three times the rate of forest loss. In contrast, human-made wetlands, largely paddy fields and reservoirs, almost doubled over this period, forming 12 percent of the wetlands. These increases have not compensated for natural wetland loss (cited from *Global Wetland Outlook: State of the World's Wetlands and their Services to People*. Gland, Switzerland: Ramsar Convention Secretariat, 2018).

India has a very large physiographic diversity (from sea level to high mountains) and high average rainfall (120 cm) marked by extreme temporal and spatial variability. Therefore, India enjoys a rich diversity of wetland types. Over millennia, humans have modified and managed most of the natural wetlands for specific purposes, such as paddy and fish. Innumerable tanks were created in the arid and semi-arid regions of western and peninsular India. More wetlands have also been created by damming rivers. Large annual fluctuations in water levels result in the exposure of extensive peripheral areas which harbour diverse aquatic vegetation and other wildlife (Gopal, 2015; Gopal and Marothia, 2017).

While several attempts have been made to document the wetlands in the country, they focused only on the better-known habitats. In 1998, the Indian Space Research Organisation, Space Application Centre Ahmedabad (SACA), conducted the first survey based on remote sensing satellite images and subsequently, in 2010, published the first *National Wetland Atlas* (Gopal and Marothia, 2017). Recently, SACA conducted a study to estimate wetland areas in the country (Patel, 2018). The study adopted the Wetland Classification System, based on the Ramsar Convention's definition of wetlands, which provides a broad framework for delineating wetlands and is acquiescent to remote sensing data. SACA's study considered all aspects of a water mass including its ecotonal area as wetland. In addition, under the Ramsar Convention's classification of wetlands, fish and shrimp ponds, saltpans, and reservoirs were also included as wetlands. Besides the Wetland Classification System, the SACA study incorporates deep-water habitats and impoundments. SACA estimated 15.260 mha of total wetland area, which is around 4.63 percent of the geographic area of the country. A total of

201,503 wetlands (>2.25 ha) have been mapped at 1:50,000 scale. In addition, 555,557 small wetlands (<2.25 ha) have also been identified. Inland-natural wetlands accounted for around 43.4 percent of the total area, while coastal-natural wetlands account for 24.3 per cent. As far as wetland units were concerned, tanks were maximum in number (122,370) as mappable units and 555,557 small wetlands (< 2.25 ha) were mapped as point features (3.64 percent). The human-made inland wetlands cover about 37 percent of the country (3941,832 ha) and the remaining 63 percent were the natural wetlands (6623,067 ha). There are also 4,140,116 ha of coastal wetlands (of which the intertidal mudflats of Kutch alone contribute about 51 percent) and 555,557 ha of wetlands smaller than 2.25 ha each (Patel, 2018). It is noteworthy that paddy fields are not included as wetlands in this inventory.

The status of wetlands in terms of open water and aquatic vegetation shows significant seasonal change, and there is a significant reduction in the extent of open water (about 32.5 percent) from post-monsoon to pre-monsoon (8.60 Mha to 5.80 Mha). This is reflected in all the inland wetland types. Small village tanks/ponds are also useful wetlands. As already noted, there are 555,557 such wetlands in the country (Patel, 2018). See also Gopal and Chauhan (2001), Gopal et al. (2008) for South Asia and tropical wetland status, and Bassi et al. (2014) for wetland status in India. (For Chhattisgarh see Indian Space Research Organisation, Space Application Center Ahmedabad, 2010.)

1.3.2 Threats to Wetlands

In a recent and comprehensive study of 189 wetland assessments, Davidson (2014; see also Davidson, 2017) estimated that global wetland losses in the 20th century were 64 to 71 percent, and, for some regions, notably Asia, even higher. He found that "losses of natural inland wetlands have been consistently greater, and have occurred at faster rates, than [those] of natural coastal wetlands." His review found that the extent of inland wetlands declined 69 to 75 percent during the 20th century, while coastal wetlands declined 62 to 63 percent. Further, 64 percent of the world's wetlands have disappeared since 1900. The loss of freshwater wetlands worldwide from 1997 to 2011 is valued at US$ 2.7 trillion per year (Briggs, 2014). India is losing wetlands at the rate of two to three percent each year (Prasher, 2018). The drivers of this decline are overfishing, agriculture, deforestation, introduced species, climate change, water drainage, land encroachment, urban development, and the absence of wetland policy and governance structures.

Despite their numerous benefits and importance in the socio-cultural life of India, the majority of wetlands are getting degraded or lost. Wetlands are directly impacted by large-scale, widespread hydrological alterations. The sources and pathways of their water supply, both as surface runoff or inflow through channels, are eliminated or blocked, or their water supply is greatly altered. This happens most commonly through urban expansion and land-use change in their catchments. Even today, human settlements continue to grow near wetlands, and most often at their expense. Those wetlands which come to lie within urban limits are the most threatened in both developed and developing countries, including India. For example, rapid loss of wetlands has occurred in Kashmir valley, Hyderabad, Bengaluru, Chennai, Delhi, Kolkata, and Guwahati, all of which have experienced disastrous floods in recent years. Wetlands are encroached upon when low water levels during dry periods are taken advantage of, or when they are drained and used them as landfill sites. Many floodplain wetlands have been eliminated by embankments and, in most cases, the flow, storage, and diversion structures upstream have greatly modified their flooding regimes. In lakes and other water bodies, water levels are regulated by the withdrawal of water for different human uses (Gopal, 2015; Gopal and Marothia, 2017; see also Euisoon et al., 1997 for the economic impacts of urban environmental problems in Asian cities, including a few Indian cities).

Wetlands are further degraded by the discharge of untreated domestic and industrial wastewaters into upstream river systems and it is not readily appreciated that such upstream wastewater discharges affect the wetlands downstream. Besides bringing in various pollutants, wastewaters also cause siltation and alter the hydrological regime (Gopal and Marothia, 2017).

Numerous wetlands are also infested with the uncontrolled growth of exotic species, particularly water hyacinth. In recent years, there has been little effort to remove and destroy them because of the misplaced perception that these plants can remove pollutants and help improve water quality. However, for these weeds to be effective they have to be selectively removed leaving young ones to multiply. Death and decay of older plants returns the nutrients and pollutants back into the water while huge quantities of undecomposed organic matter accumulate, fill in the water body, eliminate dissolved oxygen, and cause mass fish deaths (Marothia, 1997c, 2004d; Gopal, 2015; Gopal and Marothia, 2017).

The most significant factor responsible for the present state of wetlands is the absence of a clear national or state policy devoted to wetlands – natural or human-made. Wetlands receive no attention in development plans that are concerned instead with land-use changes or the development of water resources. The economic value of the goods and services of wetlands, including their contribution to livelihoods and rural economies, are never accounted for in the cost-benefit analyses of various development projects (Marothia, 2004d; Gopal, 2015; Gopal and Marothia, 2017).

1.3.3 Benefits from Wetlands

Wetlands are among the most valuable ecosystems, providing services worth US $47 trillion a year. More than one billion people rely on wetlands for their income. Nearly 40 percent of the world's species live and breed in wetlands.

Annually, about 200 new fish species are discovered in freshwater wetlands. Aquaculture is the fastest-growing food production sector, while inland fisheries alone provided 12 million tonnes of fish in 2018. Rice paddies feed 3.5 billion people annually. Coral reefs are home to 25 percent of all species. Wetlands provide protection from floods and storms, with each acre of wetland absorbing up to 1.5 million gallons of floodwater. Wetlands help regulate the climate. For example, peat lands store twice as much carbon as forests, with salt marshes, mangroves, and seagrass beds also holding vast amounts of carbon (cited from *Global Wetland Outlook: State of the World's Wetlands and their Services to People*. Gland, Switzerland: Ramsar Convention Secretariat, 2018).

Humans benefit in many ways from all kinds of wetlands. These benefits arise either directly from the use of the water, plants, animals, soils, and other components of the wetland ecosystem or indirectly from their ecosystem functions. These benefits are now called ecosystem services and categorized into provisioning, regulating, supporting and cultural services (cited from Marothia, 2021 and Gopal and Marothia, 2017, see Chapter 3 for ecosystem services derived from different types of wetlands).

Irrespective of the kinds of organisms living there, the most important function of wetlands is to regulate the hydrology of the region. Their capacity for retaining water controls flooding (especially downstream), recharges groundwater, and provides water for various human uses including irrigation. The amount and duration of water in the soil and above it governs the diversity of biota which mediates other ecosystem functions and determines various benefits. Thus, various kinds of wetlands differ in their dominant biota and the ecosystem services (Gopal and Marothia, 2017, see Chapter 3 for detailed benefits).

Wetland plants may be used directly by humans for food, fibre, fuel, medicine, or animal feed. Rice and fish, which are a staple food of humans, are major wetland products. Other wetland plants widely used for food include singhara (*Trapa bispinosa*), lotus (*Nelumbo nucifera*), makhana (*Euryale ferox*), and kodo (*Echinochloa esculenta*). Jute, reeds, cattails, *Calamus* (cane), and many grasses are used for fibre; woody plants such as *Tamarix sp.* are used for fuel, whereas species of *Echinochloa, Paspalum* and duckweeds are important animal feeds. Many plants such as vacha (*Acorus*), brahmi (*Bacopa*), motha (*Cyperus rotundus*), and bhringaraj (*Eclipta alba*), are used in medicines. Among the animals, fish, birds, amphibians, and large invertebrates (crustaceans and molluscs) are common human foods. Even some algae (*Chlorella, Spirulina, Scenedesmus, Dunaliella*) are used as health foods and medicines (Gopal, 2015; Gopal and Marothia, 2017, see Chapter 3 also for detailed benefits).

Indirectly, wetlands support high levels of biodiversity (fish, frogs, turtles, birds, molluscs, arthropods, and insects) by providing habitats and food. Plants also oxygenate water and sediments, and generally remove nutrients and pollutants to help improve water quality. Even subterranean fauna contribute significantly to the improvement of water quality as water infiltrates into shallow aquifers. Wetlands have indeed been called the 'Kidneys of the Earth'. Although some organisms in wetlands, such as mosquitoes and snails, are known disease vectors, most other wetland fauna help control pests and diseases, help in pollination, decompose wastes and sustain nutrient cycles, and support food chains. Wetlands have a moderating influence on the micro-climate and help in carbon sequestration, thereby contributing to the mitigation of climate change. Another highly important function of wetlands benefitting humans directly is related to their recreational (swimming, bathing, diving, boating, angling, fishing, bird watching) and social-cultural-religious use, besides their contribution to the aesthetics of the surroundings (Marothia, 2015, 2019; Gopal, 2015; Gopal and Marothia, 2017, see Chapter 3 and Chapter 5 for a range of benefits from wetlands).

1.3.4 Wetlands and Sustainable Development Goals (SDGs)

The Ramsar Convention's Fourth Strategic Plan (2016–2024) identified four overarching goals and 19 specific targets that directly support the achievement of both the SDGs as well as the Aichi Targets set up by the Convention on Biological Diversity. The SDGs, Aichi Targets, and the Ramsar Strategic Plan call for increased integration and synergies across existing multilateral agendas. Wetlands are implicitly or explicitly mentioned within seven sustainable goals and 27 targets (Ramsar Convention's Fourth Strategic Plan – 2016–2024).

The sustainable use of water and wetlands, by protecting the services they provide, is critical to enable society to achieve sustainable social and economic development, adapt to climate change, and improve social cohesion and economic stability. The proposed United Nations SDGs offer a universal agenda that, for the first time, recognizes the need for the restoration and management of water-related ecosystems, including wetlands, as a basis for addressing food, water scarcity and water risks. Wetlands are a solution for several key challenges around the world related to water, food, and climate, and key to meeting the SDGs. The multiple benefits and services provided by wetlands are essential to achieving the SDGs.

Most of the SDGs are relevant in some way or another to wetlands, but the following are of particular importance in the context of human-made inland wetlands: SDG 1: End poverty in all its forms everywhere, SDG 2: End hunger, achieve food security and improved nutrition, and promote sustainable agriculture, SDG 5: Achieve gender equality and empower all women and girls, SDG 6: Ensure availability and sustainable management of water and sanitation for all, SDG 8: Promote sustained, inclusive, and sustainable economic growth, full and productive employment and decent work for all, SDG 9: Build resilient infrastructure, promote inclusive and sustainable industrialization, and foster innovation, SDG 11: Make cities and human settlements inclusive, safe, resilient, and sustainable, SDG 13: Take urgent action to combat climate change and its impacts, SDG 15: Protect, restore, and promote the sustainable use of terrestrial ecosystems, sustainably manage forests, combat desertification, and halt and reverse land degradation and biodiversity loss.

1.4 Developing a Synthesis of Institutional Analysis Frameworks for Analysing Multiuse Wetlands Governance

Before discussing the relevance of a synthesis of Institutional Analysis Frameworks for analysing MWs governance, a clear understanding of the meaning of two key concepts, i.e. 'multiuse wetlands' and 'multiuse governance', is necessary. 'MWs' or 'multiple-use water systems' or 'multiuse water bodies' are the most important and valuable category of wetlands in many counties, including India and the state of Chhattisgarh. MWs include natural and human-made small water-storage bodies, village ponds, lakes, wetland complexes, sacred and temple ponds, irrigation and multipurpose tanks, riverine wetlands, reservoirs, and similar rain-harvesting structures. I have grouped these different types of wetlands as MWs. All wetlands included in this book for detailed analysis are multipurpose and multifunctional. They provide provisioning, regulating, supporting and cultural services with very high-use and non-use economic values. 'Multiuse wetlands governance' is used in this book interchangeably with 'governing multiuse wetlands' and 'managing multiuse wetlands'. The 'wetland governance' concept includes informal and formal institutional arrangements that govern wetland users' decision-making processes so as to create collective action and socially desirable patterns of interaction among users to achieve broad outcomes and evaluative criteria (economic efficiency, equity, environmental effectiveness, political legitimacy of authority and sustainability).[3] Elements of multiuse governance used here include essential components of the IUCN concept of environmental governance. The concept is equally useful for governing a wide range of CPRs.

MWs can be governed or managed sustainably under state, common, or private property regimes but are equally likely to be degraded by the same regimes (Marothia, 2004d, 2015, 2019, 2021). A property regime is a system or a set of institutional arrangements or working rules of rights and responsibilities characterizing the relationship of co-users to one another with respect to a specific natural resource (Bromley, 1989; Bromley and Cernea, 1989; Gibbs and Bromley, 1989, see also classic work of Commons (1934, 1968) on the role of institutions, property rights and collective action). In managing MWs, there are many overlaps between, and combinations of, state (public), community, and private-management systems or governance structures. In other words, these resources are managed at the interface of different property regimes. The four categories of resource governance or property rights regimes (state, private, common, and open-access regimes (Bromley, 1989a, 1991; Ostrom, 1990a; Gibbs and Bromley, 1989; Bromley and Cernea, 1989)) have been extended by Townsend and

Pooley (1995) to recognize that natural resources governance can be shared among states, communities, and private-interest groups in various ways at different decision-making levels (see also Curtis, 1991). Distributed governance involves external institutional arrangements[4] among government and local communities or resource users, as well as internal institutional arrangements[5] within local community institutions or resource users. Government, local communities, and private parties utilizing CPRs bring different interests, capabilities, and understandings to resource-management and decision-making processes (Townsend and Pooley, 1995). Alternative institutional perspectives shape the decision-making processes of government and local resource users and of members of local communities for managing a resource by converting unorganized structures into organized ones (Marothia and Phillips, 1985).

Research on institutional analysis of natural and common pool resources was initiated by John R. Commons (1924, 1931) and other institutional economists.[6] Since the late 1970s, scholars from diverse disciplines have been engaged in the development of institutional analysis frameworks to undertake a variety of empirical studies on CPRs systems. Kiser and Ostrom (1982) and Oakerson (1986, 1992) have developed institutional analysis frameworks to analyze single- and multiple-use CPRs. These frameworks have been applied by many scholars in various parts of the world to understand CPRs problems.[7] Tang (1992), Edwards (Edwards and Steins, 1996; Edwards, 1996), and a few other researchers had made attempts to extend the analytical framework of Oakerson (1986, 1992), and Ostrom (1990a) for analyzing multiple-use common pool resources under the common property resource regime.

A large number of studies in South Asia and India indicate that CPRs can be managed under different property rights regimes, including distributed and polycentric governance. My four decades of research in India has indicated that CPRs in general, and MWs ecosystems in particular, are influenced by social, ecological, cultural, economic, and political attributes. In the Indian context, I see an opportunity to synthesize diverse institutional frameworks to analyse simple and complex MWs and wetland complexes. To bring some degree of consistency between the actions of wetlands users and multiple governance agencies, a creative synthesis of various institutional frameworks is used in this book to analyse the utilization and management of MWs under different governance systems or property rights regimes. To develop a creative synthesis of institutional frameworks, I have incorporated appropriate elements of different institutional approaches to the basic framework of Oakerson (1992), and subsequently developed by Ostrom (1990a) and Tang (1992), to analyse complex, common-pool MWs ecosystems. Other components or sets of variables that emerged from Bromley's property rights and public choice analysis work (Bromley, 1989a, 1991) and Townsend and Pooley's (1995) distributed governance paradigm are also included. It is important to recognize that MWs in the Indian context also provide significant direct and indirect values (benefits) and ecosystem services (see Chapter 3). Therefore, I have added to the synthesis attributes that are related to changes in total economic values (use and non-use values) and the different categories of ecosystem services (provisioning, regulating, cultural, supporting) due to changes in other attributes of the institutional framework. A synthesized institutional framework (see Chapter 2) is used to analyse different categories of the inland MWs of Chhattisgarh, spread across different agro-climatic zones (ACZs). The synthesis of various institutional analysis approaches is also supplemented with criteria of environmental governance, concepts of total economic values and ecosystem services derived from different categories of MWs.[8]

1.5 Background and Purpose of This Book

MWs were traditionally constructed by kings, *jagirdars*, and religious bodies, who were rich and powerful in South Asian countries, including India, to create community water assets. Management, repair, and rejuvenation of these resources were governed through well-defined but informal codes for different uses and stakeholders. Control of these water bodies was with the owner (the kings, *jagirdars*, or religious bodies, for example, but in many cases ownership of MWs was transferred to village communities). MWs have traditionally been allotted for different uses such as tending cattle, washing clothes, irrigation, fishing, nutrient-rich soil and fodder-grass collection, brick making, conserving micro biodiversity, and for social and cultural rituals (funerals and worshipping, and temple or sacred ponds). MWs have also been allotted to various social groups that belong to different caste, communities, and women (Marothia, 2015, 2019).

After 1952, in many Indian states, including Chhattisgarh, most wetlands were transferred to Panchayats, municipal bodies, and irrigation departments depending on their size, water-spread areas and location. New technical and institutional arrangements, entitlements, and multiple authority systems governed the *usufruct* rights for different users and non-users. Today, MWs are administered and controlled under different property rights regimes or levels of institutional hierarchy involving Panchayats Raj Institutions at village, Janpad, and district levels, the Department of Water Resources (DWR), soil and water conservation wings of the State Department of Agriculture, State Departments of Fisheries (SDF). Some are also controlled by private owners. There is enough evidence available in Chhattisgarh and elsewhere in the country to show that, when MWs are managed under weak property rights regimes, they often degrade into open-access resources due to inadequate or poorly conceived institutional arrangements and the breakdown of local authority. Examples are also available which show that these resources degrade under an open-access system brought under a state, private or community management regime through changes in institutional arrangements.[9]

The issues related to managing MWs are complex due to the diversity of categories into which they fall, their scale, and their varying uses, such as for agricultural, fishing, domestic, and socio-cultural-religious activities, as well as due to the involvement of multiple agencies in governing them. Competition and conflicts between different parties over MWs have been observed, such as within user groups, across user groups, between institutions, between stakeholders, and between villages – particularly in the case of multipurpose irrigation tanks (Marothia, 2015). Wetlands also suffer with spatial and temporal externalities caused by pollution, congestion, in-fillings, encroachment, natural and created-use conflicts, misspecification of property rights regimes, and entitlement systems and institutionally complex hierarchies and mechanisms for adjusting the allocation of wetlands water resources (Marothia, 2004d, 2015).

MW ecosystems undeniably provide valuable environmental resources with consequently high preservation, conservation, and utilization values. Despite this, the existing evidence suggests that wetlands in many Indian states, including Chhattisgarh, are still not being managed sustainably. While a few attempts have been made in India to understand the causes and consequences of wetland degradation, little explicit research has been conducted to understand the complexities of interdependencies between economic, social, ecological, institutional, and environmental pressures and processes related to MWs. (These issues have been debated and analyzed over the years in the research work of Marothia, 1995a, 1997c, 2004d, 2015, 2019.) Further, users' perceptions have not been adequately analysed when designing wetland ecosystem policies and strategies (Marothia, 2001a, 2004d). The most challenging task in governing wetlands is to frame institutional choices – e.g. a set of institutional arrangements that users and non-users can comprehend and agree to pursue. The primary purpose of this book is to fill this gap in research and policy analysis. This is now possible because of extensive field work, the collection of required primary data, and appropriate sampling methodologies adopted for each of the MWs considered in this book. A synthesis of institutional analysis frameworks was applied to evaluate the performance of rural, peri-urban and urban MWs of Chhattisgarh – a state in eastern India that is managed under different governance structures or property rights regimes. Each MW case study reported in this book explores the kind of governance and institutional design that worked/did not work. A comparative narrative of case studies of wetlands from other regions was also drawn upon to as much as possible. Findings from these case studies provide research and policy input to address challenges and firm-up institutional choices for sustaining wetlands for inter-generational use. This book aims to provide a broader description of the issues facing researchers and policy makers working across the globe, and to improve their ability to design sustainable MW management systems.

This book aims to provide three key benefits: (i) to help wetland scholars, policy makers, and managers to understand the causes and consequences of wetland degradation and the complexities of the interdependencies between economic, social, institutional, and environmental pressures and processes, (ii) to provide a strong institutional analysis framework with which to analyse multiuse and multifunctional wetland ecosystems and other categories of multiple-use CPRs, and (iii) to provide an exhaustive analysis of rural, peri-urban, and urban, human-made wetlands to establish the relevance of institutional design and the effective role of authority in governing multiuse wetland ecosystems.

1.6 Structure of the Book

From here on, the book is organized into nine chapters. Chapter 2 presents a synthesis of institutional analysis frameworks. The synthesis framework is applied to all the case studies reported in this book to understand how institutional arrangements contribute to shaping the interests and interactions of various groups engaged in using and managing MWs. Chapter 3 presents the basic features of agro-climatic zones (ACZs); water resources; wetlands by category, size, and use; wetlands inside and outside forests; avifaunal biodiversity; Karst wetlands and ecosystems; temple and sacred ponds; wetland complexes; multipurpose irrigation tanks, *Dabries* (micro-minor irrigation tanks); wetlands constructed under the Mahatma Gandhi National Rural Employment Guarantee Act (MGNREGA); and ecosystem services derived from different categories of wetlands. Chapter 4 focuses on the current ecological health of the Ratanpur historical wetland complex and the major ecosystem services derived from it. The effectiveness of existing institutional hierarchies and proposed institutional mechanisms to restore the complex are also discussed. Chapter 5 assesses the nature and extent of water pollution in Lake Vivekananda (LV). It examines issues of resource rights regimes, estimates the willingness of users and non-users to pay for its upkeep, and takes into account the past and current efforts to restore LV. An exploration of pollution, externality, property rights, entitlements systems, the roles of different authorities, its valuation, and the willingness of various parties to pay for its upkeep are used to understand policy approaches to the interdependent activities of LV. Finally, issues and lessons learned from the restoration efforts are discussed. Recognizing the appropriateness of tanks as a traditional technology for sustaining today's farming needs, Chapter 6 attempts to examine tank irrigation and the current status of selected tanks, decision-making arrangements and their impact on patterns of interaction among farmers, and the relevance and contribution of irrigation tanks to transforming rainfed farming systems. Finally, institutional and policy interventions are suggested to sustain tanks and improve the efficiency of tank irrigation. In Chapter 7, an attempt is made to take stock of the construction of *dabries*, after the formation of the Chhattisgarh State, by block, district, and ACZs. Using the physical and technical attributes of *dabries*, the characteristics of the irrigators' community, and institutional arrangements, the chapter explores how self-organized and self-governed institutional mechanisms can sustain *dabries* and improve the income and livelihoods of farmers. Further, it explores how institutional chaos can be avoided with self-organized and self-governed, crafted institutional arrangements. In Chapter 8, looking at the importance of MWs for fishing, the governance structure of fisheries in ponds and tanks is discussed and makes conclusions based on an in-depth analysis of five Fisheries Cooperative Societies (FCSs) which cover different categories of ponds and tanks and administer them under different property rights regimes, including distributed governance. In this chapter, the performance of individual fishermen, who have leased in Panchayats and private ponds, or practised fish farming in self-owned ponds, is also examined. Finally, institutional and policy options are suggested to sustain fish culture in MWs under different property regimes. The goal of Chapter 9 is first to critically examine the current Rules of Wetland Conservation and Management (MOEFCC, GoI) in the context of the Ramsar Convention's priorities, and thereafter briefly discuss the performance of MWs under alternative governance structures. Finally, a policy, governance, and research agenda is proposed for the restoration of wetlands and getting the national and international importance of the Chhattisgarh wetlands recognized. The proposed agenda may also be applicable to other states and neighbouring countries for sustaining their inland wetland ecosystems.

Notes

1 For extent, status, benefits, and policy issues of different types of wetlands, see Gopal et al. (1992, 2008); Gopal (1995, 2003, 2002, 2013, 2018); Finlayson et al. (2005); MEA (2005); Chatterjee et al. (2008); Marothia and Gopal (2014a, 2014b); Bassi et al. (2014); Gopal and Marothia (2016, 2017); WISA (2018); Ritesh Kumar et al. (2020); Kingsford et al. (2021a, 2021b). See also McClain (2002); Huryn et al. (2003); Galbraith and Burns (2007); Indrawan et al. (2014), Sakai and Umetsu (2014); Herbert et al. (2015), Gevers et al. (2018); Bregoli et al. (2019) for ecohydrological, socio-ecological systems in transition, urban wetlands issues in the larger context. See Wood and van Halsema (2008); Marothia (2021) for agriculture-wetland-water interactions, research methodology, and policy attributes. For indigenous water-harvesting structure technology, see Mishra (1995); Agrawal and Narain (1997); Pant and Verma (2010).

2 The Ramsar Convention's definition of wetlands reflects a hydrological perspective with water as the key factor. Other scholars have stressed the link between hydrology and biology, and proposed "ecohydrological" definitions, while still others have suggested geomorphological definitions [e.g. Dugan (1990)] or agricultural (crop) definitions [e.g. FAO (1998), see also Frenken and Mahapatra (2002)]. Wetlands are "lands transitional between terrestrial and aquatic systems where the water table is usually at or near the surface or the land and is covered by shallow water" and "must have one or more of the following three attributes: (1) at least periodically, the land supports predominantly hydrophytes; (2) the substrate is predominantly undrained hydric soil; and (3) the substrate is non-soil and is saturated with water or covered by shallow water at some time during the growing season of each year" [(Cowardin et al. (1979, p. 135)]. Pittock et al. (2015, p. 579) defined

wetlands as, "Places where water is the primary factor controlling plant and animal life and the wider environment, where the water table is at or near the land surface, or where water covers the land". Wood and van Halsema (2008, p. 8) clarified further that "a wide range of wetland definitions are in circulation that are informed by the perspective taken and the primary purpose for which they were defined (e.g. nature/ecosystem conservation, small-scale agricultural development, or hydrological classification). At times, these different classifications may lead to confusion and misunderstanding as there is no clear overview of how they relate to one another. However, these different approaches to wetlands contain valuable insights and experiences of specific wetland characteristics, uses and interactions". See also Maltby (2009) for a discussion on the problem of definitions.

3 Oakerson (1992) used efficiency and equity evaluative criteria for assessing the outcome attribute of his model, Folke and Berkes (1995) added sustainability to Oakerson's outcome attribute, Adger et al. (2003) added environmental effectiveness and the political legitimacy of authority.

4 According to Townsend and Pooley (1995), an external governance structure has essentially three alternatives of management systems namely, *rights-based management* (the government grants usufruct rights to individual resource users under well-specified constraint conditions, assumes the role of monopoly over the resource base, and retains all responsibility/authority for conservation decision), *co-management* (the government and the local communities share ongoing responsibility for decision making over all or most of the resource management decisions), and *contracted management* (to transfer large part of the decision-making process to local bodies).

5 Townsend and Pooley (1995) advocated that the four alternative internal institutional arrangements, which have been closely associated with the concept of distributed governance, include: *self-organizing institutions* (institutional and organizational decisions remain with local communities and the government may use the institutional building capacity to support and gain strength from self-organization), *communal management* (to reduce the existing authority of state and vest more localized interest), *cooperative management* (membership is limited with well-defined working rules for collective governance), and *corporate* (under the corporate governance the owners and shareholders of the corporation would operate under governance rules typical of private corporations).

6 Roots of institutional analysis development frameworks can be traced in research efforts on institutions and property relations over natural and common pool resources. For seminal work in this area, see Commons (1924, 1931); Ciriacy-Wantrup (1963); Rawls (1971); Calabresi and Melamed (1972); Ciriacy-Wantrup and Bishop (1975); Ostrom (1976); Bromley (1976, 1978, 1989a, 1991, 1992); Elliott (1978); Bromley et al. (1980); Marothia (1981, 1995a); Bromley and Cernea (1989); Gibbs and Bromley (1989); Gadgil and Iyer (1989); Dasgupta and Mäler (1994); Williamson (1994); Ghezae (2009). For the diverse institutional analysis frameworks being used for analyzing CPRs, see Ostrom and Ostrom (1977); Kiser and Ostrom (1982, 1987); Marothia and Phillips (1985); Oakerson (1986, 1992); Ostrom (1986, 1988, 1990a, 1992a, 1992b, 1995, 1999a, 2007); North (1990); Arnold and Stewart (1991); Berkes and Folke (1992, 1994, 1998); Schlager and Ostrom (1992); Söderbaum (1992); Tang (1992); Ostrom and Gardner (1993); Ostrom et al. (1994); Folke and Berkes (1995); Munasinghe (1995); Townsend and Pooley (1995); Edwards and Steins (1996, 1999); McGinnis (1999); Agrawal and Ostrom (2001); Andersson and Ostrom (2006, 2008). For the role of institutional arrangements in managing diverse aquatic ecosystems under different socio-ecological systems, see Etiegni et al. (2016); Irvine et al. (2016); Radhakrishnan et al. (2017); Irvine (2018); Gevers et al. (2018); Rahman et al. (2018).

7 See studies reported in Marothia (1986, 1993b, 2005b, 2005c, 2010a, 2010b, 2012a, 2015, 2017, 2019); Arnold and Stewart (1991); Bromley (1992); Ostrom (1992); Tang (1992); Folke and Berkes (1995); Singh (1997); Singh and Ballabh (1996).
8 For conceptual frameworks of total economic values and ecosystem services and their application in valuing wetlands, see Costanza et al. (1989); Pearce and Turner (1990); Aylward and Barbier (1992); Munasinghe (1992); Berkes and Folke (1994); Smith (1996); Euisoon et al. (1997); Marothia (2001a, 2004d, 2021); De Groot et al. (2002); Adger et al. (2003); Samaraweera and Marothia (2008); Bandyopadhyay et al. (2012); Kumar and James (2012); Khai and Yabe (2014); Gopal (2015); Gopal and Marothia (2015); Gopal et al. (2016); Kumar et al. (2017).
9 For governance issues and water-sharing mechanisms among different users of MWs, see Marothia (1986, 1992a, 1992b, 1993b, 1995a, 1995b, 1997c, 2002a, 2004b, 2004c, 2004d, 2004e, 2006b, 2007, 2009b, 2010a, 2010b, 2012a, 2014, 2015, 2017, 2019).

2 A Synthesis of Institutional Approaches for Analysing Multiuse Wetlands Governance

This chapter focuses on the socio-economic and environmental aspects of MWs within the context of interdependent situations, emphasizing human interactions with respect to the conservation and management of MWs. It also explains, in brief, how the concepts of environmental governance, MW management, and interdisciplinary approaches are inseparable. The chapter attempts to develop a synthesis of institutional analysis frameworks to analyse rural, peri-urban, and urban common MWs with typical characteristics of multiple ownership, uses, diverse use and non-use values and ecosystem services under different resource regimes or combination of property rights regimes. The synthesized framework developed here provides an improved methodology to analyse multiple-use, CPRs, including MWs, in the context of India, and other South Asian, and developing countries.

2.1 Institutions, Human Interactions and Resource Governance

Studies of economic problems have their roots in the analysis of human behaviour and the relationship between institutions that shape human behaviour and society's value system. While approaches to dealing with human behaviour and controlling resources may differ from hedonic, classical, neoclassical, Marxist, and institutional economics to neo-institutional economics, the focus has been on channelling the individual's behaviour into socially beneficial expressions while preserving their freedom, security, and rights, as well as affording them justice.[1] The neoclassical economics approach to analyse the society and human interdependence in relation to natural and environmental resources utilization has been narrow in the selection of variables and simplified. The institutionalists, working from a much more complex concept of human motivation and social organization, saw conflict as the predominant driving force in economics, and, as a result, raised fundamental questions about neoclassical ideas (Leibhafsky, 1976; Randall, 1978a, 1978b). According to institutionalists, neoclassical economists have been narrow in their selection of variables for consideration, abstractive and simplistic in their approach to society, and inadequate in their assessment of

DOI: 10.4324/9781003456742-2

human interdependence and relationship with the environment. This narrow emphasis of neoclassicals has been criticized by many institutionalists.[2]

The institutionalists' approach can effectively define relationships among individuals with respect to resource utilization and other economic pursuits (Ely, 1914, p. 96). The term "institutions" as defined by Commons (1934, p. 21), "is collective action in control, liberation and expansion of individual action". One institution, namely property, is defined by Parsons (1942, page 246) as follows: "Property is a set of social relationships which ties the future to the present through expectations of stabilized behaviour regarding other persons and things". Ciriacy-Wantrup (1969, p. 1319) conceptualizes an institution as "a social decision system that provides decision rules for adjusting and accommodating, over time, conflicting demands from different interest groups in a society". Schmid (1972, p. 893, 1978, p. 6) defines institutions including property as "sets of ordered relationships among people that define their rights, exposure to the rights of others, privileges and responsibilities". The terms "institutions", "rights", and "rules of the game" are used in various definitions more or less interchangeably, though there are slight differences in the degree of aggregation and abstractness implied. The definitions of Parson, Ciriacy-Wantrup and Schmid are similar to Commons' collective action theme (see Chapter 3 of Marothia, 1981 for foundation of logical inferences in resource policy, 1989b for further details).

As societal needs change, so do our rights. The recognition that rights are subject to limits and changes (Commons, 1931, p. 648), that is, the nature and extent of rights change and individual freedom is subject to sanction, is fundamental to a clear understanding of any natural resource-policy analysis. Rights are the instruments by which any society controls and orders human interdependence and resolves the question of who gets what (Parson, 1941). In the absence of rules that govern social organization, human relations come down to the sheer play of power against power – a "war of all against all", in which individual life is "solitary, poor, nasty, brutish and short" (Long, 1953, p. 321). "Alternative rights are of interest because of their effect on economic performance and outcomes" (Schmid, 1978, p. 5). Rights define potential opportunities and the opportunities of one person are shaped by the opportunities of others (Samuels, 1972). "Thus, the opportunity sets of an individual are composed of alternative lines of action that are open because of the relative structure of rights as well as the relative capacity of the person to make use of those rights. The value of a right is also limited by available knowledge and technology. The opportunity sets of individuals interact and condition the outcome of human transactions. The components of the institutional structure are then one of the points of leverage in changing outcomes" (Schmid, 1978, p. 7).

Shifts in the structure of rights emphasize human interdependence and interactions. This interdependence can be described in terms of institutionalized cost (or transaction cost in Commons' terminology), external effects and power (Commons, 1934). Thus, property is not simply a derivation of a physical fact. It also reflects a social choice of the kind of efforts that are counted in creating an image in an individual's mind about a person's rights. A community usually defines property rights either by a common problem or by the mutual advantage of joint action (Schmid, 1978, p. 9). A basic shift in the structure of rights is evident in many problems of natural resources in general and CPRs in particular.[3]

Institutions, rules, and practice are essential components of governing environmental and natural resources, as these shape the patterns of human (users and even non-users) interaction with diverse natural and environmental resources, including MWs. Environmental issues became an important part of the international agenda for global environmental governance (United Nations Environmental Program – UNEP). The International Union for Conservation of Nature's (IUCN) concept of environmental governance includes "the rules, both formal and informal, that govern human behaviour in decision-making processes as well as the decisions themselves. Appropriate legal frameworks on the global, regional, national, and local level are a prerequisite for good environmental governance. Environmental Governance is the means by which society determines and acts on goals and priorities related to the management of natural resources" (www.iucn.org). It is important to recognize that "good environmental governance imply the principles of good governance that include: (a) Participation; (b) Rule of Law; (c) Transparency; (d) Responsiveness; (e) consensus oriented; (f) equity and inclusiveness; (g) effectiveness and efficiency; (h) accountability" (https://globalpact.informea.org). (See also Benson and Jordan, 2017 for environmental governance.)

A number of researchers in recent years have dealt with the issue of environmental decisions made by individual resource users or groups, civil society, and state and federal governments. Most of these investigations included criteria of economic efficiency, equity, environmental effectiveness, and the political legitimacy of authority. These four criteria are now widely recognized as constituting the economic, social, and environmental dimensions of sustainable development, which in turn has become an integral component of environmental governance (Adger, 2002; Adger et al., 2003).[4] Further, insights are required from different disciplines to improve these critical issues of efficiency, equity, effectiveness, and political legitimacy for analysing environmental decision arrangements. To enhance the problem-solving capacity of decision makers concerned with environmental governance and sustainable development, researchers are required to pay simultaneous attention to these four criteria (Adger et al., 2003). This is particularly important in the case of multiple-use, CPRs, which involve diverse stakeholders and multiple organizations, and which provide multiple ecosystem services and monetary and non-monetary values in the complex social-ecological context (Marothia, 1993b, 2001a, 2015, 2019; Marothia and Phillips, 1985, 1990; also see case studies reported in Chapters 3 to 8).

In addition to these four criteria in environmental decision-making, institutions (formal and informal), scale (local, regional, or global), and cultural and historical context are equally important (see Adger et al., 2003). An interdisciplinary approach ecompasses "the formulation of uniform, discipline – like terminology or common methodology" (McNeill, 1999, p. 313). Thus, "a synthesis approach that amalgamates aspects and concepts from different disciplines enable interdisciplinary research both at the individual and at the team level" (Adger et al., 2003, p. 1097).[5] The relevance of all these criteria – economic efficiency, equity, environmental effectiveness, and political legitimacy or authority – is discussed in the outcome and evaluating box-i of Figure 2.1. To this end, we discuss diverse institutional analyses and development frameworks emerging from an interdisciplinary perspective.

2.2 Emergence of Institutional Analysis and Development Framework

Researchers have developed a broad institutional framework using research in economics and economic history, political science, law, business, game theory, transaction-cost economics, and anthropology, that identifies key attributes of the typical state of affairs facing resource users, communities and decision-makers in different circumstances (Kiser and Ostrom, 1982; Ostrom, 1986, 1988, 1990a, 1990b, 1992a, 1992b, 1995, 1999a, 1999b; Oakerson, 1986, 1992; Peters, 1986; Bromley, 1989b, 1992; Schlager and Ostrom, 1992; Tang, 1992; Ostrom and Gardner, 1993; Ostrom et al., 1994; Schlager et al., 1994; Ostrom et al., 1999, 2002). Various physical, technological, community, and institutional attributes may affect outcomes directly or through collective action situations (or transaction situations) and patterns of interactions (Tang, 1992). The interrelationships between resource attributes, the attributes of a community, decision-making arrangements, and patterns of interactions outline various action situations and can ensure efficient, equitable and sustainable outcomes in managing CPRs (Oakerson, 1986, 1992).

The seminal work of Kiser and Ostrom (1982) was an attempt to integrate institutional frameworks developed by economists, political scientists, anthropologists, lawyers, and others by studying the effects of institutions on the behaviour of different categories of stakeholders in relation to resource use. Over the years, they have refined several aspects of their basic framework and applied it to the analysis of several problems concerning CPR management. They have applied the framework to a series of empirical studies and examined its relationship to theories in public choice and new institutional economics. Based on feedback from these studies, they have further added several elements at each decision-making level, namely, constitutional, collective, and operational choice. The application of this revised framework can be seen in empirical case studies carried out by Kiser and Ostrom (1982, 1987); Ostrom (1992); Tang (1992) and Marothia (2010a, 2010b).

Researchers have used the Oakerson (1986, 1992) framework and its derivative forms for analysing single- and multiple-use CPRs in various parts of the world and in the Indian context.[6] Few researchers had made attempts to extend the analytical framework of Oakerson (1986, 1992), Ostrom (1990a) and Tang (1992) for analysing multiple-use CPRs under common property resource regimes in developed countries (Edwards and Steins, 1996; Edwards, 1996). An effort was also made for the institutional mapping of meta-theories and frameworks of the Chicago and Institutional Analysis and Development (IAD) Schools (Aligica, 2006).

2.3 Synthesized Institutional Framework: Analysing MWs

I have used different institutional approaches developed by Oakerson (1992), Ostrom and her co-authors, Tang (1992), Townsend and Pooley (1995), Bromley (1978, 1989a, 1989b, 1991, 1992), Bromley et al. (1980), Gibbs and Bromley (1989) and Rawls (1971) in my earlier case studies (either a single or a combination of two or more institutional approaches) conducted during the last 30 to 35 years to assess the performance of various CPR-based development programmes implemented under different governance structures or property regimes or a combination of property rights regimes.[7] However, over the years, I have realized that in the Indian socio-economic-cultural-ecological context, a synthesis of different institutional approaches needs to be devised to analyse rural, peri-urban, and urban common MWs that have multiple ownership models, different uses, different ecosystem services, and diverse use and non-use values under different resource regimes or a combination of property rights regimes (distributed governance or/and polycentric governance).

To develop a creative synthesis of institutional frameworks, I have incorporated the appropriate elements of different institutional approaches to the basic framework of Oakerson (1992), subsequently developed by Ostrom (1990a, 1990b) and Tang (1992), to analyse complex common pool MW ecosystems. Other components or sets of variables that emerged from Bromley's property rights and public choice analysis work (Bromley, 1989a, 1989b, 1991) and Townsend and Pooley (1995) distributed governance paradigm are also included. It is important to recognize that MWs in the Indian context also provide direct and indirect values (benefits) and ecosystem services (see Chapter 3 of this book). Therefore, I have added attributes related to changes in total economic values (use and non-use values) and the different categories of ecosystem services (provisioning, regulating, cultural and supporting). Researchers have adequately documented the contribution of wetlands to total economic values and ecosystem services. Further, any change in the physical and technical parameters of a CPR, including MWs, can alter the use and non-use values and ecosystem services depending on the behaviour of users and non-users, the effectiveness of institutional arrangements, and the authority system (Marothia, 2002b, 2004d; Katiha and Marothia, 2006).

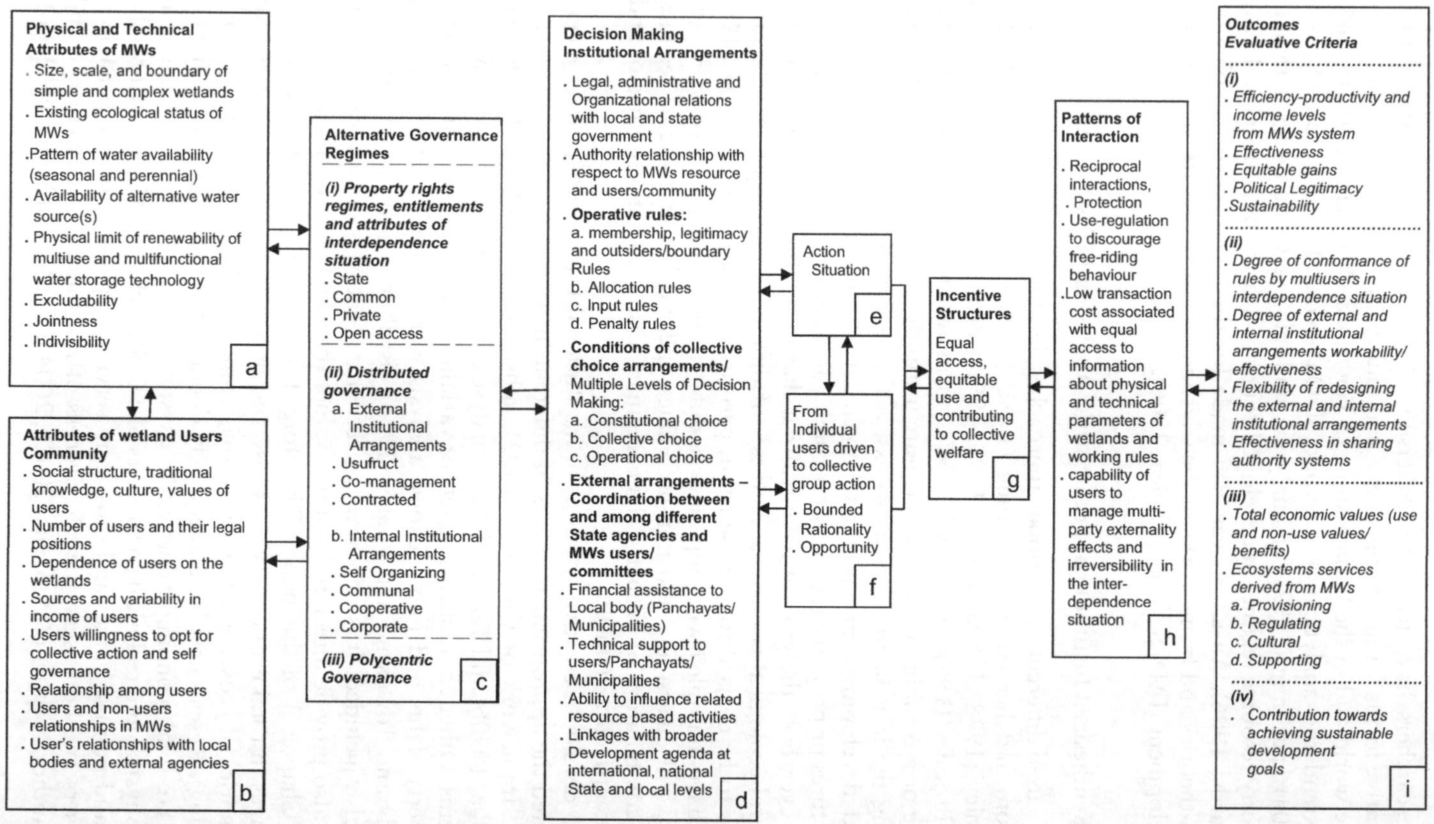

Figure 2.1 Analysing MWs Governance – A Synthesis of Institutional Approaches

Source: Devised by the author

The synthesis of institutional frameworks developed in this book, that distinguishes nine main attributes and sub-attributes (represented through the nine boxes in Figure 2.1), is used for analysing the case studies reported in Chapters 3 to 8. An attempt is being made here to connect the descriptions in each box to the case studies reported in the book. A brief mention of the key attributes of Oakerson's (1992) framework needs to be made here, as this is the base model on which the synthesis is being built. The four key attributes of Oakerson's conceptual framework are: physical and technological attributes (box-a), decision-making arrangements (box-d), patterns of interactions (box-h) and outcome (box-i). Each set of attributes relates to the others. For example, physical and technological attributes and decision-making arrangements jointly affect patterns of interactions. "The physical and technical characteristics of the commons can affect the outcome directly or through patterns of interactions. In long-run analysis institutional changes are exogenous and their effects could be iteratively assessed on interactions and outcomes" (Oakerson, 1992, pp. 41–62). The Oakerson model also has dynamic application if applied iteratively. However, Kiser and Ostrom (1982) have argued that institutional change at one level is an outcome of patterns of interaction at another level. Oakerson (1992, see note 4, page 58) has also agreed with Kiser and Ostrom (1982) that a multilevel framework is the better way to represent dynamic relationships than a recursive framework.

Several researchers, in addition to the four attributes of Oakerson's model, added social characteristics of resource users and their communities[8] along with physical attributes and institutional arrangements to their institutional analysis development framework (for a detailed interpretation of Kiser and Ostrom's (1982) attributes of institutional analysis development framework see Tang, 1992; Folke and Berkes, 1995; Edwards and Steins, 1996). In the synthesized framework, I have included characteristics of non-users of MWS, as without extracting anything directly from MWs, they derive many intangible social, cultural, religious, recreational, and educational benefits (see Chapters 3, 4 and 5).

Before discussing the attributes and sub-attributes of decision-making arrangements (box-d), an essential component of the synthesis pertaining to the resource regimes or governance regimes of multiuse CPRs, multiusers and multiple organizations engaged in management needs to be explained (see three different approaches to multiple CPRs governance and their sub-attributes in box-c). Section (i) of box-c applies Bromley's (1989a, 1989b, 1991) conceptual framework to the analysis of over the ownership of natural resources. Bromley's approach is incorporated in the synthesis of institutional frameworks. His framework deals with natural and environmental resources and problems experienced as a result of property rights, entitlements and interdependency issues (see Marothia, 1995a, 1995b, 1997c, 2004d). Based on my own and other scholars' work on property rights concerning natural resources in general, and multiple-use CPRs in particular, I am convinced that an institutional analysis framework will remain incomplete without understanding ownership of natural resources. Oakerson (1992) highlighted the role of "resource regimes" or property rights regimes (see Bromley, 1989a) in CPR management or mismanagement. CPRs can be managed or mismanaged under any of the four resource regimes: state property, private property, common property, and open access (Ciriacy-Wantrup and Bishop, 1975; Bromley, 1989a; Gibbs and Bromley, 1989; Bromley and Cernea, 1989; Ostrom, 1990a, 1990b; Feeny, 1994; Marothia, 1993b, 1996b, 2002b, 2002c, 2004e, 2006a). CPRS are also subject to degradation under distributed governance (Marothia, 2001a, 2002a, 2002b, 2005a, 2005b, 2005c, 2012a, see Chapters 6 and 8). Property rights regimes are part of the institutional arrangements through which resource users convert natural resources and environmental services into human-made capital or inputs of production (Folke and Berkes, 1995). The nature of institutional arrangements defines the extent of property regime over land, water, MWs and related resources. In the same vein, Ostrom (1992a, 1992b) argued that crafting rules, enforcement, dispute-management and decision-making in general are all matters pertaining to property rights, and they are also part of institutional inquiries. Folke and Berkes (1995, page 130), while working on mechanisms that link property rights to ecological systems using Oakerson and Ostrom's framework, highlighted that traditional and local knowledge of environmental resources are "embedded" in local institutions and property rights are "embedded" in institutions (see Ostrom, 1990a, 1990b, 1992a, 1992b).

The basic requirement for any property rights regime (or a combination of property rights regimes) is an authority system that can guarantee the interests of the property rights holders. When the authority system breaks down, a particular resource regime degenerates. Under such circumstances, new institutional arrangements need to be made to protect the interests of those holding the rights under a particular regime (Bromley and Cernea, 1989; Gibbs and Bromley, 1989; Marothia, 1989b, 1993b, 1997a, 1997b, 2010a, 2010b). The findings of case studies reported in this book confirm the importance of an authority system in protecting MWs and the rights of users.

The critical issues are the nature and extent of the rights of individuals or groups of individual resource users as compared to the rights of society over natural and environmental resources and collective actions for their management (Bromley, 1976). Environmental and resource-management problems are defined by property, property rights, property regimes, multi-party externalities, and institutional arrangements governing the interdependence between users of a multipurpose and multi-user resource (Bromley, 1989a, 1989b; also see Chapters 4 and 5 of this book which deal with such issues in cases of wetland complexes and lake ecosystems). Due to unrealistic assumptions about the two-party externalities model, its direct utility for policy regarding pollution problems in MWs is limited. Bromley (1978, 1993) filled this theoretical gap in the literature on environmental economics by using the taxonomy of entitlements by property rules, liability rules, and inalienability rules developed by Calabresi and Melamed (1972). I have used Calabresi and Melamed's (1972) and Bromley's (1978) Property Rights Framework to understand human interference in the interdependent and complex situation of Lake Vivekananda (see Marothia, 1995a, 1995b, 1997c, 2004d, see also Chapter 5 for a discussion about the application of the taxonomy of entitlements by property rules, liability rules, and inalienability rules, and designing institutional arrangements based on Bromley's theoretical institutional framework (Bromley, 1978, 1991, Chapter 3, pp. 41–58). These concepts are related to section (i) of box-c of Figure 2.1.)

Section (ii) of box-c of Figure 2.1 deals with the distributed or shared management of multiple-use CPRs, including MWs. The four categories of resource governance or property rights regimes e.g. state, private, common and open-access regimes (Bromley, 1989a, 1989b, 1991; Ostrom, 1990a, 1990b; Gibbs and Bromley, 1989; Bromley and Cernea, 1989) have been extended by Townsend and Pooley (1995) to recognize that natural resources governance can be shared among states, communities and private-interest groups in various ways at different decision-making levels (see also Curtis, 1991). Distributed governance involves the external institutional arrangements[9] among government and local communities or resource users as well as internal institutional arrangements[10] within local community institutions or resource users. Government, local communities and private parties utilizing CPRs bring different interests, capabilities and understanding to the resource-management and decision-making process (Townsend and Pooley, 1995). I have used Distributed Governance in analysing tank irrigation, *dabries*, wetland complex and culture fisheries in MWs and results are reported in Chapter 4, 6, 7 and 8. Alternative institutional perspectives shape the decision-making process among government and local resource users' communities and within members of local community for managing a resource by converting unorganized structures into organized ones (Marothia and Phillips, 1985).

Section (iii) of box-c of Figure 2.1 deals with the Polycentric Governance Regime developed by Ostrom et al. (1994) and Andersson and Ostrom (2008). "It is a system that seeks to unleash ingenuity and stimulates the creativity of political entrepreneurs. It is a system that is structured so that actors within the system are given opportunities for institutional innovation and adaptation through experimentation and learning" (Ostrom et al., 1994, p. 2). "A key aspect of all proposals for increased polycentricity (as opposed to *just* centralization or *just* decentralization) is the effort to enable institutions of multiple scales to more effectively blend local, indigenous knowledge with scientific knowledge" (quoted from Andersson and Ostrom, 2008; see also Ostrom and Ostrom, 1977, p. 3; Berkes and Folke, 1998; McGinnis, 1999). Along with other attributes of the synthesis framework, the polycentric approach was used to understand how resource users create opportunities for crafting institutional arrangements for self-organized and self-governed small and simple MWs (see Chapter 7) that possess a number of specific institutional attributes capable of providing and producing essential collective goods and services to the resource users in a polycentric regime.

Oakerson's (1992) model of the three sub-attributes of decision-making institutional arrangements are listed in box-d of Figure 2.1. Under the decision-making arrangements (box-d of Figure 2.1), among other sub-attributes, the conditions of collective action have three levels. If a government is willing to devolve property rights over CPRs that comply at the operational, choice, and constitutional levels, then meaningful decentralized governance of MWs can take place. At the operational level, users interact with each other to use or withdraw resource units from a CPR (including MWs). At the collective-choice level, rules are established and decisions are taken by existing and potential resource users to define the operational, institutional, and technical arrangements for managing CPRs. In constitutional situations, decisions are taken to determine who has the authority to create rules for collective-choice situations. The outcomes of decentralization at different levels are largely affected by the types of property rights structures assigned to the resource users (Agrawal and Ostrom, 2001). It is not necessary that the three levels correspond to the three actual levels of authority or the institutional arrangements in the political or legislative system. It is entirely possible that the same political or administrative body designs and implements rules for the decentralized management of CPRs at all three levels (Ostrom, 1992a, 1999b; Andersson and Ostrom, 2006).

For example, in Chhattisgarh, the DWR empowers the Water Users Associations (WUAs) under Participatory Irrigation Management (PIM), and at the same time also designs institutional arrangements at the operational, collective, and constitutional levels to manage the canal irrigation system within a distributed governance framework. The physical and technical attributes of canal irrigation put water resources under the category of CPRs which are used by numerous farmers under the private property regime. The efficient and equitable distribution of canal water among different categories of farmers depends upon technical and institutional arrangements (Chambers, 1988; Vaidyanathan, 1999; Marothia, 2002b, 2005a, 2010a, 2016; see Chapter 6). Multiple-level decision-making arrangements have also been functioning in the case of the cooperative management of non-timber forest products in the Central Indian States (see Marothia, 1996a, 1996b, 1997d, 2001b).

Action situation (Box-e) is the crucial point of the IAD framework (Kiser and Ostrom, 1982, see also Edwards and Steins, 1996; Aligica, 2006). Tang (1992, page 19) added action situations, mode of individuals in terms of bounded rationality and opportunism, and incentives to the Oakerson model. Tang (1992) argued that transaction-cost economics adopts transactions, which are similar to action situations as the essential unit of analysis. However, "Action situation is a broader concept than transaction because the former deals with all kinds of social situations and the later concerns specific exchange between or among people" (Tang, 1992, page 139).

Action situation is affected by external variables (the biophysical and technical condition of a resource, the attributes of resource users, the community rules in use (Ostrom, 2005), or the decision-making institutional arrangements – represented in box-a, b, and box-d of Figure 2.1 under alternative governance structures (box-c). In action situations, individual resource users or groups of users adopt actions or strategies. Different outcomes may result from interactions between users, depending on the number of users, the choices available to them (box-f), their ability to process information, and the incentives in front of them (box-g). (See Tang, 1992 for a detailed description of the attributes listed in box – e, f and g.) Thus, individual user rationality is bounded (see Tang, 1992). This is particularly true in the case of different types of MWs. For example, in the case of tank irrigation, to craft mutually beneficial institutional arrangements, all farmers located at the head, middle, and tail ends of a water-carrying canal need to have all the information pertaining to the availability of water at normal and critical crop-growth stages (see Chapter 6). Further, equal access of information related to the physical and technical parameters of MWs helps users, groups of users, and authorities to make decisions regarding maintenance and water allocation under rotational patterns (see Chapters 6 and 7, for example). Opportunism is another important attribute that affects the collective action process in simple and complex MWs. Since MWs' resource users and community are characterized by bounded rationality and opportunism (refer box-e,f), they respond according to the incentives and constraints (see box-g) inbuilt in the situations they are confronted with (see Tang, 1992; Ostrom, 2005 for further details on action situations and associated variables). Action arenas are also linked to the operational level, the collective-choice level, and the constitutional or constitutive decisions level. Three levels of rules cumulatively affect the actions, interaction patterns, and outcomes in any setting (Ostrom and Kiser, 1982; Ostrom, 1986; Ostrom and Crawford, 1995; see also Aligica, 2006; Marothia, 2005a, 2010a for further elaboration).

The sub-attributes of patterns of interaction of the Oakerson model include: reciprocity – individual co-users contribute to each other's welfare, and free-riding behaviour – degrades reciprocity, breeds destructive competition and conflicts, and ultimately leads to the over-exploitation of a resource (see box-h of Figure 2.1). Edwards and Steins (1996) connected "context bound" factors to the patterns of interaction attributes of Oakerson (1992), Tang (1992), and Ostrom (2005). In the case of users, groups of users, and communities dependent on MWs, "context bound" factors like social, cultural, economic, political, historical, and institutional environment are part of box-b of Figure 2.1. These factors influence patterns of interaction among users irrespective of resource regimes (see Edwards and Steins, 1996, 1999 for the important role of these factors in patterns of interaction). Other attributes that influence the behaviour of users of MWs also include low transaction costs associated with equal access of information about the physical and technical parameters of wetlands and working rules, the capability of users to manage multi-party externality effects, and irreversibility in an interdependence situation (see Bromley, 1991, pages 52–54, for an interpretation of policy-relevant variables of interdependence to deal with various types of interference to manage natural and environmental resources). For example, given the resource attributes and decision-making arrangements of governing water use from multiuse village ponds/tanks, it is important to identify the patterns of interactions that characterize the behaviour of fishermen, farmers, fisheries, and irrigation Panchayats collectively concerned with community village ponds/tanks management. In the case of fish and irrigation use, fishermen and farmers contribute to each other's welfare through reciprocity which eventually leads to collective action. Due to well-defined institutional rules and their effective implementation by the FCSs, and irrigation Panchayats, free riding behaviour could not breed among users (see Chapter 8). But in another example of multiuse irrigation tanks (Chapter 6 and Dulhara irrigation tank study reported in Chapter 4), due to the ineffective implementation of decision-making arrangements by the defunct WUAs, free-rider behaviour spreads, and in turn yields individualistic patterns of interaction. Further, conflicts in water allocation between farmers and fisheries surfaced frequently. The collapse of decision-making arrangements, individual users extracting the maximum possible quantity of water, and the absence of an authority system resulted in the degradation of irrigation tanks (see Chapter 6). It is interesting to note from a case study of self-organized and self-governed multiuse *dabri* irrigation systems (Chapter 7), that internal governance is fairly effective even without formal institutional arrangements. This was possible perhaps due to *dabri* users being small in number, the physical size of *dabries* being manageable, and users of *dabri* were capable of protecting the boundaries of multiuse *dabries* from non-members. The findings of this case study are fairly similar to those of Swallow and Bromley's (1994) work pertaining to co-management or no management, in which they examined the prospects for internal governance of common property regimes through dynamic contracts.

Box-i of the synthesized framework (Figure 2.1) deals with outcome and evaluative criteria. Oakerson's (1992) outcomes attributes have two sub-attributes to with which analyse CPRs under common property regimes: (i) economic outcomes – evaluated using concepts of efficiency (overall use rates of a resource), and (ii) distributive outcomes – evaluated in terms of equity (for example, equity criteria can be assessed in terms of fair share to co-users on their contribution to a collective choice, effectiveness of a management system to exclude non-users enforcement of rules). A few researchers have incorporated sustainability criteria to the outcome attribute of Oakerson's framework, just after the inception of his framework (see Marothia, 1993b; Folke and Berkes, 1995). In addition to efficiency and equity criteria, I have added environmental effectiveness and political legitimacy criteria (section-i of box-i, Figure 2.1) to relate decision-making arrangements to the concept of environmental governance (see Adger et al., 2003 for the importance of these four criteria in relation to environmental governance within an interdisciplinary framework to deal with broad categories of natural and environmental resources, including multiple-use commons). Effectiveness criteria deal with the level of understanding of policy makers to address the question of how to integrate multiple wetlands values in decision making (see also Kumar et al., 2017 for the relevance of multiple wetland values in decision making). These four criteria are particularly important to assess a MWs' ecosystem. For example, under the Smart City Mission, beautification work that was carried out on Lake Vivekananda focused on urban recreation, jointly by Raipur Smart City Limited (RSCL) and Raipur Municipal Corporation (RMC), without considering the irreversible loss of wetland biodiversity and the loss of provisioning, regulating, cultural, and supporting ecosystems services (represented by direct-use values, indirect use values, option, bequest, and existence values). Furthermore, the political legitimacy (RSCL and RMC) factor completely ignored the participation of different segments of society while reaching the decision to convert the historical common-pool lake as a recreational and tourist spot (see Chapter 5). A similar story is seen in Chapter 9, dealing with peri-urban multiuse wetlands users and non-users, effectiveness.

In the specific context of self-governance irrigation system, Tang (1992, page 36) added adequacy of water supply, degree of rule conformance, maintenance, and distribution of benefits and costs, and other criteria listed in section-ii of box-i. Both Oakerson and Tang have assessed the outcome and impact of three attributes i.e. physical, technological, and community and institutional arrangements in terms of efficiency and equity. It is important to recognize that MWs in the Indian context also provide direct and indirect values and ecosystem services. Therefore, I have added changes in total economic values (use and non-use values) and the different categories of ecosystem services (provisioning, regulating, cultural, supporting) due to changes in other attributes in the outcome and impact box-i (section iii) of Figure 2.1. Researchers have adequately documented the contribution of wetlands in general and MWs in particular to total economic value and ecosystem services.[11] Further, any change in the physical and technical parameters of a CPR, including MWs, will alter the use and non-use values and ecosystem services depending on the behaviour of users and non-users, the effectiveness of institutional arrangements, and the authority system.[12] The contribution of different MWs reported in this book to SDGs is also assessed. The nine boxes in Figure 2.1 may be considered components of a MWs ecological and social system and are in dynamic interrelationship with one another. Feedback loops among the nine components are indicated with arrows (see Folke and Berkes, 1995 for the application of Oakerson and Ostrom institutional frameworks).

The synthesized institutional framework depicted in Figure 2.1 is used to analyse simple and complex MWs and the results of this application are reported in this book (the synthesized framework is an improved version of Marothia (2021) and his earlier work). The focus in this synthesis is on the manner in which a particular institution or set of institutions (working rules) affects individual freedom, securities, rights and individual views of efficiency, equality, and output, which dominate when individuals' interests are in conflict in an interdependent situation. The synthesized institutional framework suggests possibilities for analysing major, complex, natural, and environmental resources, and multiple-use problems where property rights, efficiency, equity, resource use, and human interaction are interwoven in an interdependent world.

Notes

1 Refer to Marothia (1981, 1989b); Marothia and Phillips (1985) for a critical review of different schools of thoughts on resource use and human interactions.
2 For criticism, refer to: Commons (1931, 1934, 1950); Parson (1941, 1942, 1974); Mukherjee (1942); Ayers (1944); Boulding (1957, 1975); Clark (1957); Myrdal (1960, 1978); Robinson (1963, 1971); Dorfman (1968); Gruchy (1969, 1972); Shaffer (1969); Schmid (1972, 1978); Ward (1972); Galbraith (1973); Kelso (1977); Elliott (1978).
3 For additional literature on institutions, arrangements, rules, property rights institutions, and cultural and social capital, see Dewey (1922); Tugwell (1922); Commons (1931, 1934, 1950); Downs (1957); Black (1958); Coase (1960); Buchanan and Tullock (1962); Johnson (1962); Ricker (1962); Ciriacy-Wantrup (1963, 1969, 1971); Tullock (1964); Demsetz (1967); Dolbear (1967); Krutilla (1967); Kapp (1968); Schultz (1968); Davis and North (1970); Gonce (1971); Mishan (1971); Olson (1971); Randall (1971, 1974, 1975); Sen (1971); Barkley and Seckler (1972); Furubotn and Pejovich (1972); Timmons (1972); Alchian and Demsetz (1973); Goldberg (1974); Kanel (1974); Solo (1974); Ciriacy-Wantrup and Bishop (1975); Ostrom (1975, 1976); Philips et al. (1975); Coelho (1976); Phillips (1976); Schmid (1976); Bishop (1978); Binswanger and Ruttan and others (1978); Castle (1978); Wilber and Harrison (1978); Myrdal (1978); Veeman (1978); Ruge and Bromley (1979); Dasgupta (1982, 1993); Walzer (1983); Sugden (1984); Marothia and Phillips (1985, 1990); Gupta (1986); Matthews (1986); Peters (1986); Söderbaum (1986, 1992); Uphoff (1986); Livingston (1987); Coleman (1988); Savory (1988), Bromley (1989a, 1989b, 1991, 2007); Marothia (1989a, 1992a, 1992b, 2002b, 2002c, 2003b, 2003c, 2004c, 2004d, 2004e, 2006b, 2007, 2010a, 2010b); North (1990); Beetham (1991); Ostrom (1990a, 1990b, 1992a, 1992b); Berkes and Folke (1992, 1994, 1998); Anderson (1993); Dasgupta and Mäler (1994); Saleth (1994); Williamson (1994); Folke and Berkes (1995); Weimer (1995); Drobak and Nye (1997), Mehta et al. (1999); Cleaver (2000); Agrawal (2001); Berkes (2002); Saleth and Dinar (2004); Ostrom (2010); Gopal (2015); Kumar Ritesh et al. (2020).
4 See Marothia (2002a); Ballabh (2008); Bandyopadhyay et al. (2012) for natural and environmental resources governance.
5 Adger et al. (2003) reviewed the research work of many scholars on the appropriateness of the interdisciplinary framework for the analysis of environmental governance (see particularly the work of Savory (1988); Parsons (1995); Hanna et al. (1996); Harvey (1996); Owens (1997); Low and Gleeson (1998); Dore (1999); McNeill (1999); O'Riordan and Jordan (1999); Jordan (2001); Bromley and Paavola (2002); Holland (2002); Paavola (2002); Young (2002). See also Guba (1990), Guba and Lincoln (1994) for the theoretical strength of combining perspectives from a number of disciplines and the deficiencies of a narrow, single

discipline-based approach. Refer to Marothia (2002c) for the application of the interdisciplinary framework for institutionalizing CPRs.

6 Refer to endnote 7 of Chapter 1.
7 Refer to endnote 7 of Chapter 1.
8 See Kiser and Ostrom (1982), Tang (1992); Marothia (1993b, 2001b, 2002a, pp. 701–716, 2004c, 2004e, 2005, 2012a); Singh (1994); Singh and Ballabh (1996).
9 Refer to endnote 4 of Chapter 1.
10 Refer to endnote 5 of Chapter 1.
11 Refer to Costanza et al. (1989); Aylward and Barbier (1992); Munasinghe (1992); Smith (1996); Costanza et al. (1997); Marothia (2001a, 2004d, 2021); De Groot et al. (2002); Gopal (2015); Gopal and Marothia (2015, 2016, 2017); Gopal et al. (2016); Kumar et al. (2017).
12 See Marothia (2002b, 2004d); Katiha and Marothia (2006); Samaraweera and Marothia (2008).

3 Wetland Diversity, Water-Wetland-Agriculture Interactions and Ecosystem Services

Since all the case studies in this book were carried out in the different ACZs of Chhattisgarh, it is important to document all the categories of wetlands existing in the state and how they have sustained ecosystem services over the centuries, despite changing socio-ecological situations. Chhattisgarh is gifted with a rich cultural and historical heritage. It is home to some of India's most ancient caves and Karst wetlands, Buddhist sites, and rock paintings. The landscape of the state is dotted with hill plateaus, waterfalls, wetland complexes, river basins, community ponds, temples, and temple ponds, and its varied ecosystems support an abundance of wildlife and a rich endemic biodiversity. It is home to traditional farming, and indigenous people with rich ecological knowledge of art, folk music, weaponry, and diverse languages. Historically, Chhattisgarh shared its boundaries with many kingdoms.[1] Wetland construction in Chhattisgarh is an old-age practice to harvest rainwater for multiple uses. This traditional technology of water storage is still in use in Chhattisgarh and other parts of India. This chapter presents details of the state and ACZs; water resources; categories, size, and use of wetlands; wetlands both within and outside forests; avifaunal biodiversity; Karst wetlands and ecosystems; temples and sacred ponds; wetland complexes; riverbed or floodplains systems; multipurpose irrigation tanks; *Dabrires* (micro-minor irrigation tanks); wetlands constructed under MGNREGA; wetlands-Agriculture Interactions (WAIs); and ecosystem services derived from these different categories of wetlands.

3.1 Chhattisgarh State: Basic Features

Chhattisgarh came into existence on 1 November 2000 by the partitioning of the south-eastern districts of Madhya Pradesh, with Raipur as its capital city. The state covers a total geographical area of 13.79 million ha, situated between 17°46' to 24°06' north and 80°15' to 84°20' east. Chhattisgarh borders the states of Madhya Pradesh in the north-west, Maharashtra in the west, Telangana and Andhra Pradesh in the south, Odisha in the east, Jharkhand in the north-east and Uttar Pradesh in the north. It is the 10th

DOI: 10.4324/9781003456742-3

largest state of India with 27 administration districts (Table 3.1). The total population of Chhattisgarh (as per the 2011 Census) is 25.55 million, of which males and females are 50.24 and 49.76 percent respectively. The sex ratio is 991 females per 1000 males, and its population density is 189 people per sq. km. The literacy rate is 70.28 percent with male literacy at 80.27 percent and female literacy at 59.58 percent. The majority of the population of Chhattisgarh lives in rural areas (76.76 percent), in 20,532 villages, as compared to the urban population (23.24 percent). Nearly 12.81 percent of the population belongs to the scheduled caste (SC). The total scheduled tribe (ST) population of the State is 33.13 percent, which is distributed among all the districts of the state except Raipur. There are 43 tribal communities concentrated in the undivided districts of Bastar, Bilaspur, and Surguja, and different tribal groups have very strong social, cultural, ethical, and moral institutional traditions that govern their natural resources and environments (Marothia, 2004a).

A brief description of agriculture and allied activities in the state is needed to understand the interactions between the state's wetland ecosystems and its agricultural practices. Of the state's total geographical area, 47 percent is under forest cover. The net cropped area of the state is 33.79 percent of its geographical area. Nearly 7.50 percent of Chhattisgarh's total area is not available for agriculture while 2.65 percent is cultivable barren land, which can be made cultivable by minor changes or improvements. Eighty percent of the land holdings belong to small and marginal farmers (up to two hectares of land) and cover 42 percent of the cultivated land in Chhattisgarh. The average land holding of the state is 1.36 ha. The total land utilization pattern can be seen in Table 3.1.

In 2018–2019, agriculture contributed 22.15 per cent of the Gross State Domestic Product (GSDP) of Chhattisgarh. The share of agriculture and allied sectors in the GSDP of the state from 2011–12 to 2018–19, at current prices, has been static at around 17.21 percent, though at constant prices the trend is declining. It is evident, then, that GSDP from the agriculture and allied sector is growing at a rate lower than the growth of total GSDP. The compound annual growth rate (CAGR) of GSDP in Chhattisgarh was 7.21 percent per annum while the CAGR of agriculture was 3.67 percent during the period of 2011–12 to 2016–17 (cited from Chapter 1 of Economic Survey, Chhattisgarh, GoCG, 2018–19).

Rice is the prime crop of Chhattisgarh, and is sowed over 71.13 percent of total cropped area (55.40 lakh ha) (2017–18 TE). Other than rice, gram, lathyrus, maize, and soybeans are the other major crops of the state. These crops, including rice, are grown on 85 percent of the gross cropped area. Gram is grown in 3.22 million ha and the cropping areas of maize and soybean are 122,000 ha and 100,000 ha respectively. Wheat, Kodo-Kutki (millets), black gram lentils, Tur (pigeon pea), ramtil, rapeseed, mustard, linseeds, and groundnuts are all also grown.

Table 3.1 Main Attributes of ACZs in Chhattisgarh (Area in lakh ha.)

Agro-Climatic Zone	*Districts*	*Total Geo. Area*	*Forest*#	*Barren and Unculturable Land*	*Permanent Pasture and Other Grazing Land*	*Culturable Waste Land*	*Net Sown Area*	*Share of Paddy in Net Crop Area*	*Area under Reservoirs Tanks/ Ponds/ Dams*	*Total Irrigated Area*	*Other Important Crops in Net Crop Area*	*Soil Type***	*Annual Rainfall (mm)*	*Mean Temperature (°C)*	
														Max	*Min*
Central Plains	Raipur, Gariyaband, Balodabazar, Mahasamund, Dhamtari, Durg, Balod, Bemetara, Rajnandgaon, Kabirdham, Bilaspur, Mungeli, Korba, Janjgir, Raigarh & a part of Kanker Districts (Narharpur & Kanker Block)	76.80 (55.70)	27.90 (44.17)	1.31 (45.45)	5.43 (61.18)	1.60 (45.52)	32.40 (69.62)	29.37 (73.83)	1.84 (64.62)	17.48 (95.09)	Lathyrus, Wheat, Gram, Pigeonpea, Soybean, Fruits, Vegetable, and other crops	Lateritic (Bhata), Sandy loam (Matasi), Clay loam (Dorsa), Clay (Kanhar), Sandy (Tikra), Alluvial (Kachar)	950-1200 (54)***	32.4	20.2
Bastar Plateau	Jagdalpur, Narayanpur, Bijapur, Kondagaon, Dantewada, Sukma and the remaining part of Kanker Districts (Antagarh, Bhanupratappur, Charama, Durgukondal and Koilebeda)	32.62 (23.66)	21.29 (33.71)	0.83 (28.87)	0.81 (9.13)	1.82 (51.99)	6.07 (13.05)	4.79 (12.04)	0.46 (16.29)	0.21 (1.14)	Kodo-Kutki, Maize, Black Gram, Fruits, Vegetable, and other crops	Coarse sandy, gravely (Marhan), Gravely with coarse sand (Tikra), Sandy loam (Mal), Clay & clay loam (Gabhar)	1160-1380 (63)***	31.1	18.7
Northern Hills	Surguja, Surajpur, Balrampur, Koria, Jashpur & Dharamjaigarh Tehsil of Raigarh Districts	28.47 (20.65)	13.97 (22.12)	0.74 (25.68)	2.63 (29.69)	0.09 (2.50)	8.07 (17.34)	5.62 (14.13)	0.54 (19.09)	0.69 (3.77)	Maize, Back Gram, Pigenpea, Minor millets, Fruits, Vegetable, and other crops	Sandy (Hilly, Tikra/Dand), Lateritic (Goda Chawar), Silty loam (Bahara)	1250-1410 (59)***	29.8	16.6
Total		137.90 (100)	63.16 (100)	2.88 (100)	8.87 (100)	3.51 (100)	46.53 (100)	39.78 (100)	2.84 (100)	18.38 (100)			1190 (58)***	31.1	18.5

Source: Agriculture Statistics 2017–18, Land Revenue Commissioner, Government of Chhattisgarh, compiled by the author

Note: 2017–18, Figures in brackets are a percentage of the respective parameters. # Reserved Forest, Protected Forest, Unclassified Forest, and Revenue Forest.
** Local soils name inside brackets. *** Figures in bracket are Annual Average Rainy days.

Agricultural productivity is the prime concern of the state, due to the wide gap between the average productivity of Chhattisgarh and nation-wide average for prime crops. The productivity of rice for the year 2017–18 for Chhattisgarh was 1,639 kg/ha, against an all-India average of 2490.66 kg/ha. In the case of maize, for which the area is gradually increasing, the productivity of Chhattisgarh is less than the all-India average. In the case of pulses and oilseeds, productivity is better, and is nearer to the national average. The gap in the consumption of fertilizers between Chhattisgarh and the all-India average is also wide. Data shows that in 2017–18, the fertilizer consumption of Chhattisgarh was 98.09 kg/ha, against the national average of 199.68 kg/ha.

Chhattisgarh is reasonably rich in horticulture, with crops like fruits, vegetables, spices, and medicinal and aromatic crops grown over 827,000 ha (2017–18). Vegetables and fruits use 55.38 percent and 30.22 percent of total horticulture area respectively. The total production of horticulture crops in the state in 2017–18 was 9.727 million metric tonnes.

Chhattisgarh is rich in livestock wealth. The state has about 15.04 million livestock as per the 19th livestock census of 2012. The livestock population increased by 4.31 percent over the previous census of 2007. In Chhattisgarh, 65 percent of livestock are cows, 22 percent are goats and nine percent are buffalo. At the national level, 37 percent are cows, 21 percent are buffalo, 21 percent are goats, and 13 percent are sheep.

In an agrarian state like Chhattisgarh, fish farming occupies pride of place in the rural economy and as a source of occupation. Fish production in the state is being taken up in various water sources like tanks, ponds, reservoirs, and rivers. There is an enormous potential for the development of fisheries and the climate and meteorological conditions are favourable for this. The total area devoted to fisheries is 194,000 ha, which is 95.24 percent of the total available area for fisheries. In the state, fish production mainly comes from rural ponds. Fish production from reservoirs and rivers is nominal. In 2018–19, the total fish production of the state was 469,000 tonnes. Of this, only 15,000 tonnes and 4,000 tonnes were from reservoirs and rivers respectively. The remaining 450,000 tonnes of production were from rural ponds (cited from Chapter 7, Economic Survey: Chhattisgarh, GoCG, 2018–19).

3.1.1 Agro-Climatic Zones

Three ACZs or regions are recognized in Chhattisgarh,[2] namely, the Northern Hills, the Central Plains, and the Bastar Plateau (Map 3.1). The main features of each of these three ACZs are given next, and also listed in Table 3.1.

Central Plains: This region is spread across 15 of the 27 districts of the state. Mahanadi is the primary source of irrigation and water for domestic use. The fertile Central Plains region hosts numerous indigenous aromatic and wild species of rice (Marothia, 2003a) and is also known as the 'rice bowl' of Central India. This region has a distinctive place in the state, with the 'pond

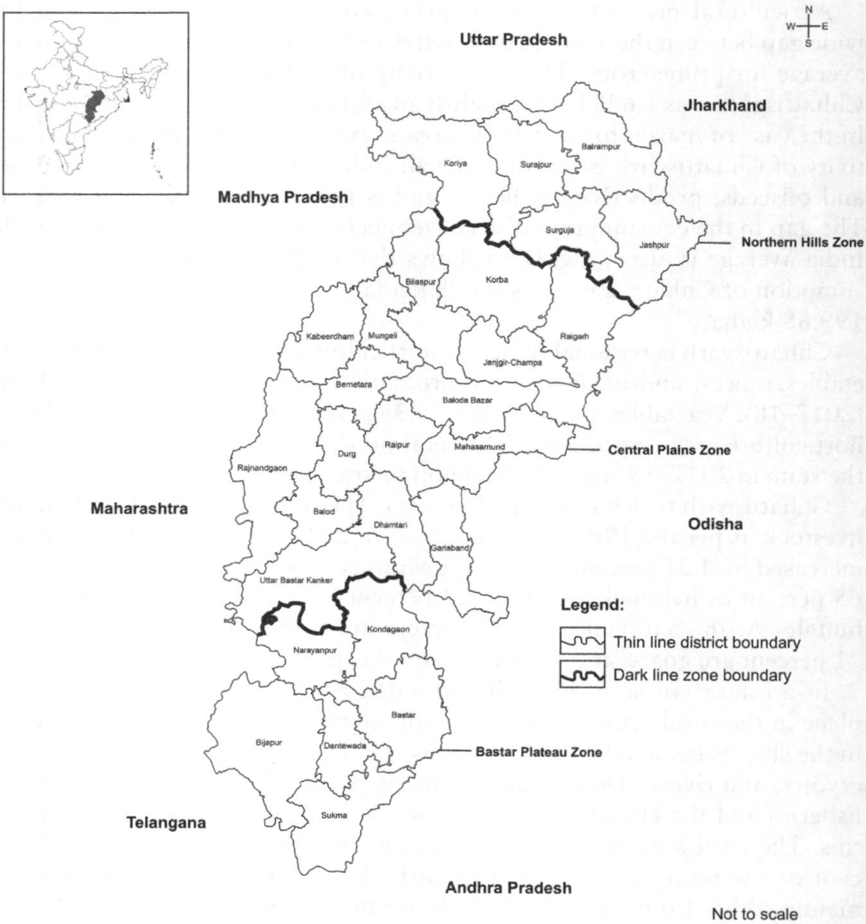

Map 3.1 Map of India and Chhattisgarh State (devised by the author)

town' of Ratanpur in Bilaspur district. The region is well-known for traditional silk cloth and rural crafts. A large number of *arjuna* trees still grow in the forest of Janjgir – Champa district – and are used for rearing silkworms. The region is densely populated, with Raipur and Durg accounting for almost half the total urban population of Chhattisgarh.

Bastar Plateau: This region covers the southern districts of the state. These districts are known for varied and rich forests, and a diverse tribal population with a unique culture. These districts border the states of Maharashtra, Telangana, Andhra Pradesh, and Orissa. The people of the region are dependent on traditional agriculture and forests (non-timber forest produce, aromatic and medicinal herbs and shrubs) for their livelihood. Bastar terracotta and metal artwork has earned international recognition. The Bastar Plateau is rich in mineral resources. The area around Bailadila has huge deposits of iron ore and the NMDC is exploiting these deposits. Other important minerals are bauxite, limestone, dolomite, corundum, tin ore, lepidolite, granite, quartz, copper, feldspar, magnetite, gold, graphite, galena, mica, and sillimanite (Marothia, 1990, 2018; Phillips and Marothia, 1982). The Bailadila iron ore mines and the more recent Nagarnar Steel Plant in Dantewada district represent the limited industrial activity in the region. Bastar Plateau has several sacred hills, mountains, ponds, and river ecosystems. Districts of this region have national parks and wildlife sanctuaries with historical Karsts and cave wetlands. Tribal peoples worship these sacred commons and the natural resources are an inseparable part of their life. The *Gonds* are the numerically dominant tribal group of the state, followed by *Kawars, Halbas, Bhataras, Oraons, Binjhuars, Baigas, Sewars, Bharias,* and *Agharias* (Marothia, 1990, 1994, 1993a; Mandal et al., 2002; Singh Deo, 1990).

Northern Hills: The districts in this region harbour dense forests, hills, and water reservoirs. It is home to several indigenous tribal communities such as *Pahadikorba*, and *Pando*, and other sub-groups. Governed largely by tribal customs, cultures, and traditions in the rural areas of the region, people are dependent largely on agriculture and non-timber forest products (NTFPs). The region has historical and archaeological caves and Karst wetlands which have religious, spiritual, and cultural significance. The level of migration from this region is comparatively limited. There are coal mines in Surguja and Korea districts.

The main features of the ACZs are briefly mentioned in Table 3.1. It would be meaningful to have a look at the land-use data across the different ACZs. Dry deciduous forests cover about 44 percent of the state. Broadly, forests (excluding national parks and wildlife sanctuaries, and with limited access to reserve forests and greater access to protected forests and village forests), cultivable wastelands, grazing lands, and fallows, other than current fallows, could be considered as common-pool land resources used by local communities under usufruct rights (Marothia, 2010a, 2010b).

A large number of natural and human-made wetlands are located both inside and outside the forests, meeting the requirements of wildlife and the population living in the fringe forests. Barren and Uncultivable Land, Permanent Pasture and Other Grazing Land, and Culturable Waste Land have been used to create community water-harvesting ponds under MGNREGA.

Land not available for agriculture largely covers waterlogged areas, areas under social forests, reservoirs tanks/ponds/dams, and buildings and roads. Reservoirs, tanks/ponds/dams, and buildings and roads cover the largest area in the Chhattisgarh Plains, followed by the Northern Hills and Bastar Plateau. Waterlogged area is mainly found in the Central Plains (526 ha), while in the other two zones it is negligible. Similarly, social forestry attempts have largely been concentrated in the Central Plains.

Paddy fields are an important category of wetlands in the state, occupying more than 76 percent of the total cropped area. The remaining area of the total cropped land is occupied by the crops mentioned in Table 3.1. Aquatic crops, medicinal and aromatic herbs and plants, vegetables, and plantation crops are also grown in Chhattisgarh. More than 55 percent of the area cultivated in Chhattisgarh is under canal and human-made wetlands.

The climate of Chhattisgarh is tropical. It is hot and humid because of its proximity to the Tropic of Cancer, and it is dependent on the monsoons for rains. The monsoon season is from late June to October. Chhattisgarh receives an average of 1,292 millimetres of rain. The winter months are from November to January, with low temperatures and low humidity.

The temperature varies between 30°C and 47°C in summer, and between 5°C and 25°C during winter. However, extremes in temperature can be observed, falling to less than 0°C in winter and touching 49°C in summer. The summer months in Chhattisgarh extend from April to June and temperatures. Variations in soil type, rainfall, rainy days, area covered under water structures, and irrigated areas across the state are reflected in Table 3.1.

3.2 Water Resources Status

Water and wetlands are inseparable and essential for sustaining lives. The major source of water in the state is rainfall, which drains through streams and rivers and is collected in surface depressions forming tanks and ponds. Due to its undulating topography, in the hills, Bastar and middle ranges, these tanks and ponds were traditionally used for irrigation, drinking, fisheries, raising animals, and for domestic purposes. As per the State Water Policy 2012 (DWR, GoCG), the priority for water allocation is for household use and drinking, fisheries, agriculture and irrigation, and industry, in that order. The current allocation is to the tune of ten percent for household and drinking (allowable up to 55 percent, with current use being 22 percent), 80 percent for irrigation and agriculture, and ten percent for industry.

Water quality continues to deteriorate in the rivers, canals, tanks, village ponds, streams, and groundwater in many districts of the state. The main sources of water pollution are domestic sewage, industrial effluents, and agricultural runoff. There are many rural areas in the state where lack of drinking water is becoming an issue. Arsenic, fluoride, sulphates, iron, and even a cocktail of contaminants have emerged as a big problem in a sizable number of districts in the state. For example, out of 27 districts, 12 are contaminated by fluoride, 20 by iron, and four by sulphates. Villages of Rajnandgaon are suffering from arsenic contamination. The worst case reported in recent years was the sad story of Supabeda village of Deobhog block of Gariyaband district (Bodhankar and Sinha, 2018) where about 70 villagers died due to water-contamination-induced kidney disease. The rest of the villagers from Supabeda and nine adjoining villages marched barefoot to the Chief

Minister's house to grant them their death wish (prime story covered by INH and IBC 24 News Channels, 11 February 2021). The rapid depletion of water quality is the major factor affecting the health of the human and livestock population in Supabeda village and is displacing a large number of people, reducing their status to that of environmental refugees.

The rural-urban water divide in the state is stark and will further intensify with the rapid development of industries and as urbanization continues. Municipalities also contribute to surface and groundwater pollution due to improper urbanization, the expansion of industries in peri-urban areas, recycling, and the absence of civic amenities. Efforts should be directed towards understanding the links between groundwater and drinking water, and quantifying the socio-economic impact of arsenic, fluoride, and other chemical contamination of groundwater. Alternative technologies for wastewater treatment, economic incentive structures to minimize pollution to the prescribed standards, the role of central and state groundwater boards and state and central pollution control boards, and water-pollution control laws need a hard look.

An understanding of the water-food-energy nexus in the wider context of natural ecosystems and the services they provide is essential to develop strategies for achieving water, food, and energy security in a climate-changing world. The implications of the water-food-energy-climate change nexus extend to the entire domain of the United Nations' *Transforming Our World: The 2030 Agenda for Sustainable Development*.

These issues need to be taken into account in water governance and management. Current approaches to managing these resources are fragmentary and there are overlapping claims and counterclaims on them. In the absence of programmatic policy and strategy, if the current approach continues, it would escalate the conflict over already deteriorating and depleting natural resources in general and water resources in particular. Understanding of wetland-water interactions can provide crucial information for making a comprehensive integrated water policy. To this end, different types of water resource are discussed next.

3.2.1 Rainwater Balance

Chhattisgarh receives an average of 1,292 millimetres of rain during the monsoon season, with wide interregional variation. The state's rainwater balance is given in Table 3.2. Of the total rainfall, 9.4 Mha-m infiltrates the groundwater but deep percolation is only 2.2 Mha-m. Due to geohydrological reasons, a large part of infiltrated groundwater (77 percent) is retained in the soil. Thus, groundwater yield varies across different regions and districts of the state. In-situ water storage in the form of tanks and ponds has been an integral part of water resource management in the state. Subsurface streams are found in many areas and these have been sustaining groundwater recharge as well as working as drought-proofing through protective irrigation.

Table 3.2 Rainwater Balance in Chhattisgarh

S.No.	*Particular*	*Amount in Mha-m*
1.	Water received as rainfall	17.55 (100)
2.	Lost through evaporation	2.8 (17.0)
3.	Infiltration in ground • Available soil moisture 7.2 • Deep percolation 2.2	9.4 (58.0)
4.	Surface	4.1 (25.0)

Source: https://cgwrd.in/water-resource/wr.html, 2018–19

Note: Figures in parentheses are percentages

Table 3.3 Usable Surface and Groundwater Resources and Existing Use in Chhattisgarh State

S.No.	*Water categories*	*In BCM*	*Percentage*
1.	Usable Surface Water	48.9	100
2.	Current Surface Water used	9.2	22
3.	Usable Groundwater	14.54	100
4.	Current Groundwater used	2.79	20.4

Source: https://cgwrd.in/water-resource/groundwater-status.html
https://cgwrd.in/water-resource/wr.html, 2018–19

3.2.2 Status of Surface and Groundwater

The total usable amount of surface and groundwater and its existing use in Chhattisgarh is given in Table 3.3. It may be seen from this table that only 22 percent of surface water and 20 percent of groundwater is being used. Thus, one finds that almost 75 percent of usable water remains unused. In fact, one of the estimates suggests that, from groundwater alone, 75 percent of cultural land could be irrigated. One of the redeeming features of groundwater in the state is that only one of the 146 blocks is over-exploited, only 21 blocks are in the critical zone, and the remaining 124 blocks are in the safe zone. Less than 30 percent of the gross cropped area in the state is irrigated. Chhattisgarh's Central Plains have 41.8 percent gross cropped area under irrigation. All in all, the state has ample water resources to harness productivity without threatening its sustainability.

3.2.3 River Basin Network

Located in Chhattisgarh's Central Plain region is the upper Mahanadi River basin. The basin lies at an elevation of about 800 to 950 feet (250 metres to 300 metres) above sea level. Hillocks, undulating interfluves (areas between adjacent watercourses), and valleys flanked by belts of clay soils are characteristic of the region. The upper Mahanadi basin is separated from the upper

Narmada basin in the west by the Maikal Hill (part of the Satpuras) and from the plains of Odisha in the east by ranges of hills. The southern part of the state lies on the Godavari River and its tributary, the Indrāvati River. The Mahanadi is the chief river of the state. The other major rivers are Hasdo (a tributary of the Mahanadi), Rihand, Indrāvati, Jonk, Arpa and Shivnath (https://cgwrd.in/water-resource/wr.html).

The state is located on the oldest rocks of the Indian subcontinent. One can find the Archaean granites and gneisses and the Gondwana formations dating back 250–300 million years, bearing coal deposits. The Deccan Trap can also be found, formed from volcanic basalt flows capped with laterite that dates back 65 million years. About 100 miles (160 km) wide, the Chhattisgarh Central Plain is bounded by the Chota Nagpur plateau in the north, the Maikal range in the west, the hill of Raigarh in the north-east, the Raipur upland in the south-east, and the Bastar plateaus in the south. These highlands comprise mostly erosional plateaus reaching an elevation of more than 2,300 feet (700 meters) in the Maikal range and the Dandakaranya hills. The Maikal hills coincide with the Vindhyas and the Satpura, extending over 1430 sq km. Its elevation is highest at Lafagarh, standing tall at 1067 m, 75 km north of Bilaspur. To the north-east are the Raipur uplands, and the south-east lies the Raigarh Hills. The plains run out in the former princely state of Kanker to the south, beyond which is the Keshkal ghats. In the Kutumsar and Dandak caves in Bastar District, the remains of plant and burnt charcoal have been discovered. Cave paintings are found in Kanker, Raigarh, and Surguja districts, which prove that the region was inhabited from early times (Ball, 1877; Das et al., 1992; Dutt, 1964; GSI, 2005; Murti, 1987; Ramakrishnan and Vaidyanadhan, 2008).

The rivers Ganga, Godavari, Mahanadi, Brahmani, and Narmada form major basins crisscrossing the state, though they are also part of other states (Table 3.4). Mahanadi, Sheonath, Indrāvati, Arpa, Hasdeo, Kelo, Son, Rehar, and Kanhar are some of the main rivers providing a lifeline to agriculture and the socio-economic development of the state. Most of the rivers in the state have a torrential regime characterized by a high flow of water for three to four months during the monsoon (June to September), during which nearly 80 percent of the annual runoff flows (https://cgwrd.in/water-resource/wr.html).

In the north lies the edge of the great Indo Gangetic Plain (IGP). A small part (1.7 percent) of the Ganga basin, namely the river Son sub-basin, falls within the state of Chhattisgarh. Spread over 18,406 sq km, it drains the districts of Koriya, Surajpur, Balrampur, Surguja, and Jashpur. The backwaters of the Rihand dam reservoir (in the state of Uttar Pradesh) stretch into parts of Chhattisgarh. The eastern end of the Satpura range and the western edge of the Chota Nagpur plateau form an east-west belt of hills that divide the Mahanadi River basin from the IGP. Geologically, low hills in the northern and southern parts of the state enclose a central fertile plain which is formed by the headwater streams of the Mahanadi. The northern and southern parts of the state are hilly, while the central part is a fertile plain where rice is grown extensively (https://

Table 3.4 Area Under Different River Basins in Chhattisgarh

Basin	*Place of Origin of Main River*	*Basin Area in State (Mha)*	*Area of State (percent)*	*Major Tributaries*
Ganga	Gangotri Uttar kashi District (Uttarakhand)	1.9	13.63	Rihand & Banas
Godavari	Nasik district (Maharashtra)	3.9	28.64	Sabri & Indrāvati
Mahanadi	Sihawa hills, Dhamtari district (CG)	7.6	56.15	Seonath, Hasdeo, Tel, Mand & Kelo
Brahmani	Chhotanagpur plateau (Jharkhand)	0.13	1.03	Sankh
Narmada	Amarkantak hill ranges, Shahdol (MP)	0.07	0.55	Banjar
	Chhattisgarh	13.6	100	

Source: https://cgwrd.in/water-resource/wr.html, 2018–19

cgwrd.in/water-resource/wr.html). As these rivers and basins are part of other states, Chhattisgarh would have to deal with sharing water with co-riparian states. Conflicts over water sharing between Chhattisgarh and Orissa have recently arisen in the case of Mahanadi and Indirawati rivers (Marothia, 2017).

3.2.4 Aquifers Corresponding to River Basins

Using a map of major aquifers (Plate IV), aquifers corresponding to river basins have been worked out and delineated in Map 3.2 and Table 3.5.

3.2.5 Reservoirs of Chhattisgarh

Out of five major reservoirs, four (Ravishankar Sagar, Dudhawa Reservoir, Sondur Reservoir, and Minimata Bango) were constructed in the Mahanadi basin, and the Tandula Reservoir was completed on the confluence of the Sukha Nala and Tandula rivers. All five reservoirs are multipurpose in nature, with their waters being used for irrigation, hydro-electric power generation, fisheries, industrial applications, and recreation. A brief description of these reservoirs is given in the Indian Space Research Organisation (ISRO), Space Applications Centre (2010), *National Wetland Atlas: Chhattisgarh*, pp. 99–119. Most of the reservoirs are experiencing a loss in capacity due to sedimentation, and all of them need treatment in the upper reaches of the catchment (description drawn from Indian Space Research Organisation (ISRO), Space Applications Centre (2010), *National Wetland Atlas: Chhattisgarh*, pp. 99–119).

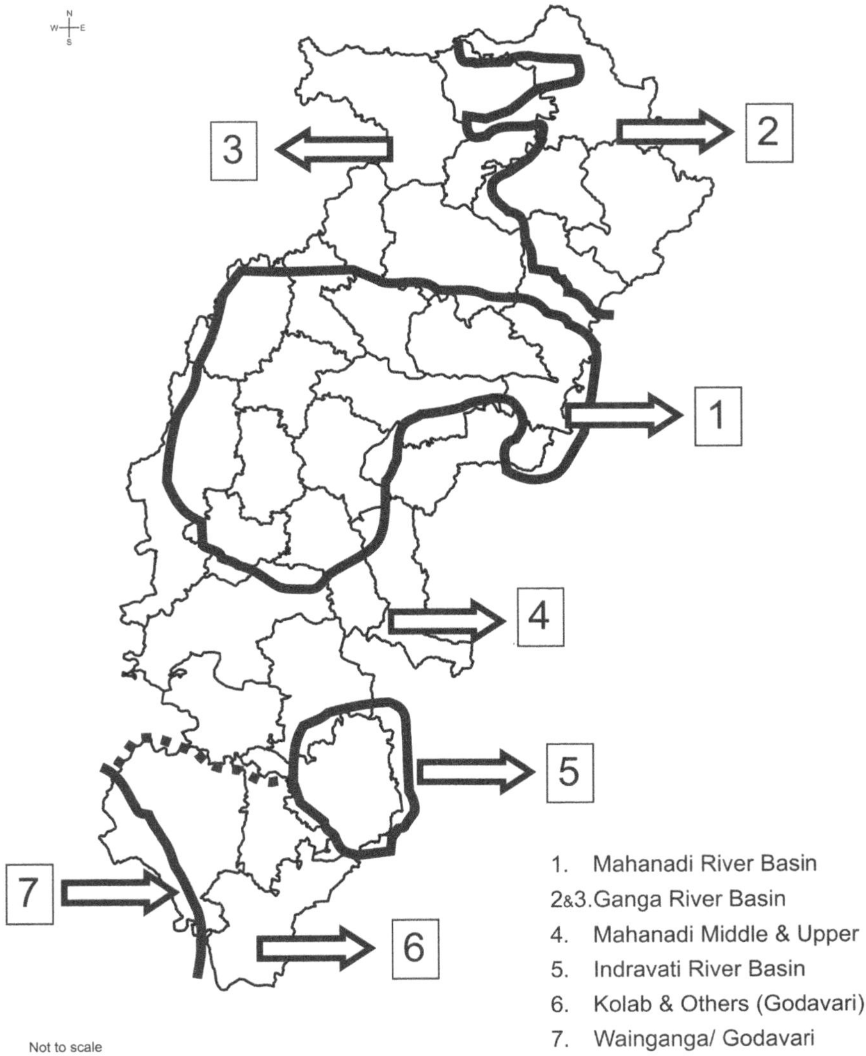

Map 3.2 Aquifers Corresponding to River Basins (identified 1 to 7 aquifers)

Note: Aquifers delineated using a map of the aquifer systems[3] of Chhattisgarh, Central Groundwater Board North Central Chhattisgarh Region, Ministry of Water Resources Government of India, Plate IV.

3.2.6 Source-wise Irrigated Area

In 2017–18, 32 percent of the state was irrigated. If we look at the latest data available, in 2018–19, the net irrigated area was 1.564 million ha (27.82 %), whereas the total irrigated area was just over two million ha (35.68 %) of the total cropped area of 5.6 million ha. In Chhattisgarh, canals are the main source

Table 3.5 Aquifers Corresponding to River Basin[3]

Code	River Basin	Geological Unit	Major Aquifer Litho Units	Hydrogeological Properties	Hydrochemistry	Percentage of Total Wetland Number	Percentage of Total Wetland Area
1	Mahanadi	Rocks of Chhattisgarh Supergroup	Sandstone, Shale, and Limestone	Sandstone: Hard and compact; lacks primary porosity Shale: yellow in colour, occurs near surface below laterite, commonly known as *chhui mitti* and is known for water-retention capacity, which makes it favourable for construction of surface water-harvesting structures Limestone: black & pink varieties encountered. The black or Charmuria limestone is extensively used for as building material while the pink variety, depending on the CaO%, is used for cement production	Potable and extensively used for human consumption through tubewells Anthropogenic effects are evident in urban areas, and the water needs to be filtered before consumption	51.85	40.39
2	Ganga River Basin	Banded Gneissic Complex	Gneiss and Granite	Shallow and deep groundwater is tapped through dugwell and borewell, but due to less population density the stress on groundwater is less	At places fluoride occurrences are reported in groundwater	4.90	8.08
3	Ganga River Basin	Sedimentary rocks of Gondwana Supergroup	The Mahanadi Rift Zone is known to be occupied by Gondwana Supergroup rocks, which are sedimentary in nature with occurrences of massive coal deposits	Gondwana Supergroup forms the major aquifer which is a reliable source for domestic water supply	Anthropogenic effects on groundwater quality evident in populated area, otherwise the water quality is reliable	14.61	20.90

4	Mahanadi Middle & Upper	Bastar Craton	Banded Gneissic Complex and Granite	Shallow and deep groundwater is tapped through dugwell and borewell	In places, fluoride occurrences are reported in groundwater	24.29	21.79
5	Indrāvati River Basin	Rocks of Indrawati Supergroup	Sandstone, shale, and limestone	Similar to the rocks of Chhattisgarh Supergroup, as discussed in 1	Potable and extensively used for human consumption through tubewells Anthropogenic effects are evident in urban areas	1.43	1.30
6	Kolab & Others (Godavari)	Bastar Craton	Gneiss and granite of Southern Bastar	Shallow and deep groundwater is tapped through dugwell and borewell, but due to less population density the stress on groundwater is less	At places, occurrences of fluoride and iron is reported in groundwater	2.33	6.04
7	Wainganga/ Godavari	Sedimentary rocks of Gondwana Supergroup	The Godavari Rift Zone is known to be occupied by Gondwana Supergroup rocks, which are sedimentary in nature with occurrences of massive coal deposits	Shallow and deep groundwater is tapped through dugwell and borewell, but due to less population density the stress on groundwater is less	Potable	0.58	1.51

[3] Technical support in preparing details received from Professor Ninad Bodhankar

Table 3.6 Irrigated Area by Different Sources in Chhattisgarh, 2018–19

S.No.	*Sources*	*Number*	*Net Irrigated Area Lakh ha*
1.	Canal (2051 km length)	1137	8.98 (57.4)*
2.	Tank/pond	35211	0.29876 (1.9)
3.	Well	124314	0.15807 (1.0)
4.	Tube well and other sources	196293	6.21159 (39.7)
Total		356955	15.64842 (100)

Source: Economic Survey Chhattisgarh, 2019–20, p. 154

* Figures in parentheses are percentages in respective columns

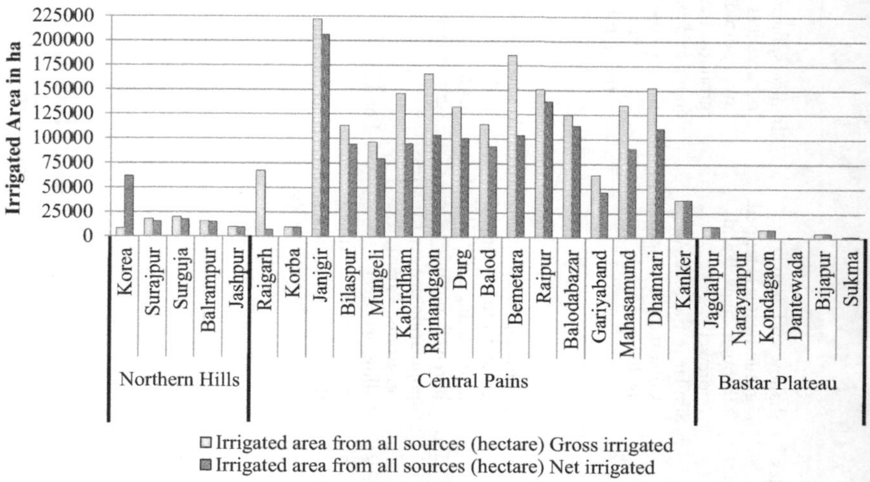

Figure 3.1 Irrigated Area in District and ACZs-Wise

Source: Derived from Economic Survey Chhattisgarh, 2019–20, p. 154

of irrigation followed by tube wells and other sources (Table 3.6). A little over 90 percent of the gross irrigated area is available in the Central Plains region. Very poor irrigation facilities were developed in the Northern Hills (3.48 percent) and the Bastar Plateau (1.23 percent) regions (Economic Survey Chhattisgarh, 2019–20, p. 154). Districts and ACZs wise irrigated area is given in Figure 3.1.

3.3 Wetlands: Categories, Size, and Use

This section gives the details about the wetlands in the state: the number, the area under different categories, and the wetlands inside and outside forested areas.

3.3.1 Wetland Distribution in the State

Although Chhattisgarh is exceptionally rich in water resources and agricultural wetlands, there has been hardly any effort to document or investigate the state's wetland wealth. Almost a decade ago, the National Wetland Inventory prepared by the Indian Space Research Organisation (ISRO), Space Applications Centre (2010) identified 7,711 wetlands (mapped at 1:50,000 scale), besides 27,823 small wetlands (<2.25 ha) covering an area of 337,966 ha (2.5 percent of the geographic area) in the state. Rivers and streams account for about 53 percent of the wetlands (179,088 ha), reservoirs and barrages cover 90,389 ha, and tanks and ponds have a total area of 40,226 ha. The small wetlands (<2.5ha) account for about 8.2 percent, assuming that each is of one hectare. Out of the total wetland area, the extent of open water is 243,814 ha post-monsoon and 173,678 ha in the pre-monsoon period (Table 3.7).

Table 3.7 Wetland Categories in Chhattisgarh (Area in ha)

S. No.	*Wetland Category*	*Number of Wetlands*	*Total Wetland Area*	*Percent of Wetland Area*	*Open Water (Area)*	
					Post-Monsoon	*Pre-Monsoon*
Inland Wetlands – Natural						
1	Lakes/ponds	-	-	-	-	-
2	Ox-bow lakes/ cut-off meanders	6	26	0.01	26	13
4	Riverine wetlands	8	174	0.05	83	76
5	Waterlogged	-	-	-	-	-
6	River/stream	156	179088	52.99	124712	93095
Inland wetlands – man-made						
7	Reservoirs/ barrages	604	90389	26.74	85148	54012
8	Tanks/ponds	6906	40226	11.90	33671	26366
9	Waterlogged	31	240	0.07	174	116
Sub-Total		7711	310143	91.77	243814	173678
Wetlands (<2.25 ha), mainly tanks*		27823	27823	8.23	-	-
	Total	**35,534**	**337,966**	**100.00**	**243,814**	**173,678**
	Area under Aquatic Vegetation				2123	19600
	Area under turbidity levels					
	Low				28985	79103
	Moderate				183025	85841
	High				31804	8734

Source: Indian Space Research Organisation, Space Applications Centre (2010). *National Wetland Atlas: Chhattisgarh*, p. 19

* Assuming that each wetland under this category is of one ha

There is a significant reduction in the extent of open water from post-monsoon to pre-monsoon. This is reflected in all categories of wetlands. Turbidity is mostly moderate post-monsoon (183,025 ha out of 243,814 ha of open water), followed by high turbidity (31,804 ha), and low turbidity (28,985 ha) (Table 3.7).

The aquatic vegetation in Chhattisgarh accounts for between 0.6 percent and 5.8 percent of total wetland area post-monsoon (2,123 ha) and pre-monsoon (19,600 ha) (*National Wetland Atlas: Chhattisgarh*, p. 19). Wetland distribution in districts, and zones and the ACZs of Chhattisgarh is given in Table 3.8 (see Map 3.3 for Chhattisgarh's share of wetlands in India, in terms of number and area. The number and area of wetlands in the three regions of Chhattisgarh are also depicted in Figure 3.2).

3.3.2 Wetlands in Forest Areas[4]

The State Forest Department has identified wetlands larger than 5 ha, the majority of which lie within the forested areas (Table 3.9). In view of the large difference between Forest Department and SAC estimates, wetlands outside forest areas still need to be inventoried and mapped. The number of wetlands inside forest areas account for nearly 89 percent of the total wetlands in the state, covering an area of about 17 percent of the state. Interestingly, the number of wetlands outside forest area is only 11.4 6 percent of the state total, but they cover 83 percent of the state's area (Table 3.9). Out of the total number of wetlands inside forest areas, about 78.52 percent fall between 5 ha and 25 ha. The remaining 22.48 percent of wetlands cover an area above 25 ha, and up to 200 ha (Marothia, 2009a).

The picture of wetlands outside forest areas is quite different. About 36 percent of wetlands fall into the category of 10–25 ha, and 29 percent above 100 ha. The remaining 35 percent of wetlands have an area between 25 ha and 100 ha. The number of wetlands inside protected areas is highest in Bilaspur and Jagdalpur wildlife circles. Although the number of wetlands is slightly lower in the Raipur wildlife circle, the area of wetland is almost four times higher than in the Surguja wildlife circle. It is interesting to note that most of the animal-human conflicts, particularly elephant-human, have been observed in the Sarguja wildlife circle. Human life, cattle, crops, and property have been lost due to shrinking wetlands and biomass areas (see Marothia, 2018, for minimizing human-wild life conflicts). The average wetland size in protected areas is 29.5 ha. The wetlands range from 6 ha to 1,700 ha. The highest number and area fall under the Bilaspur protected area, followed by Raipur, Surguja, Durg, and Kanker respectively. Jagdalpur protected area has the lowest number and area of wetlands. One wildlife sanctuary and three national parks cover 34 percent of the area, whereas six tiger reserves have 64 percent of the total wetland area. (See Forest Management Information System (FMIS), 2019–20.)

Table 3.8 Wetland Distribution in Districts and ACZs of Chhattisgarh

District Code	ACZs	District	Geographic area*	No. of Wetland	percent of Total number of Wetlands	Wetland area	Percent of total wetland area	Area per wetland	Percent of district geographic area	Open Water		Aquatic Vegetation		Turbidity (Post-monsoon)			Turbidity (Pre-monsoon)		
										Post-monsoon	Pre-monsoon	Post-monsoon	Pre-monsoon	Low	Moderate	High	Low	Moderate	High
			(sq. km)			(ha)		(ha)		(ha)	(ha)	(ha)	(ha)	(ha)	(ha)	(ha)	(ha)	(ha)	(ha)
1	Northern	Koriya	6604	793	2.22	14515	4.29	18.30	2.20	9666	7486	60	155	736	8064	866	1053	6284	149
2	Hills	Surguja	15733	1880	5.26	28794	8.52	15.32	1.83	20650	16658	119	519	298	17953	2399	2418	12514	1726
3		Jashpur	5838	215	0.60	7585	2.24	35.28	1.30	6314	4007	36	87	-	6264	50	188	3753	66
4		Raigarh	7086	1985	5.56	17719	5.24	8.93	2.50	12951	8236	113	457	860	11422	669	5258	2414	564
5		Korba	9010	1648	4.61	28624	8.47	17.37	3.18	25026	17708	70	705	420	9781	14825	7743	9615	350
6		Janjgir-Champa	3672	4333	12.13	24886	7.36	5.74	6.78	13452	11035	218	1206	855	10718	1879	5399	5004	632
7		Bilaspur	7215	3103	8.69	19443	5.75	6.27	2.69	15069	8838	50	730	1441	13551	77	4051	4321	466
8	Central	Kawardha	4223	760	2.13	6899	2.04	9.08	1.63	4623	2959	111	321	696	3635	292	995	1736	228
9	Plains	Rajnandgaon	6904	2606	7.30	18149	5.37	6.96	2.63	13272	10054	116	2195	330	11222	1720	4122	4706	1226
10		Durg	8537	4348	12.17	30291	8.96	6.97	3.55	25958	18187	358	3772	9553	15753	652	11475	6193	519
11		Raipur	12507	5777	16.18	44211	13.08	7.65	3.53	26783	18739	181	2130	3649	20524	2610	10899	6743	1097
12		Mahasamund	4702	2475	6.93	16557	4.90	6.69	3.52	11468	6857	508	2312	3096	7216	1156	4161	1801	895
13		Dhamtari	4049	1325	3.71	26909	7.96	20.31	6.65	19676	11891	52	2981	4778	14210	688	9174	2670	47
14		Kanker	6506	1723	4.82	13219	3.91	7.67	2.03	9854	6833	15	1061	1751	6505	1598	1380	5200	253
15	Bastar	Bastar	14974	1703	4.77	14662	4.34	8.61	0.98	10732	9976	4	460	315	9216	1201	1533	8111	332
16	Plateau	Dantewada	17634	1041	2.91	25503	7.55	24.50	1.45	18320	14214	112	509	207	16991	1122	9254	4776	184
Total			135194	35715	100	337966	100.00	9.46	2.50	243814	173678	2123	19600	27951	183025	31804	79103	85841	8734

Source: Indian Space Research Organisation, Space Applications Centre (2010). *National Wetland Atlas: Chhattisgarh*, Ahmedabad. Compiled by the author

Note: * Currently there are 27 districts

A. Raipur includes Gariyaband and Balodabazar
B. Durg includes Balod and Bemetara
C. Bilaspur includes Mungeli
D. Bastar includes Narayanpur and Kondagaon
E. Dantewada includes Bijapur and Sukma
F. Surguja includes Balrampur and Surajpur

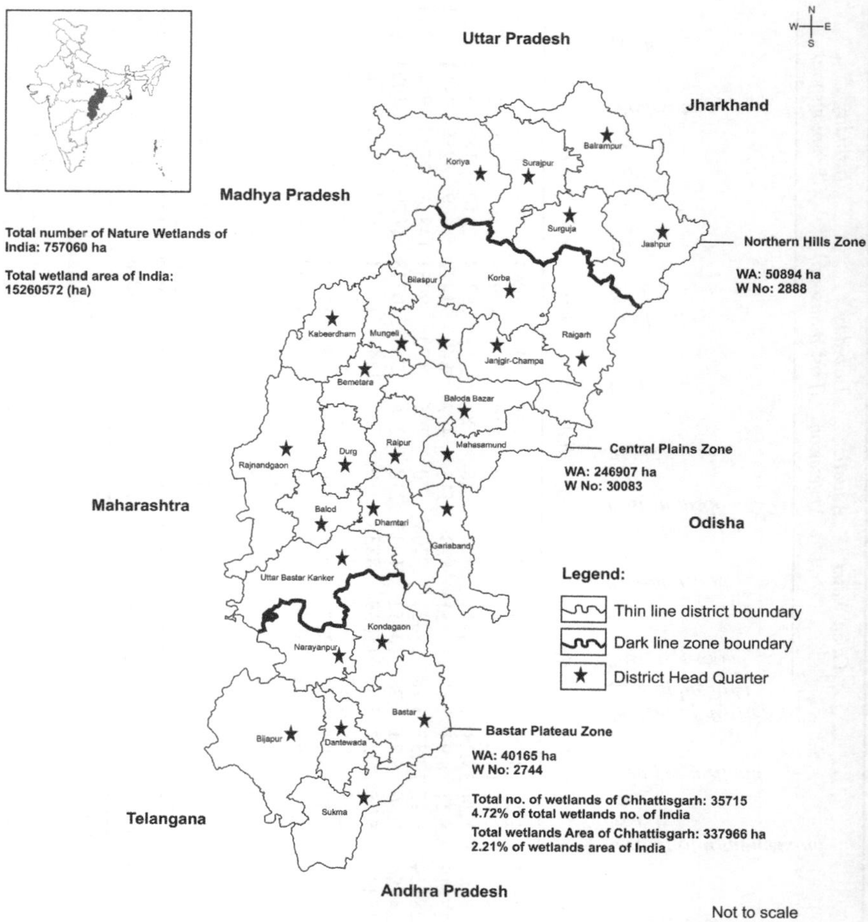

Map 3.3 Wetland Number and Area in ACZs of Chhattisgarh

Source: Devised by the author

Table 3.9 Wetland in Forest Areas (5.00 ha. and above)

S.No.	*Wetland*	*No.*	*Area in ha.*
1.	Inside forest	1159 (88.54)	24531.27 (17.06)
2.	Outside forest	150 (11.46)	119236.75 (82.94)
Total		1309 (100)	143768.02 (100)

Source: Forest Management Information System (FMIS), 2019–20. Aranya Bhawan, Atal Nagar, Nava Raipur, Chhattisgarh

Note: Figures in parentheses are percentages of total number and area under wetlands

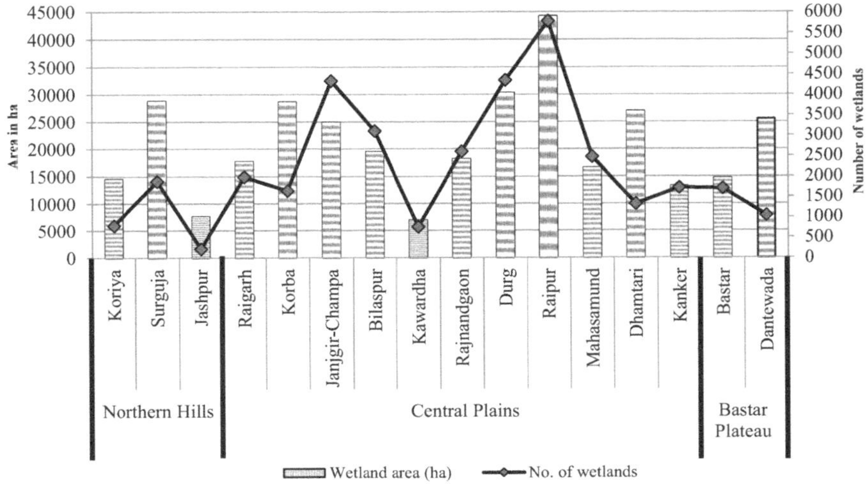

Figure 3.2 Wetland Distribution in Districts and ACZs of Chhattisgarh

Source: Drawn by the author using data of Indian Space Research Organisation, Space Applications Centre (2010). *National Wetland Atlas: Chhattisgarh*, Ahmedabad

Different types of wetlands and their associated ecosystem services are depicted in Photo 3.1 (a to l).

Photo 3.1 (a) Temple pond and lotus cultivation (*source:* Marothia assisted by Prafull Katre)

Photo 3.1 (b) Riverbed/ floodplain cultivation in the Mahanadi River Basin (*source:* Marothia assisted by Navneet Rana) (c) Foxnut cultivation in rural wetlands (*source:* S.K. Patil) (Continued)

Photo 3.1 (d) Chestnut and agro-forestry cultivation in rural wetlands (*source:* Marothia), (e) Fish cultivation in rural and peri-urban wetlands (*source:* Marothia) (Continued)

Photo 3.1 (f) Cattle tending in rural and peri-urban wetlands (source: Marothia), (g) Meeting domestic water requirements in peri-urban & rural wetlands (*source:* Marothia) (Continued)

Photo 3.1 (h) Recreation in lotus pond (*source:* Marothia and Sukhdev Prasad Kashyap) (i) Water-centric recreation in urban wetland (*source:* Ajit Degwekar and Marothia) (Continued)

Photo 3.1 (j) Avifaunal diversity in wetlands and wetlands complex (*source:* Moiz Ahmed, Nova Nature) (k) Wetlands inside forest area (*source:* Moiz Ahmed, Nova Nature) (Continued)

Photo 3.1 (l) Kotumsar cave and Karst wetlands (*source:* Pinturam Kadeyari)

3.4 Avifaunal Diversity

The Bombay Natural History Society has identified four Important Bird Areas (IBAs) in the state for conservation: Barnawapara and Gomarda wildlife sanctuaries in Raipur district, Udanti and Sitanadi sanctuaries in Raipur and Dhamtari, and Indrāvati tiger reserve in Dantewada (Bastar). Ratanpur (Chhattisgarh), known as a pilgrimage and wetland town, has also been the favourite home of several summer and winter migratory birds, besides being rich in local bird diversity. Avifaunal surveys have been carried out in wetlands located in different parts of Chhattisgarh, but only a few studies have been conducted in Ratanpur. The findings of these studies covering domestic and migratory birds are summarized in Table 3.10. Several other areas remain yet to be explored. These IBAs include wetlands which are used by migratory and globally threatened bird species, besides many other fauna.

3.5 Karst Wetlands

Karsts are areas characterized by the presence of soluble limestone rocks, sinkholes, caves, and underground drainage systems through which energy and water move (Gunn, 2004). Given that karsts have complex geological and hydrological characteristics, they are among the most fragile and vulnerable environments in the world (Brinkmann and Parise, 2012). Cave biodiversity is extremely unique and susceptible to natural and human-made disturbances. In addition to their valuable biodiversity, caves are known for their archaeological/

Table 3.10 Avifaunal Diversity in Wetlands of Chhattisgarh

Wetlands	*Bird Diversity*		*Duration of Survey*	*Author/* **Organization**
	Domestic	*Migratory*		
Bairagban, Bhedimuda, Bikma, Dulahara, Gireejaban, Jagannath, Kaira, Krishnarjuni, Maharaiya, Ratneshwar-all wetlands of Ratanpur town		WV – Gadwall, Ferruginous Pochard, Lesser Whistling Duck, Cotton Pygmy Goose, Pheasant-tailed Jacana, Little Grebe SV – Black-winged Stilt MV – Greater Painted-snipe, Asian Openbill Stork VC – Bronze-winged Jacana, Indian Pond Heron, Cattle Egret, Little Cormorant, NR – Red-wattled Lapwing, Green Sandpiper, Black-Crowned Night- heron, Common Coot, Common Moorhen R – Purple Heron, White-Breasted Waterhen, Great Cormorant C – Great White Egret, Little Egret, Chestnut Bittern, Intermediate Egret, Common Kingfisher, Pied Kingfisher, White – breasted Kingfisher, Purple Swamp Hen	Jan. 2013 to Dec. 2014	Porte & Gupta (2018)
Bandhwa, Bikma, E. Nawagaon, S. Nawagaon, one by Chami Rd, one by Highway, one near Ratanpur Temple, Champi and Khutaghat Dam- all wetlands of Ratanpur town	Purple Swamp Hen, Bronzed Wing Jacana, Red Vented Bulbul, Lesser Whistling Duck, Black Ibis, Little Cormorant, Little Egret, Cattle Egret, Common Kingfisher, White-Throated Kingfisher, Asian Pied Kingfisher, Black-Winged Stilt, Purple Heron, Paddy Field Pipit, Rufous-Tailed Lark, Indian Roller, Jungle Crow, House Crow, Pheasant-Tailed Jacana, Blue Cheeked Bee-eater, Common Swift, Black Bitten.	Winter visitor from abroad: Great Egret, Red-Rumped Swallow, Common Redshank, Little Bittern, Ruddy Shelduck, Black-Crowned Night Heron, Barn Swallow, Citrine Wagtail, Common Sandpiper, Common Teal Eurasian Wigeon, Great Gadwall, Great Bittern, Green Sandpiper, Grey Wagtail, Marsh Sandpiper, Northern Pintail, Red-Crested Pochard, White-Browed Wagtail, Tufted Duck, White Wagtail, Yellow Wagtail. Local Migratory: Common Hawk Cuckoo, Eurasian Golden Oriole, House Swift, Common Quail, Common Coot, Asian Openbill Stork, Common Moorhen, Oriental White Ibis Threskiornis Melanocephalus.	January 2017	Dey (Feb 2017)

Ratanpur water bodies near Kaka Pahad	Dendrocygna Javanica, Nettapus Coromandelianus, Tachybaptus Ruficollis, Microcarbo Niger, Anastomus Oscitans, Ardea Purpurea, Bubulcus Ibis, Egretta Garzetta, Ardeola Grayii, Nycticorax Nycticorax, Francolinus Pondicerianus, Ixobrichus Cinnamomeusm, Amaurornis Phoenicurus, Porphyrio Porphyrio, Gallinula Chloropus, Fulica Atra, Hydrophasianus Chirurgus, Metropidius Indicus, Vanellus Indicus, Columba Livia, Streptopelia Decaocto, Spilopelia Senegalensis, Spilopelia Suratensis, Elanus Caeruleus, Butastur Teesa, Accipitus Badius, Milvus Migrans, Perdicula Asiatica, Centropus Sinensis, Athena Brama, Eudynamys Scolopaceus, Clamator Jacobinus, Cacomantis Passerines, Hierococcyx Varius, Coracias Benghalensis, Psilopogon Haemacephalus, Psittacula Krameri, Tephrodornis Pondicerianus, Alcedo Atthis, Halcyon Gularis, Ceryle Rudis, Merops Orientalis, Dicrurus Macrocercus, Corvus Macrorhynchos, Pycnonotus Luteolus, Pycnonotus Cafer, Dendrocitta Vagabunda, Corvu Splendens, Aegithina Tiphia, Lanius Schach, Oriolus Oriolus, Mirafa Erythroptera, Orthotomus Sutorius, Prinis Sylvatica, Prinia Inornata, Prinia Socialis, Prinia Hodgsonii, Turdoides/ Argya Caudate, Turdoides Malcolmi, Turdoides Striata, Dumetia Hyperythra, Chrysomma Sinense, Saxicola Caprata, Saxicoloides Fulicatus, Copsychus Saularis, Apus Affinis, Acridotheres Tristis, Gracupica Contra, Sturnia Pagodarum, Zosterops Palpebrosus, Cinnyris Asiaticus, Leptocoma Zeylonica, Dicaeum Erythrohynchos, Dicaeum Agile, Anthus Rufulus, Motacilla Maderapatensis, Passer Domesticus, Passer Domesticus, Ploceous Philippinus, Amandava Amandava, Euodice Malabarica, Lonchura Punctulata.	Summer visitor: Terpsiphone Paradise Winter visitor: Phylloscopus Trochiloides, Acrocephalus Dumentorum, Acrocephalus Stentoreus, Hirundo Rustica.	25th June to 15th Oct. 2017	Preliminary survey funded by CGSPC (2017)* Survey conducted by Mohit Sahu

(*Continued*)

Table 3.10 (Continued)

Wetlands	*Bird Diversity*		*Duration of Survey*	*Author/* **Organization**
	Domestic	*Migratory*		
Gidhwa	Lesser Whistling Duck, Cotton Pygmy Goose, Common Kingfisher, White-Throated Kingfisher, Pied Kingfisher, White-Breasted Waterhen, Purple Swamphen, Common Moorhen, Pheasant-Tailed Jacana, Bronze-Winged Jacana, Little Grebe, Red-Crested Pochard, Common Pochard, Gadwall, Ferruginous Pochard, Northern Shoveller, Northern Pintail, Gargeny, Black Tailed Godwit, Marsh Sandpiper, Common Sandpiper, Green Sandpiper, Common Greenshank, Temminck's Stint, Common Ringed Plover, Little Ringed Plover, Kentish Plover, Yellow Bittern, Grey Francolin, Spotted Owlet, Red Collared Dove, Oriental Honey Buzzard, Short-Toed Snake Eagle, Eurasian Marsh Harrier, White Eye Buzzard, Common Kestrel, Brown Shrike, Blue Throat, Variable Wheatear, Ashy Prinia, Clamorous Reed Warbler, Little Cormorant, Great Cormorant, Grey Heron, Purple Heron, Indian Pond Heron, Great Egret, Little Egret, Cattle Egret, Black-Crowned Night Heron, Common Moorhen, Purple Swamphen, Pheasant-Tailed Jacana, Red-Wattled, River Lapwing, Spotted Dove, Laughing Dove, Oriental Turtle Dove, White-Throated, Green Bee-eater, Indian Roller, Ashy-Crowned Sparrow Lark, Jerdons Bush Lark, Peddy Field Pipit, Red-Vented Bulbul, Indian Robbin, Plain Prinia, Purple Sunbird, Red Avadavat, Indian Silverbill, Yellow-Throated Sparrow, Baya Weaver, Common Myna, Lesser Pied Myna, Brahminy Starling, Black Drongo, Ashy Drongo, House Crow, Jungle Crow, Common Hoopoe, Lesser Whistling Duck, Greater Spotted Eagle, Steppe Eagle, Brown Crake, Pied Kingfisher, Black Kite, Plum-Headed Parakeet, Little Bittern, House Sparrow, Indian Courser.	Great Crested Grebe, Asian Openbill, Woolly-Necked Stork, Eurasian Wigeon, Cotton Teal, Red-Crested Pochard, Common Pochard, Tufted Pochard, Gadwall, Eurasian Coot, Common Sandpiper, Marsh Sandpiper, Grey Wagtail, Yellow Wagtail, White-Browed Fantail, Bluethroat, Rosy Starling, Black Bittern, White Wagtail, Northern Pintale, Northern Shoveller,	January 2013	Yadav (2017)
Parsada Lake, Khuteri Lake I, Khuteri Lake II, Khanduwa Dam, Bandha Dam, Chherkapur		Common Greenshank, Gadwall, Green Sandpiper, Red-Crested Pochard, Wood Sandpiper, Bluethroat, Common Pochard, Marsh Sandpiper, Tufted Duck, Asian Openbill Stork, Black-Crowned Night Heron, Black-Headed Ibis, Common Hoopoe, Common Sandpiper, Western Yellow Wagtail, Red Start, Great Cormorent, Grey Heron, Western Reef Egret, Great Cristed Grebe	September, 2015	George et al. (2015)

Uttar Bastar-Kanker, Kondagaon, Narayanpur and Dakshin Bastar-Dantewada, Bijapur, Bastar and Sukma	Little Grebe, Little Cormorant, Indian Shag, Darter, Little Egret, Cattle Egret, Median Egret, Large Egret, Chestnut Bittern, Black Bittern, Purple Heron, Black-crowned Night Heron, Little Green Heron, Indian Pond Heron, White-necked Stork, Asian Openbill- Stork, Oriental White Ibis, Black Ibis, Cotton Teal, Lesser Whistling-Duck, Crested Serpent Eagle, Short-Toed Snake Eagle, Changeable Hawk-Eagle, Indian White-backed Vulture, Egyptian Vulture, Long-billed Vulture, Oriental Honey Buzzard, White-eyed Buzzard, Black Kite, Jerdon's Baza, Black –shouldered Kite, Black Baza, Besra Sparrowhawk, Crested Goshawk, Shikra, Painted Francolin, Grey Francolin, Rock Bush-Quail, Painted Bush-Quail, Rain Quail, Indian Peafowl, Red Junglefowl, Yellow-legged Buttonquail, Common Buttonquail, Slaty-legged Crake, Purple Moorhen, Common Moorhen, Blue-breasted Rail, Brown Crake, White-breasted Waterhen, Common Coot, Watercock, Great Indian Bustard, Bronze-winged Jacana, Pheasant-tailed Jacana, Yellow-wattled Lapwing, Red-wattled Lapwing, Little Ringed Plover, Black-winged Stilt, Greater Painted-Snipe, Stone Curlew, Small Pratincole, Indian Courser, River Tern, Little Brown Dove, Red Collared Dove, Oriental Turtle Dove, Emerald Dove, Purple Wood-Pigeon, Blue Rock Pigeon, Green Imperial-Pigeon, Spotted Dove, Eurasian Collared Dove, Orange-breasted Green-Pigeon, Yellow-legged Green-Pigeon, Indian Hanging-Parrot, Alexandrine Parakeet, Rose-ringed Parakeet, Plum-headed Parakeet, Brainfever Bird, Indian Plaintive Cuckoo, Banded Bay Cuckoo, Rufous-bellied Cuckoo, Indian Cuckoo, Greater Coucal, Asian Koel, Large Green-billed Malkoha, Small Green-billed Malkoha, Sirkeer Malkoha, Spotted Owlet, Jungle Owlet, Brown Wood-Owl, Brown Fish-Owl, Collared Scops-Owl, Eurasian Scops-Owl, Dusky Eagle-Owl, Eurasian Eagle-Owl, Brown Hawk-Owl, Barn Owl, European Nightjar, Franklin's Nightjar, Indian Jungle Nightjar, Large-tailed Nightjar, Common Indian Nightjar, Asian Palm-Swift, White-rumped Needletail-Swift, House Swift, Alpine Swift, Crested Tree-Swift, Malabar Trogon, Black-capped Kingfisher, Blue-eared Kingfisher, Small Blue Kingfisher, Lesser Pied Kingfisher, White-breasted Kingfisher, Small Bee-eater, Blue-bearded Bee-eater, Indian Roller, Malabar Pied Hornbill, Indian Grey Hornbill, Common Hoopoe, Brown-headed Barbet, Coppersmith Barbet, Yellow-fronted Pied Woodpecker, Lesser Golden-backed Woodpecker, Heart-spotted Woodpecker, Speckled Piculet, Brown-capped Pygmy Woodpecker, Fulvous-breasted	Great Crested Grebe, Great Cormorant, Spot-billed Pelican, Grey Heron, Little Bittern, Northern Pintail, Gadwall, Northern Shoveller, Eurasian Wigeon, Garganey, Bar-headed Goose, Brahminy Shelduck, Grey lag Goose, Common Teal, Red-crested Pochard, Common Pochard, Black Eagle, Greater Grey-headed Fish-Eagle, Lesser spotted Eagle, Booted Eagle, Bonelli's Eagle, Montagu's Harrier, Western Marsh Harrier, Brahminy Kite, Pied Harrier, Osprey, Common Kestrel, Common Quail, Baillon's Crake, Pacific Golden-Plover, White-tailed Lapwing, Grey-headed Lapwing, Common Greenshank, Common Redshank, Spotted Redshank, Green Sandpiper, Common Sandpiper, Wood Sandpiper, Common Snipe, Temminck's Stint, Black-headed Gull, Pied Crested Cuckoo, Chestnut-headed Bee-eater, Blue-tailed Bee-eater, Eurasian Wryneck, Indian Pitta, Common Swallow, Grey Wagtail, White Wagtail, Citrine Wagtail, Yellow Wagtail, Forest Wagtail, Oriental Tree Pipit, Brown Shrike, Rufous-backed Shrike, Bay-backed Shrike, Siberian Rubythroat, Bluethroat, Blue-headed Rock-Thrush, Blue Rock-Thrush, Dark-throated Thrush, Black Redstart, Blyth's Reed Warbler, Common Chiffchaff, Greenish Leaf-Warbler, Orphean Warbler, Common Lesser Whitethroat, Verditer Flycatcher, Tickell's Blue-Flycatcher, Blue-throated Flycatcher, Kashmir Flycatcher, Asian Brown Flycatcher, Red-throated Flycatcher, Ultramarine Flycatcher, Black-lored Yellow Tit, Black-headed Bunting, Common Rosefinch, Rosy Starling, Grey-headed Starling, Black-naped Oriole, Ashy Drongo	July, 2011 to December, 2014	Chandra et al. (2015)

(Continued)

Table 3.10 (Continued)

Wetlands	*Bird Diversity*		*Duration of Survey*	*Author/ Organization*
	Domestic	*Migratory*		
	Woodpecker, Grey-capped Pygmy Woodpecker, Rufous Woodpecker, Great Black Woodpecker, Little Scaly-bellied Green Woodpecker, Small Yellow-naped Woodpecker, Greater Golden-backed Woodpecker, Black-shouldered Woodpecker, Singing Bush Lark, Ashy-crowned Sparrow Lark, Bengal Bush Lark, Red-winged Bush Lark, Rufous-Tailed Finch Lark, Eastern Skylark, Dusky Crag-Martin, Streak-throated Swallow, Wire-tailed Swallow, Red-rumped Swallow, Large Pied Wagtail, Paddyfield Pipit, Large Cuckoo-Shrike, Black-headed Cuckoo-Shrike, Common Woodshrike, Pied Flycatcher-Shrike, Scarlet Minivet, Rosy Minivet, Small Minivet, Black-crested Bulbul, Red-vented Bulbul, Red-whiskered Bulbul, Gold-fronted Chloropsis, Jerdon's Chloropsis, Common Iora, White-rumped Shama, Tickell's Thrush, India Robin, Pied Bushchat, Oriental Magpie-Robin, Orange-headed Thrush, Yellow-breasted Babbler, Quaker Tit-Babbler, Spotted Babbler, Yellow-eyed Babbler, Indian Scimitar Babbler, Jungle Babbler, Common Babbler, Large Grey Babbler, Streaked Fantail-Warbler, Jungle Prinia, Ashy Prinia, Rufous-fronted Prinia, Plain Prinia, Common Tailorbird, Olivaceous Leaf-Warbler, Grey-headed Flycatcher, Asian Paradise-Flycatcher, Black-naped Monarch-Flycatcher, White-throated Fantail-Flycatcher, Great Tit, Velvet-fronted Nuthatch, Chestnut-bellied Nuthatch, Spotted Creeper, Tickells's Flowerpecker, Thick-billed Flowerpecker, Purple-rumped Sunbird, Crimson Sunbird, Ruby-cheeked Sunbird, Purple Sunbird, Oriental white-eye, Red Munia, White-rumped Munia, Black-headed Munia, Green Munia, White-throated Munia, Spotted Munia, Yellow-throated Sparrow, House Sparrow, Black-breasted Weaver, Baya Weaver, Common Myna, Common Hill- Myna, Asian Pied Starling, Brahminy Starling, Jungle Myna, Eurasian Golden Oriole, Black-headed Oriole, Black Drongo, Spangled Drongo, Bronzed Drongo, White-bellied Drongo, Greater Racket-tailed Drongo, Ashy Woodswallow, Jungle Crow, House Crow, Indian Treepie.			

Dulhara, Ratanpur	Cotton Pigmi Goose, Copper Smith, Hoopoe, Blue Jay, White-Breasted King Fisher, Small Blue King Fisher, Pied King Fisher, Koel, Indian Cocoo, Crow Pheasant, Rose Ring Parakete, Padaki, White Pegeon, Moor Hen, Purple moor hen, Pheasant-Tailed Jassana, Bronze-Winged Jassana, Shikra, Cormorant, Cheast nut Beeten, Cattle Egret, Little Egret, Paddy Bird, Large White Heron, Open Bill Storke, Mag Pied Robin, Drongo, Bulbul, Bush Chat, Indian Robbin, Crow, Pied Myna, Common Myna, Brahmini Myna, Jungle Babler, Common Babler, Black Bellied Finch Lark, Red Winged Bush Lark, Plain Prinia, Ashy Wren Warbler, Red Munia, Silver Throated Munia, Spotted Munia, Purple-Rumped Sun Bird, Purple sun Bird, Pipit (paddy field), Yellow-eyed Babler.	Lesser Whistling Teel, Green Bee-eater, Crusted pied Cocoo, Gray Shrike, Roufous backed Shrike, Roufous Tailed Lark, Yellow Wagtail, Black Red Start.	April 2008 to March 2010	Rahalkar and Tiwari (2014)
Dalpatsagar	Lesser Whistling Duck, Bronze-Winged Jacana, Pheasant-Tailed Jacana, Eurasian Coot, Darter, Cattle Egret, White-Breasted Waterhen, Nail Eater, Asian Openbill, Indian Pond Heron Or Paddybird, Common Moorhen, Cattle Egret, Purple Swamphen, Little Cormorant.		2nd February, 2017	Pradhan (2017)

Note: I=IUCN status; M=Migration status; LC= Least concern, NT= Near threatened, WV=Winter visitor, SV= Summer visitor, MV= Monsoon visitor, VC= Very common, C= Common, NR= Not rare, R= Rare

Note: Complied by the author * The survey was funded by CGSPC and carried out by Mohi Sahu, Ornithologist and Wetland Surveyor. The survey was monitored by Marothia, Chair TF on Ag & Allied sector and assessed by Brij Gopal, Chair, SWG, CGSPC on Wetlands (2017)

paleontological, historic, cultural, spiritual, recreational, aesthetic, educational, research, and economic values, which, when not sustainably managed, can compromise their ecological integrity (Centre for Karst Gospi, Croatia, 2009; Andreo et al., 2010; van Beynen, 2011; Bachmann, 2016).

Some wetlands in Chhattisgarh also exist below ground, where water infiltrating through the ground dissolves carbonate minerals in limestone rocks over thousands of years and forms caves (Gopal and Marothia, 2017). So far, 21 karsts/caves have been identified, listed next, in different agro-climatic/ecological zones of Chhattisgarh (http://tourism.cg.gov.in). For the location of caves

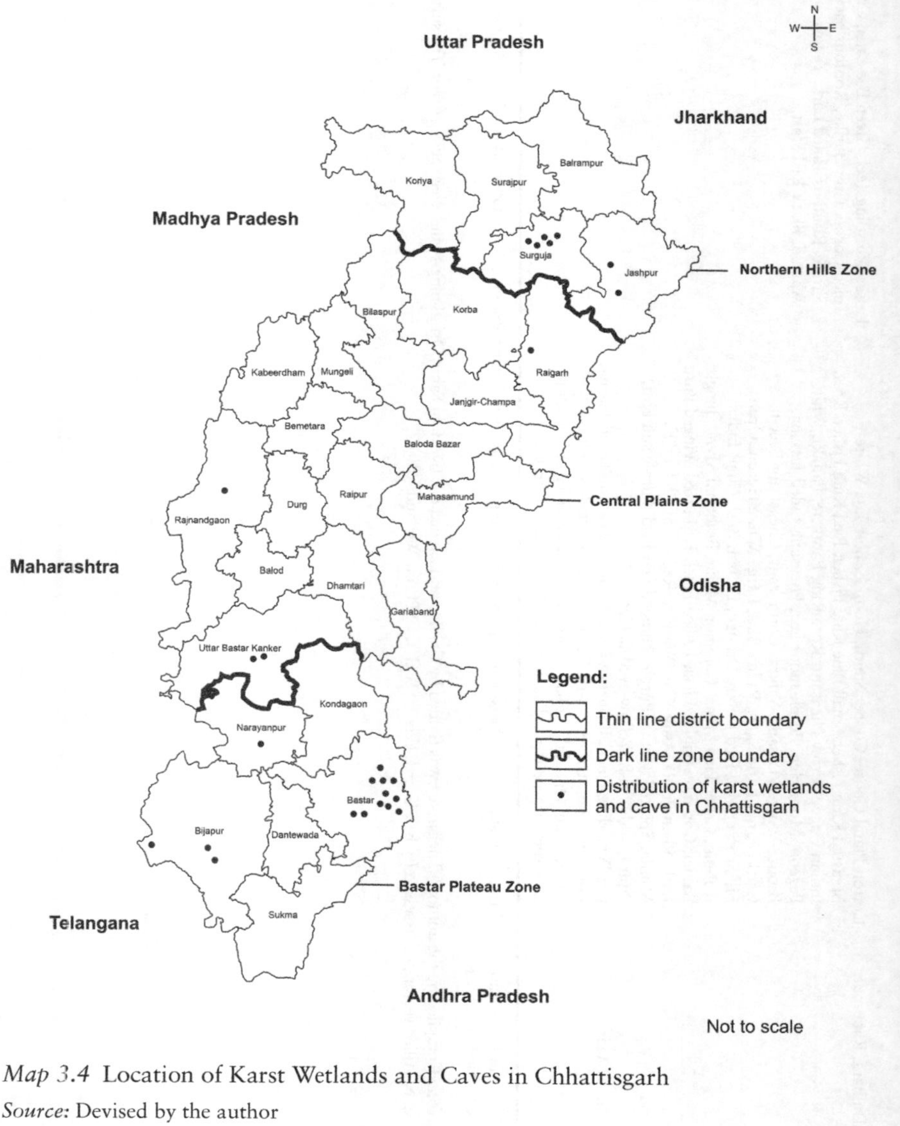

Map 3.4 Location of Karst Wetlands and Caves in Chhattisgarh

Source: Devised by the author

and Karst wetlands, see Map 3.4. For a brief description of caves located in different parts of the state, see Gupta (1973), Sankhyan (2011), and Alung (2014).

3.5.1 Caves and Karst Wetlands in Bastar Plateau

The Kanger Valley National Park in Bastar District Host Nine Caves/Karsts

Aranyak Cave, Dandak Cave (length: 362 m (1,188 ft)), Kailash Caves (length: 253 m (830 ft)) Devgiri Cave, Jhumar Cave, Kanak Cave (unexplored), Kotumsar Cave (length: 660 m (2,170 ft)), Mendhkamaari Cave, Sheet Cave (length: 25 m (82 ft)) (http://tourism.cg.gov.in).

There are three more caves identified in Bastar District: Kapatdwar Cave of Tokapal, Bastar (length: 30 m (98 ft)), Rani Cave of Tokapal, Bastar (length: 50 m or 160 ft)), Tular Cave, Narayanpur District (unexplored) (http://tourism.cg.gov.in).

Kanker district has two caves: Jogi Cave of Gadiya Mountain Kanker (length: 50 m (160 ft)), Sondayi Cave, Kanker (length: 40 m or130 ft)) (http://tourism.cg.gov.in).

Bijapur District has three caves, namely, Usur Cave of Usur Block, Shankarpalli Caves, Shakal Narayan Cave of Bhopalpatnam (unexplored) (http://tourism.cg.gov.in).

3.5.2 Caves and Karst Wetlands in Northern Hills

Jashpur district has Kailash Cave and Khuriarani Cave of Bagicha Village (length: 27 m (89 ft)). Khuriaraniki cave is one of the most important historical places in Jashpur Nagar (http://tourism.cg.gov.in).

Surguja District has the Sita Bengra Cave, which is the most mythological and historically important tourist destination of Ramgarh. It is known as the residence of the Goddess Sita. Located on a north-eastern slope of the hillock, the Sita Bengra Cave is 14 m long, 4.2 m wide, with a height of 2 m in the front, which lessens towards the back. Outside the cave there are many circular gates and benches carved out of ancient stone. Lord Hanuman footprints are also present, which add to the mystical aura and are said to stand testimony to the fact that the Goddess Sita lived in the cave during her forest sojourn (http://tourism.cg.gov.in).

Jogimara Caves are located in the Surguja District in the state of Chhattisgarh. These caves are around 10x6x6 feet in dimension and age back to 300 BC. There are many paintings of animals, human beings, birds, and flowers in these caves. Each painting is of white base plaster with a red outline (http://tourism.cg.gov.in).

Here, if someone speaks in the theatre-like space inside the cave, his/her sound can be heard far away. Ramgarh Cave is the most ancient among the historical places of Surguja. It is also called Ramgiri. Ramgarh is a centre of interest due to the legend of Lord Rama and Kalidas (one of the most famous poets in the world in Sanskrit). According to an ancient belief, Lord Rama, his brother Lakshman, and his wife Sita had resided here during the Vanasthan period (http://tourism.cg.gov.in).

Kailash Caves are situated at Sabarbar, which is 60 km to the east of Ambikapur. These caves were created by Saint Rameshwar Gahira Guruji. The temple of Lord Shiva-Parvati and the temples of different gods and goddesses can be found in the caves. On the occasion of Mahashivratri, a huge festival is organized here. The major attractions are the temple of Lord Shiva-Parvati, the Yagya Mandap (sacrificial altar), the Sanskrit school, the Gahira Guru Ashram, Badgdrat Beir, and Badh mada (http://tourism.cg.gov.in).

3.5.3 Caves and Karst Wetlands in Central Plains

Rajanandgaon has Mandhip Khol, Gandai (length: 430 m (1,410 ft)).

Singhanpur is most popular for its ancient caves. Singhanpur is situated 20 km from Raigarh and has the distinction of hosting the oldest sculptures on Earth. These sculptures are claimed to have been created in 30,000 BC and look very similar to Mexican and Spanish sculptures (http://tourism.cg.gov.in).

Kotumsar and Dandak caves have been explored and researched, but there are many others with nearly unexplored wetlands. These wetlands, which cannot be identified by remote sensing images, also need investigation, conservation, and protection against anthropogenic pressures. The Kotumsar Cave of Chhattisgarh is one of the best-explored caves of India, and biologists have researched several dimensions of this subterranean cave ecosystem (Pradhan and Agarwal, 1984; Biswas et al., 1990; Biswas, 1992; Reddy, 2006; Messouli et al., 2007; Reddy and Defaye, 2009).

3.6 Temples and Sacred Ponds

The worship of nature is an integral part of human society. In historic times, it was common in the cultures, religions, and the social norms of traditional societies. Nature worship can focus on various natural objects like stones, rocks, mountains, ponds, rivers, forests, and plants and animals, which are considered to have divine value (Ray and Ramachandra, 2010; Ray et al., 2014). Sacred groves, a form of nature worship, includes forests, lakes, ponds, streams, rivers, and hills, which are considered as "Sacred Natural Sites" by the IUCN (Oviedo et al., 2005). Cultural values related to them have a very important place in the life of villagers (Deb and Malhotra, 1997; Malhotra, 2002).

Ponds are considered by many to be sacred centres in the life of the village. Villagers often have some beliefs and spirits (Vidyarthi, 1963; Vidyarthi et al., 1979; Jha, 1997; Tripathi and Panigrahi, 2017) related to ponds. In most of the villages, towns, and big cities across states in India, Hindu temples customarily have ponds in their surroundings which are centres of worship and described as temple ponds. In most places, ponds are also found inside temples, while some ponds are located at the fringe of the temples. Like in other parts of the country, one or more temple ponds are situated in almost every village of Chhattisgarh, some of which are centuries old. Temple ponds with daily sacramental functions provide very important cultural and spiritual services in the total estimation of pond ecosystem services. Temple ponds have been historically managed by the temple trustee, or priest, or a traditional priest family, and these age-old institutional arrangements continue in many villages. Temple management bodies generally make every effort to keep holy ponds pollution-free, using soft institutional arrangements, or working rules, like imposing restrictions over the misuse of these holy ponds, and so they remain comparatively clean. Devotees of temples use the holy water for washing their limbs and sometimes take a dip in the holy waters, as they believe that the waters will wash their sins away. However, temple ponds located outside temples are used by people for normal bathing and even washing clothes.

The incremental flow of devotees performing rituals within and around temple premises has polluted many such ponds. For example, in Ratanpur and other highly religion-centric temple ponds, poor maintenance has led to their eutrophication. This is due to the disposal of ritual offerings that are rich in organic matter, which poses a serious threat to the existence of temple-pond ecosystems. A few studies have adequately estimated the level of pollution associated with this dumping of ritual materials, washing, and bathing in such ponds (Anithakumari and Aziz, 1989; Maya, 2002; Nayar, 2019).

Sacred ponds have existed for centuries in almost every village of Chhattisgarh. Sacred ponds, even during modern urbanization, remain the epicentre for cultural practices, religious beliefs, life cycle rituals, and customs related to sacred matters. In most of tribal-dominated villages, as in the case of other communities, all sacred acts, like sacrifice, ancestral worship, festivals, and other social and cultural activities, are performed in sacred ponds by the person dealing with the spiritual well-being of the community. Relationships between people, ponds and spirits is still a complex issue in the tradition of village life of Chhattisgarh (Vidyarthi, 1963; Pandey, 2000), and is important to the interaction between nature, man, and spirit (see Chapter 4 for more on the people-pond-spirit relationship).

3.7 Wetland Complex

Like eastern and southern states of India, several villages and towns in Chhattisgarh have wetlands chains, or interconnected, or complex systems of wetland (particularly in the case of village ponds, locally known as *talabs* and *dabries*, and irrigation tanks). Such complexes, some of which are centuries old, were generally designed to store rainwater or connect wetland systems with rivers, perennial streams, or water sources to meet the multiuse demands of rural and semi-rural communities. One example is the Chhattisgarh sacred wetland complex of Ratanpur town, which can be considered a unique socio-cultural-religious-ecological-archaeological site (details are given in Chapter 4).

The Chandkhuri Kausalya temple pond is connected with seven other ponds, all of which are used for a variety of activities including growing aquatic crops (mainly lotus cultivation) and for irrigation see Photo 3.1 (a). Lake Vivekanand (*Budha talab*) was also connected with seven ponds (including temple ponds and small *dabri or pethu* i.e., ponds). Another wetland complex in the state is the Dalpat Sagar wetland of Jagdalpur town, Bastar District. Dalpat Sagar was constructed by King Dalpat Dev around 1772 to meet requirements for domestic use in Jagdalpur and irrigation in nearby areas. An earthen bund was constructed on the northern side of a low-lying area, connecting an enlarged area of three ponds near Dalpat Dev palace to create the 148 ha lake. In the centre of the lake is a beautiful temple of Lord Shiva, which still stands after almost five centuries. Dalpat Sagar has an approximately 1,500 ha catchment area. Ratanpur and Dalpat Sagar are potential choices for Ramsar sites in the state (see Chapter 9).

Similarly, in Kurra Village of Dharsiwa block in Raipur district, there are 29 multipurpose ponds of different sizes. Some of these are linked with canals while others are seasonal, but all of them are interconnected (Marothia, 1993b). It is worth mentioning here that wetland chain systems, though variable in number and size, are still operative in several towns and villages in Chhattisgarh, other Indian states, and South Asian countries.

3.8 Riverbed or Floodplains Cultivation

Rivers have been central to the development of human civilisations and cultures globally. Rivers have provided water and fish ever since humans evolved. Rivers supported food production, first on floodplains and, later, away from them by the use of irrigation. Chhattisgarh has an extensive network of riverine wetlands. Riverbeds or floodplains have been used for centuries for cultivating cucurbitaceous vegetables. Largely mixed cultivation has been a common system in riverbed areas. Riverbeds represent a distinctive farming system. Floodplains are available in the summer months. Muskmelon, watermelon, bottle gourd, cucumber, summer squash, bitter gourd, Indian squash or round melon, long melon, and tomatoes are common crops grown in the riverbed systems of the state (see Photo 3.1b). Watermelons from the state have historically been exported through traders of Maharashtra to Arabian countries. Village Panchayats lease out floodplains to landless, agricultural labourers and marginal farmers with some financial support, who cultivators pay after harvesting the crops. Such an arrangement is institutionalized between cultivators and the Gram Samiti of the village Panchayat. Marketing has never been a problem due to high seasonal demand from local and neighbouring towns and cities. Riverbed cultivation has been profitable for the farmers, so the repayment of loans is not seen as a major issue.

During the last decade, riverbed cultivation suffered a big setback as a result of the construction of a chain of stop dams along most of the rivers of the state. This had three major implications: first, the intensity of annual flooding in riverbed areas reduced significantly; second, surplus water from stop dams is sometimes released during the summer season resulting in the washing away, or causing damage to, the cucurbits crops; and, third, loss of income for poor landless and marginal farmers, trapping them in debt. Further, seasonally dry riverbeds have been declining and in the process affecting the income, food, and nutritional security of under privileged people. Furthermore, the whole of society is deprived of high-quality summer vegetables and fruits (for field cases of riverbed farming, see Kumari et al. (2018), Ramjan et al. (2018) and Mishra (2021).

3.9 Irrigation Tanks

Tank irrigation is a centuries-old established practice for rainwater harvesting in Chhattisgarh. Thousands of irrigation tanks were constructed before the British regime, during the British period, and after independence (see details in Chapter 6). A large number of reservoirs and tanks are under the ownership of water resources authorities and other departments (agriculture, Panchayas and rural development agencies), and some water structures are also under private ownership in districts and ACZs (Marothia, 1986, 1992b, 2004b; Marothia et al., 2010).

3.10 *Dabries* Irrigation System

In Chhattisgarh, *dabri* (small or micro-minor irrigation tanks) have been created by the State Department of Agriculture to provide life-saving irrigation to sustain a rainfed-based cropping pattern, particularly paddy fields under the *Dabri Yojana*. An institutional analysis of the Dabries Programme was carried out and is reported in Chapter 7 along with details related to district, ACZs, and state-level construction of *dabries* (see also Marothia, 1992b, 2004b).

3.11 MGNREGA: Move Towards Wetland Centric Livelihoods[5]

The Mahatma Gandhi National Rural Employment Guarantee Act (MGNREGA) is a "demand-driven" programme, whereby any adult who is willing to do unskilled manual work at the minimum wage is entitled to be employed for local public works within 15 days of applying, with a guarantee of 100 days employment per household per year (for a detailed analysis of MGNREGA, see Desai et al., 2015). MGNREGA has in recent years moved towards creating rural assets on private and village common lands to induce a multiplier effect to the ongoing rural transformation process.

To create social and economic assets, several works and schemes have been aligned under MGNREGA in all three ACZs in Chhattisgarh. These include the creation of assets under Bharat Nirman Rajeev Gandhi Sewa Kendra, drought proofing, fisheries development, flood control, building irrigation canals and irrigation facilities for SC/ST/IAY/LR, land development, renovation of traditional water bodies, rural connectivity, rural drinking water, rural sanitation, water conservation and water harvesting. Figure 3.3 provides details of constructed wetlands across the three ACZs and also gives the aggregate picture of the state. A comparative picture of beneficiaries along with the construction of different types of rural water systems is also provided in Figure 3.3.

Given the lack of electricity, the most reliable energy source for pumping is diesel fuel, which has many drawbacks (environmental pollution and ever-increasing costs, among others). To mitigate this, under "*Saur Sujala Yojana*", CREDA provides solar irrigation pumps with subsidized rates to small and marginal farmers. One of the most obvious advantages of a solar power irrigation system (SPIS) is that its operation does not produce any greenhouse gas emissions, minimizes the spatial and temporal externalities associated with groundwater, and can substantially reduce water-extraction costs and overuse or misuse (Gopal and Marothia, 2016; Marothia, 2017). By reducing costs, SPIS can improve people's access to water. The solar-powered water pump, running on photovoltaic energy, is appropriate in Chhattisgarh

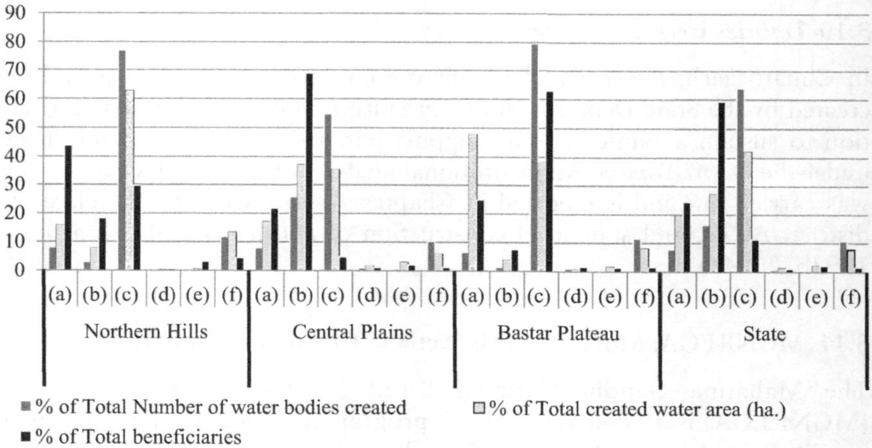

Figure 3.3 Water Conservation Works Under MGNREGA (Water Conservation Works FY 15–16 to 17–18)

Source: Derived from the data from Commissioner MGNREGA, Department of Panchayat, and Rural Development, GoCG

Note: (a) Community ponds (new), (b) Community ponds renovation, (c) Farm pond (Dabri), (d) Flood-protection works, (e) Stopdam, (f) Under FRA: Farm Pond+Fisheries+Crop

as plenty of sunlight is available and it is quite intensive, more than compared to many other states during nearly eight to nine months of the year. SPIS uses solar energy to pump up water from the source to an elevated storage tank, or directly to fields from small ponds or seasonal/perennial streams (direct feed). Once water in the storage tank or pond is needed for irrigation, it is released gravitationally at a certain pressure dependent on the height difference from the tank to the irrigated area, which can be regulated by pipe diameter, length and the type of emitters employed.

After the construction of the wetlands under MGNREGA, the Department of Rural Development and Panchayat Raj developed institutional mechanisms to sustain them. Maintenance of individual farm ponds is the responsibility of the beneficiaries of those ponds. Wetlands created for self-help groups and Panchayats have had ownership transferred to them and they are responsible for managing them. Individual beneficiaries use wetlands for growing vegetables and operating fisheries. Most of the bunds of wetlands are used for growing fruit, flowers, and vegetables. Self-help groups are engaged in the diversification of agriculture and allied activities. Panchayats have been leasing out wetlands for fisheries, water chestnut production, and lotus cultivation. Other common wetlands are used for providing life-saving irrigation to the kharif and rabi crops.

The construction of small check dams has helped farmers to increase cropping intensity. Many wells, small farm ponds, and streams are connected with solar panel pumps for irrigation. Several ponds are linked by installing solar panels to energize small pumps. This not only saves energy from the grid, but may also be able to feed the grid. The use of micro-irrigation helps reduce water consumption and optimally uses the harvested water. Fish can also be raised in ponds where the water stays for several months. Once the ownership of farm ponds and their benefits are assured, farmers will readily adopt new measures. The construction and restoration of tanks and village ponds under MGNREGA can be effectively converged with RRR and PMKSY and other schemes of the Ministry of Water Resources for sprinklers and drip irrigation (Gopal and Marothia, 2016; Marothia, 2017).

3.12 Wetlands-Agriculture Interactions (WAIs)

The Ramsar Convention adopted Resolution VIII in 2002 to enhance the management of agriculture, wetlands, and water resources, and subsequently to support the UN International Year of Family Farming. Ramsar's theme for World Water Day 2014 was "Wetland and Agriculture". In the Indian context, WAIs have not received even scant attention over the decades, with wetland scholars focusing on different disciplines.

It is now well-recognized that there exists an intimate relationship between wetlands and the landscape of which they are an integral part, and that neither of them can be managed independently. The findings of the Millennium Ecosystem Assessment (MEA, 2005) and the Comprehensive Assessment of water management in agriculture (CA) indicate that ecosystem services tend to be skewed towards over-exploiting provisioning services at the expense of regulating and supporting services This is due to rapid and invisible changes in wetland ecosystems and market demands, as well as substantive failures of government policies. Common WAIs practices that operate in major rural MWs (community wetlands, wetland complexes, riverine wetlands, multi-purpose irrigation tanks, *Dabries*, private ponds, *Munda*, *Tarbandha*, paddy fields converted for lotus cultivation) are discussed in different chapters of this book. A brief mention of *Munda*, *Tarbandha*, paddy fields converted for lotus cultivation is included here, and other MWs are covered in different chapters.

Tar Bandhs are traditional rainwater-harvesting structures designed to provide life-saving irrigation to local, long-duration paddy crops during dry conditions. These are constructed to retain maximum water by using mini earthen-dam structures. *Tar bundhs* require maintenance before every rainfall. Small fishes are also reared in their seasonal waters. *Tar Bundhs* have been part of indigenous water-harvesting techniques in Bastar and Surguja areas. In the Central Plains region, *tar bunds* have been constructed by the community and Panchayats. There are several types of traditional design of *tar bundhs*, and *mundas*, another such technology, can still be seen in remote tribal villages and hamlets.

Mundas are an age-old technology practiced by tribal communities of Bastar District to create common water sources on private land through collective action. *Mundas* are very small ponds constructed at the centre of four or five farm plots for supplementary irrigation and fish production. Mundas have traditionally been dug out by the local population with an average size of about 0.57 ha. The average size of *mundas* range from 0.02 to 0.6 hectares (Marothia, 1994, see also Chapter 8).

There are several types of traditional design of *tar bunds* and *mundas*. These structures are managed under private and collective, or common property regimes, depending on the size and number of stakeholders using the water resource for single or multiple activities, e.g., for irrigating crops and small fish cultivation. In the second phase of MGNREGA, thousands of wetlands have been created for irrigating crops, fisheries, and for domestic use. Many wetlands have been connected with solar panels to improve water efficiency.

Many shallow wetlands have been used for cultivating aquatic crops, and flooded fields are extensively used for the same purpose. Recently, many farmers have converted paddy fields into fields for lotus cultivation to enhance their income and reduce the risks associated with rice production, e.g. MSP, increasing input cost and market hazards.[6] In centuries-old wetlands that are still functional, the Panchayats, the Soil and Conservation wing of the State Agriculture Department, the Irrigation Department, and the Fisheries Department have constructed thousands of wetlands, some of which are for irrigation and fisheries purposes, while others are multipurpose.

Table 3.11 Attributes of Wetland-Agriculture Interactions in MWs: A Few Examples

Attributes of WAIs in MWs	*Multiuse Wetlands Categories*						
	Community Wetland (1)	*Wetland Complex (2)*	*Riverine Wetlands (3)*	*Irrigation Tanks (4)*	*Dabri Irrigation System (Micro-Minor Irrigation Tanks) (5)*	*Private and Community Ponds for Foxnut (6)*	*Paddy field converted for Aquatic Crops (7)*
WAI Combinations:	Fish + paddy	(i) Sole fish, (ii) Fish + lotus, (iii) Fish + water chestnut	(i) Watermelon (ii) Muskmelon (iii) Cucumber	(i) Paddy (ii) Fish	Paddy	Fox nut	Lotus
(i) *Main uses*	Food and nutritional security, ornamental fish for aquarium medicinal. few indigenous rice varieties have high zinc value	Fish: Food and nutritional security, ornamental fish for aquarium medicinal. Lotus+ water chestnut: Food, socio-cultural, religious medicinal, cosmetic, leaves of lotus vital for water purification of wetlands	Food and nutritional	Food and nutritional security, ornamental fish for aquarium medicinal. Basically, food security, few indigenous rice varieties have high zinc value	Basically, food security, few indigenous rice varieties have high zinc value	Dry fruit, sweet dishes, medicinal, starch use for silk and cotton polishing, Foxnut straw for as a poultry birds & cattle feed	Food, socio-cultural, religious medicinal, cosmetic, leaves of lotus vital for water purification of wetlands

(*Continued*)

Table 3.11 (Continued)

Attributes of WAIs in MWs	*Multiuse Wetlands Categories*						
	Community Wetland (1)	*Wetland Complex (2)*	*Riverine Wetlands (3)*	*Irrigation Tanks (4)*	*Dabri Irrigation System (Micro-Minor Irrigation Tanks) (5)*	*Private and Community Ponds for Foxnut (6)*	*Paddy field converted for Aquatic Crops (7)*
(ii) Yield (q/ha)	554 (Fish) 38 (Paddy)	3400 (sole fish) Fish + lotus: 2000 (Fish), 13500 (Lotus flower), 6000 (Rhizome), 700 (Pods nos.), 12 (Seeds), Fish + Water chestnut: 2000 (Fish), 25000 (Chestnut)	800 (Watermelon) 730 (Muskmelon) 600 (Cucumber)	Perennial Tank: 276.8 (Paddy), 450 (Fish), Seasonal Tank: 255 (Paddy), 330 (Fish)	20 (Paddy)	275 (Fox nut)	1000 (Lotus flower) 50 (Rhizome) 500 (Pods nos.) 5 (Seeds)
(iii) Return per rupee invested	1:3.99 (Fish) 1:1.26 (Paddy)	1:1.92 (Sole Fish); Fish + Lotus: 1:4.79 (Fish), 1:3.17 (Lotus); Fish + Water chestnut:1:5.00 (Fish), 1:2.86 (Chestnut)	1:4.50 (Watermelon) 1:5.60 (Muskmelon) 1:5.30 (Cucumber)	Perennial Tank: 1:1:74 (Paddy), 1:1.39 (Fish) Seasonal Tank: 1:1:75 (Paddy), 1:1.16 (Fish)	1:1.30 (Paddy)	1:1.80 (Fox nut)	1:1.98 (Lotus)

(iv) Price determining mechanism	Largely fish price fixed by producers Paddy at MSP	Largely price fixed by producers Price of lotus flower, pods very during normal (Rs. 1–5) and festival season (Rs. 30–50) Bulk of harvest of flowers done during festival season producer	Largely price fixed by producers	Largely price fixed by producers Paddy to Mandi at MSP, direct to open market	At MSP, direct to open market	Largely price fixed by producers	Largely price fixed by producers Price of lotus flower pods vary during normal (Rs. 1–5) and festival season (Rs. 30–50) Bulk of harvest of flowers done during festival season producer
(v) Marketing channels	Producer to retailer, producer to wholesaler, Producer to Mandi	Producer to retailer, retailer to consumer, producer to wholesaler	Producer to wholesaler, retailer to consumer, producer to consumer	Producer to retailer, producer to wholesaler, Producer to Mandi	Producer to Mandi	Producer to wholesaler	Producer to retailer, producer to wholesaler
(vi) Export opportunity	Domestic markets	Domestic markets	Domestic markets	Domestic markets	Not applicable	High export opportunity for fox nut pop	Opportunity in South-East Asia
(vii) *Value addition* (by products)	Low to medium Limited byproducts	Low to medium Medium to high (direct consumption, roasted and by products after grinding the dry fruits	Direct consumption and fruit juices	Low to medium Limited byproducts	Limited by products	A range of products add values	A range of byproducts add values

Source: Author's analysis based on field work

Note: Nutritional levels of aquatic crops listed are very high in terms of calories, sodium, potassium, calceium, tcarbohydrates, fibre, suger, protein, and water content (see Marothia, 2021)

For illustrative purposes, common WAIs practices in operation in major rural MWs (community wetlands, wetland complexes, riverine wetlands, multipurpose irrigation tanks, *Dabries*, private ponds, paddy fields converted for lotus cultivation) are briefly shown in Table 3.11. The production, prices, and marketing mechanism value addition and the export potential of different wetland farming products (fish, lotus, water chestnut, fox nut, paddy) are briefly described in Table 3.11. (For detailed analysis of WAIs see Marothia, 2021.)

Photographs of WAIs pertaining to paddy and fish, fish and lotus, fish, and water chestnut, foxnut, and lotus in MWs are shown in Photo 3.1 c, d, e. Production of watermelon, muskmelon, and cucumbers in riverine wetlands are shown in Photo 3.1b.

WAIs, in the majority of cases, are not a major cause of wetland degradation and loss. However, in a number of cases, declining fish yields have been reported due to the intensification of paddy production and increased demand for domestic use of water under water-stress situations. Weak institutional arrangements governing access to water have also been responsible for conflicts among stakeholders. With effective enforcement of institutional arrangements, provisioning services can be improved for agriculture and allied activities without affecting regulating and supporting services (see Chapter 6). In fact, some wetlands in the state have degraded and have vanished due to pollution, the dumping of rural and peri-urban wastes, the large-scale conversion of physical infrastructure, encroachment by powerful people and people suffering from chronic poverty, and government policy failures or the absence of a wetland policy. This needs a restructuring of existing institutional arrangements (this is discussed throughout this book).

The previously mentioned factors have increased pressure on the remaining wetlands and their ecosystem services. Therefore, there is a need to explore how to improve the nature of WAIs in order to strike an appropriate balance in ecosystem services for sustainably supplying all services, and not only provisioning services. In view of the above facts, there is an urgent need to enhance the options for valuing and understanding non-provisioning ecosystem services whose economic benefits can accrue within the local context, especially in rural economies in Chhattisgarh. To this end, ecosystem services of different categories of wetlands are summarized in Table 3.12 and discussed briefly next. Photographs in Photo 3.1 a to l also depict the ecosystem services derived from multiuse WAI.

3.13 Ecosystem Services of Different Categories of Wetlands

Wetlands provide human society with essential goods, services, and many direct and indirect benefits; option, bequest and existence non-use values (see Munasinghe, 1992, for components of total economic values concepts, see Chapters 4 and 5 for user and non-user assessment of values of wetland). Humans benefit in many ways from all kinds of wetlands. These benefits

Table 3.12 Comparative Extent of Ecosystem Services Derived From Different Categories of Wetlands in Chhattisgarh

Ecosystem services	*Shallow Lakes*	*Deep Lakes*	*Flood plains*	*Marshes/Swamp*	*Urban Wetlands*	*Reservoirs*	*Irrigation Tanks*	*Temple Tanks/sacred Ponds*	*Fish Ponds*	*Paddy fields*	*Village Ponds*	*Paddy Field converted for Aquatic Crops*	*Caves/Karst*	*Wetland Town*	*Water Bodies MGNREGA*	*Wetlands Inside Forest*
Provisioning Services																
Food	◆◆◆	◆	◆◆	◆◆	◆	◆	◆	◆	◆◆◆	◆◆◆	◆	◆◆◆	◊	◆◆◆	◆◆	◆◆
Fodder	◆	◊	◆◆◆	◆◆◆	◊	◊	◊	◊	◊	◆	◊	◊	◊	◆◆	◆◆	◆
Fuel	◆◆	◊	◆◆◆	◆◆	◊	◊	◊	◊	◊	◊	◊	◊	◊	◆◆	◆◆	◆
Fibre	◆◆	◊	◆◆◆	◆◆◆	◊	◊	◊	◊	◊	◆	◊	◆	◊	◆◆	◆◆	◆
Medicinal	◆◆	◊	◆◆◆	◆◆◆	◊	◊	◊	◊	◊	◆	◊	◆	◊	◆◆◆	◊	◆
Bio-chemicals	◆◆	◊	◆◆	◆◆	◊	◊	◊	◊	◊	◆◆	◊	◆	◊	◆◆	◊	◆
Genetic materials	◆◆	◊	◆	◆◆	◊	◊	◊	◊	◊	◊	◊	◆◆	◆◆◆	◆◆◆	◊	◆
Water storage	◆◆	◆◆◆	◆◆	◆◆	◆◆◆	◆◆◆	◆◆	◆◆	◆◆	◆	◆◆◆	◆◆	◆◆	◆◆◆	◆◆	◆◆◆
Domestic requirement of local community*	◆◆◆	◆◆	◆◆	◆◆	◆◆◆	◆◆	◆◆◆	◆	◆	◊	◆◆◆	◊	◊	◆◆◆	◆	◆**
Regulating Services																
Sediments	◆◆◆	◆◆	◆◆◆	◆◆◆	◆	◆◆	◆◆	◆◆	◆	◊	◆	◊	◊	◆◆◆	◆	◆◆
Heavy metals	◆◆	◆	◆◆◆	◆◆◆	◆	◊	◊	◊	◊	◊	◆	◊	◊	◆	◊	◊
Toxics	◆◆	◆	◆◆◆	◆◆◆	◆	◊	◊	◊	◊	◊	◆	◊	◊	◆	◊	◊
C- Sequestration	◆◆◆	◆	◆◆◆	◆◆◆	◊	◆	◊	◊	◆	◆	◊	◆	◊	◆◆◆	◆◆	◆◆
GHG emission	◆◆	◆	◆	◆◆	◊	◆◆	◊	◊	◊	◆◆	◊	◊	◊	◊	◊	◊
Erosion control	◆◆	◊	◆◆◆	◆◆◆	◊	◊	◊	◊	◊	◊	◊	◆◆	◊	◊	◊	◊

(*Continued*)

Table 3.12 (Continued)

Ecosystem services	*Shallow Lakes*	*Deep Lakes*	*Flood plains*	*Marshes/Swamp*	*Urban Wetlands*	*Reservoirs*	*Irrigation Tanks*	*Temple Tanks/sacred Ponds*	*Fish Ponds*	*Paddy fields*	*Village Ponds*	*Paddy Field converted for Aquatic Crops*	*Caves/Karst*	*Wetland Town*	*Water Bodies MGNREGA*	*Wetlands Inside Forest*
Disease vectors	◆◆	◊	◆◆	◆◆	◆◆	◊	◆	◊	◆	◆	◆◆	◊	◊	◆	◊	◊
Pollination	◆◆	◊	◆◆	◆◆	◊	◊	◊	◊	◊	◊	◊	◆	◊	◆◆	◆	◆◆
Cultural and Social Services																
Recreation	◆◆◆	◆	◆◆◆	◆◆◆	◆	◆	◆◆	◆	◊	◊	◆	◆	◆◆◆	◆◆◆	◆	◆◆
Aesthetics	◆◆◆	◆	◆◆◆	◆◆◆	◆	◆	◆	◆	◆	◊	◆	◆	◆◆◆	◆◆◆	◆	◆◆
Spiritual	◆◆	◊	◆	◆	◆	◊	◆	◆◆	◊	◊	◆	◆◆	◆◆◆	◆◆◆	◊	◊
Religious	◆◆	◊	◆	◆	◆	◊	◆	◆◆	◊	◊	◆	◆◆	◆◆◆	◆◆◆	◊	◊
Education R&D	◆◆◆	◆◆◆	◆◆◆	◆◆◆	◆◆◆	◆◆◆	◆◆◆	◆◆◆	◆◆◆	◆◆◆	◆◆◆	◆◆◆	◆◆◆	◆◆◆	◆◆	◆◆
Supporting Services																
Soil Formation	◆◆◆	◊	◆◆◆	◆◆◆	◆	◊	◊	◊	◊	◊	◆	◆	◆	◆◆◆	◊	◆
Groundwater Recharge	◆◆◆	◆	◆◆◆	◆◆◆	◆◆	◆	◆	◆	◆	◆	◆◆	◆◆	◆◆	◆◆	◆◆	◆◆
Nutrient cycling	◆◆◆	◆	◆◆◆	◆◆◆	◆	◆	◆	◆	◆	◆	◆	◆◆	◆	◆◆	◆◆	◆◆
Water cycling	◆◆	◆◆◆	◆◆	◆	◆	◆◆◆	◆◆	◆◆	◆	◆	◆	◆◆	◆	◆◆	◆	◆◆
Photosynthesis	◆◆◆	◆◆	◆◆◆	◆◆◆	◆◆◆	◆◆	◆◆	◆◆◆	◆◆◆	◆◆◆	◆◆◆	◆◆◆	◊	◆◆◆	◆◆	◆
Biodiversity																
Microphyte	◆◆	◆◆◆	◆◆	◆◆	◆◆	◆◆◆	◆◆	◆	◆◆◆	◆	◆◆	◆	◊	◆◆	◆◆	◆◆
Macrophyte	◆◆◆	◊	◆◆	◆◆◆	◆	◊	◆◆	◆	◆	◆	◆	◆	◊	◆◆	◆◆	◆
Fish	◆◆◆	◆◆	◆◆	◆◆	◆	◆◆	◆◆	◆	◆◆◆	◆	◆	◆	◆	◆◆◆	◆◆	◆
Zooplankton	◆◆◆	◆	◆◆	◆◆◆	◆◆	◆	◆◆	◆	◆◆	◆	◆◆	◆	◊	◆◆	◆◆	◆

Benthos	◆◆◆	◆	◆◆	◆◆◆	◆	◆	◆◆	◆	◆◆	◆◆	◆	◆	◊	◆	◆◆	◆
Birds	◆◆◆	◆	◆◆	◆◆◆	◆	◆	◆◆	◆	◆	◆	◆	◆◆	◊	◆◆◆	◆◆◆	◆◆◆
Herpetofauna	◆◆◆	◆	◆	◆◆◆	◆	◊	◆◆	◆	◆	◆	◆	◆	◊	◆	◆◆	◆
Lotus diversity	◆	◊	◆	◆◆◆	◆◆◆	◊	◊	◆◆	◆◆	◊	◆◆◆	◆◆◆	◊	◆◆◆	◆◆	◆
Aquatic crops diversity (water chestnut, water spinach/Makhana (*Euryale ferox*), watermelon, seasonal vegetables)	◆	◊	◆◆	◆◆	◆◆	◊	◊	◆◆	◆◆	◊	◆◆	◆◆	◊	◆◆	◆◆	◊
Floral diversity	◆	◊	◆◆	◆◆	◆◆	◆	◊	◆◆	◆◆	◊	◆◆	◆◆	◊	◆◆	◆	◆◆
Screeds vegetation	◆◆◆	◆◆◆	◆◆◆	◆◆	◆◆◆	◆◆	◆◆	◆◆◆	◆◆	◆	◆◆◆	◆◆	◆	◆◆◆	◆	◆◆

Source: Author's analysis based on field work

Benefits: – ◊no benefits; ◆ little; ◆◆ medium, ◆◆◆ large

*Bathing, washing utensils, clothes, tending cattle, drinking, and sanitation.

** Indigenous population living inside and at forest fringe using wetland water for domestic activities.

Note: Option value reflects future direct and indirect uses of goods and services mentioned. Non-use values include: existence, bequest and altruistic value of wetland habitats and species; traditional/cultural knowledge and traditions.

arise either directly from the use of the water, plants, animals, soils, and other components of the wetland ecosystem, or indirectly from their ecosystem functions. These benefits are now called ecosystem services and categorized into provisioning, regulating, supporting, and cultural services (see MEA, 2005, for categorization of ecosystem services and examples of related services, see Gopal, 2015; Gopal and Marothia, 2017, for extended ecosystem services of wetlands in India). Ecosystem services differ from wetland to wetland, even within the same category of MWs under different agro-climatic environments. Ecosystem goods and services play an important role in delivering the benefits from different types of wetlands. A comparative picture of the extent of ecosystem services derived from different categories of wetlands[7] [8] in Chhattisgarh is presented in Table 3.12.

Notes

1 The exact history of the state cannot be outlined, but it has been traced that Lord Rama stayed here during his days of exile and the region was referred to as Dakshin Koshala in ancient times. The history of Chhattisgarh or South Koshala can be traced back to the 2nd-4th century AD. The first rulers of the state were the Satavahanas, who ruled during the 2nd-4th century AD. Between the 6th and 12th centuries AD, the Sarabhpurias, Panduvanshi, Somvanshi, Kalchuri, and Nagvanshi rulers dominated this region. The Satavahana king Mahendra was defeated by Samudragupta in the 4th century AD. The next ruler who ruled the region was the Panduvanshi, whose king, Mahashiva Gupta Balarjuna, is famous for freeing his country. It was during his reign, 595–655 AD, that Hieun Tsang visited Dakshin Koshala. In the early 11th century, the Chedis came into power. King Ratnadeva founded his capital at Ratanpur and from there they started calling themselves Haihaya. The dynasty split in the 14th century and the elder brother remained at Ratanpur while the younger brother settled at Raipur. In the medieval period, the region came to be known as Gondwana and became a part of the kingdom of the Kalchuris, who ruled the region until the end of the 18th century AD. The Muslim chroniclers of the 14th century AD have described well the dynasties that ruled over the region. The region also came under the suzerainty of the Mughal Empire in the 16th century and later under the Marathas in 1745. By the year 1758, the whole region of Chhattisgarh was annexed by the Marathas, who plundered its natural resources mercilessly. The word 'Chhattisgarh' was popularized during the Maratha period and was first used in an official document in 1795. With the British entry in the early 19th century, much of the territory was subsumed into the Central Provinces. From 1854 onwards the British administered the region, with a deputy commissionership headquartered at Raipur. Chhattisgarh took part in the 1857 Revolution. The region was mentioned in the *Imperial Gazetteer* of India, Vol. X, 1908, as one "being far-removed from the routes of armies and protected from invasion or disturbances by the precipitous ranges which fringe the plain on three sides. (The details of this note have been drawn from Gupt 1973; Sharma, 2008).

2 Bhuyan and Marothia (1990); Marothia (2018); Garg (2017); Chaudhary et al. (2015; Chaudhary, Das, Unjan et al., 2018; Chaudhary, Das, Patel et al., 2018).

3 Technical support and insights received from Professor Ninad Bodhankar in creating the aquifer delineation map is gratefully acknowledged.

4 B.P. Nonhare, Additional Principal Chief Conservator of Forest (Monitoring and Evaluation, GoCG) provided information related to wetlands inside forest areas.

Moiz Ahmed, Nova Nature, provided photographs of wetlands inside forest and protected areas. Thanks to them.

5 The temple of Kaushalya (mother of Lord Rama) is the rarest, similar to the ancient temple of Brahma Ji in Pushkar, and is located near Raipur. Chandkhuri (ancient name Chandrapuri) is a village about 25 km from Raipur, the state capital. The village, which has about 126 ponds, has an Eesa temple on the ancient island in the middle of the Jalsan Talab surrounded by seven ponds, where the statue of Lord Shri Rama's mother Kaushalya is installed. It is believed that Chandkhuri is the birthplace of Mata Kaushalya. It is said that Sushyan/Sukhain Vaidya also came from Lanka with Lord Rama after the end of Ravana and he gave up his life at Chandkuri. The state government has been working to develop Chandkuri as a religious tourism centre.

6 P.C. Mishra, Commissioner, MGNREGA, Department of Panchayat and Rural Development, GoCG provided information related to water-centric activities. Thanks to him.

7 See Sahu and Chandravanshi (2017a, 2017b); Sahu and Gopal (2017) for agronomic aspects of foxnut and converted paddy fields for lotus cultivation. I gratefully acknowledge the assistance of S.S. Chandravanshi, Senior Scientist and Head, KVK, Dhamtari, for data collection on economic parameters for foxnut on private and community ponds, and lotus crops on paddy-converted fields for my research.

8 Ecosystem Services developed in the table is based on the research framework developed in Costanza et al. (1989); Barbier et al. (1997); Aylward and Barbier (1992); Munasinghe (1992); Marothia (2001a, 2001b, 2004d, 2021); De Groot et al. (2002); Gopal and Marothia (2015, 2017); Gopal et al. 2016). An alternative classification of ecosystem services was also suggested by Wallace (2007).

4 Ratanpur Wetland Complex

The Existential Crisis

This chapter is divided into 11 sections. Section one provides a brief profile of the ancient and sacred temples and ponds of Ratanpur in Chhattisgarh state. Section two gives brief profile of agro-climatic and socio-economic characteristics of Ratanpur. Sections three and four describe the unique attributes of wetland complexes and historical Dulhara *sarovar* (multiuse water bodies) respectively. Activities and the current status of ecological health of ponds within wetland complexes are discussed in section five. Evidence of the poor ecological health of the wetland complex and institutional failures at different hierarchy levels is presented in section six. A synthesis of institutional analysis frameworks, developed in Chapter 2, is used here to analyse the performance of ponds and tanks of wetland complex in section seven. Sections eight and nine provide facts about a wide range of ecosystem services, financial gains, income, and employment opportunities for different stakeholders deriving from diverse activities undertaken in ponds of wetland complexes. WTP for restoration, conserving, and managing temple ponds and wetland complexes, perceptions from direct users and non-users are presented in section ten. Workable institutional arrangements are suggested in the final section to reverse ecological degradation and facilitate the process of getting Ratanpur classed as a Ramsar site under Ramsar Wetland City Accreditation.

4.1 The Sacred Historical Town

The historical timeline of Ratanpur runs over millennia. Many illustrious references are made to a Ratanpur Kingdom in the epics and *Punanic granth* (ancient literature), like *Ramayana, Mahabharata, Jamini*, and the *Rewa Khand of Skanda Purana*. There are many mythological stories connected with Ratanpur.[1] Ratanpur, originally known as Ratnapura, was established by Kalingaraja's grandson Ratnadev I in 1050 AD. Ratanpur continued as the capital of Haihaiavansi Kalachuri Kingdom until the 18th century.[2]

Ratanpur, full of temples and ponds, is an ancient town of immense socio-economic, cultural, archaeological, architectural, and historical value, and is important for hydrological science and ecological heritage. Ratanpur town is popular as a religious centre and a million or more Hindu devotees travel

DOI: 10.4324/9781003456742-4

there twice a year to offer their prayers and seek the blessings at one of the many temples: the Mahamaya Temple (one of the *Sakti pithas*), Baba Kaal Bhairavnath Temple, Bhuddeshwar Shiva Temple, Ekbira Temple (Mahalakhmi or Lakhani devi), Ratneswar Mahadev Temple, Ramtekri temple, Kanthi Devel Mahadev Temple, Bairagban, Bees Duvariya Temple, Bhuvaneshvar Mahadev Temple, Pali ka Shiv temple, Girijabandh Hanuman temple, Khandoba temple, Jagannath temple, Kedarnath Temple, and the Dusdus Mahadev Temple. All these temples have ponds attached to them and each of these temple ponds, and other ponds, have ancient and historical backgrounds. That is why Ratanpur is popularly known as 'Temple town' as well as 'Town of Ponds'. During their long rule, the Kalachuri left various powerful legacies in the area that can be found in and around Ratanpur and other parts of Chhattisgarh.[3]

There are also places of historical significance for other religions, like Kabir Panthis, Jainism, Buddhism, and Islam. The town is surrounded by five hills, namely, Kaka Pahad, Ghughus Pahad, Gopal Pahari, Bedarchua Pahad, and Chhotak bandha-Ghodbandha. Ratanpur has a unique endemic ecosystem of plants and organisms indigenous to this historical town. Because of these unique geographical environs, it is claimed by some historians that King Ratandev and his legatees made Ratanpur the capital of their Kalachuri kingdom. These hills have been, and are still, considered as sacred common land. Even just 30 years ago crocodiles, turtles, frogs, and ornamental fish were some rare species that made their homes in many of these wetlands. After commercial fisheries and lotus cultivation became established, crocodiles, and to a great extent, turtles, disappeared over the years. Even so, turtles are still occasionally seen.

Ratanpur has many natural and human-made MWs. These play an important role in life-saving irrigation, freshwater aquaculture, aquatic-crop production (particularly large-scale lotus cultivation for religious purposes), domestic use, and to serve as homes for resident and migratory birds.

4.2 Ratanpur: Agro-Climatic and Socio-Economic Profile

Ratanpur is situated in the Kota block of Bilaspur district, 25 km away from Bilaspur city, at 22.28° N latitude, 82.15° longitude, and at an elevation of 306 above mean sea level (see Map 4.1). The total geographical area of Ratanpur is 4,426.26 ha, of which 28.10 percent is covered by forest. Net cropped area is 87.12 percent of the total cropped area of 1,987.251 ha. Of the total irrigated area of 1256.075 ha, kharif accounts for about 79.10 percent, the remaining area under rabi season crops. Tank and tube well water irrigates nearly 90 percent of the area. Paddy is the prime crop and is sown in 99.70 percent of total cropped area. Vegetables and pulses are also cultivated. As a result, the cropping intensity has been hovering around 144 percent. Fish, lotus, and water chestnut are the main produce of the ponds of the complex. NTFPs and useful herbs and shrubs are available in the town,

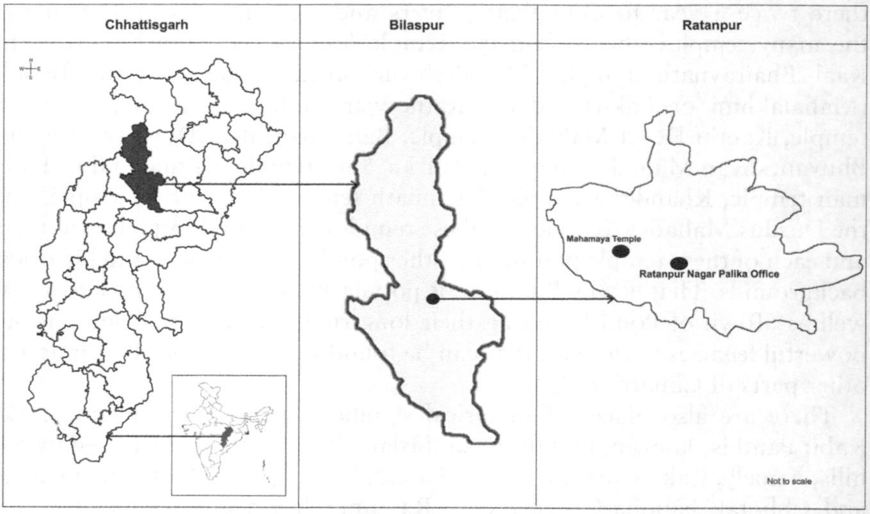

Map 4.1 Location Map of Ratanpur Town
Source: Devised by the author

and their production is critical to the livelihoods of tribal and other under-privileged people.

Ratanpur is representative of the agro-climatic and socio-economic conditions of the Central Plains of Chhattisgarh. The climate of Ratanpur is sub-tropical. The average annual rainfall varies between 1,187 mm and 1,200 mm. The monsoon season extends from mid-June to early October, and more than 90 percent of the total precipitation occurs during the months of July and August. The winter season (November to February) is relatively warm with a mean temperature of 25°C to 30°C between December and March, followed by very hot and dry weather in summer from May to June, with mean temperatures between 40°C and 45°C. As in the case of the other parts of the state, five types of soils are found in Ratanpur, namely, clayey (>2mm), fine, fine loamy (0.25–01mm), fine silty (0.25–0.1mm), and loamy skeletal (2mm). Of the area of 4,418.20 ha, the fine silty soils and the loamy skeletal soils account for about 62.04 percent and 14.84 percent respectively. The remaining three soils occupy 2.65 percent of the total area. As per the census of 2011, the total population of Ratanpur Nagar Palika (RNP) is 24,636, representing all social groups (Table 4.1). The main workforce is engaged in fisheries, lotus and other aquatic-crop cultivation, livestock rearing, and the collection of NTFPs.

Table 4.1 Profile of Ratanpur Town

S.N.	*Rural Resources*	*Extent of Availability*
1.	Geographical area (ha)	4426.262
	(i) Forest area (percent)	28.10
	(ii) Fallow land cultivation (percent)	4.61
	(iii) Pastureland (percent)	7.50
2.	Total cropped area (ha)	1987.251
	(i) Net cropped area (percent)	87.12
	(ii) Double-cropped area (percent)	12.88
	(iii) Area under kharif (percent)	87.12
	(iv) Area under rabi (percent)	12.88
3.	Cropping intensity (per cent)	144.00
4.	Total irrigated area (ha)	1256.075
	(i) Irrigated under kharif (percent)	79.10
	(ii) Irrigated under rabi (percent)	20.90
5.	Total irrigated area (ha)	993.558
	(i) Canal (percent)	15.16
	(ii) Tanks (percent)	42.02
	(iii) Tube wells (percent)	42.82
7.	Total farmers (on the basis of caste) (nos)	2202.00
	(i) Schedule caste (percent)	14.08
	(ii) Schedule tribes (percent)	13.53
	(iii) Other caste (percent)	72.39
8.	Total farmers (category wise) (nos)	2202.00
	(i) Marginal (percent)	72.52
	(ii) Small (percent)	16.35
	(iii) Large (percent)	11.13
9.	Total land owned by farmers (ha)	2102.534
	(i) Schedule caste (percent)	11.42
	(ii) Schedule tribes (percent)	13.61
	(iii) Other caste (per cent)	74.97
10.	Total land owned by farmer (percent)	1902.534
	(i) Marginal (percent)	23.80
	(ii) Small (percent)	26.12
	(iii) Large (percent)	50.08
11.	Total population (Census 2011) (nos)	24636.00
	(i) Male (percent)	51.28
	(ii) Female (percent)	48.72
	(iii) Male literacy (percent)	73.60
	(iv) Female literacy (percent)	56.92
12.	Total caste population (Census 2011) (nos)	24636.00
	(i) Schedule caste (percent)	20.59
	(ii) Schedule tribes (percent)	14.86
	(iii) General (percent)	64.55
13.	Total cropped area kharif (ha)	1731.221
	(i) Paddy (percent)	99.70
	(ii) Maize (percent)	0.14
	(iii) Pigeon pea (percent)	0.12
	(iv) Niger (percent)	0.05

(Continued)

Table 4.1 (Continued)

S.N.	*Rural Resources*	*Extent of Availability*
14.	Total cropped area rabi (ha)	256.00
	(i) Wheat (percent)	31.25
	(ii) Chickpea (percent)	0.78
	(iii) Pea (percent)	23.44
	(iv) Lathyrus (percent)	3.12
	(v) Mustard (percent)	4.69
	(vi) Kusum (percent)	1.17
	(vii) Sugar Cane (percent)	0.78
	(viii) Other Vegetables (percent)	34.77
15.	Local institutions (FCSs, self-help groups, Mahamaya Temple Trust, Primary Minor Forest Produce Cooperative Society, bank institutions and the Agriculture Cooperative Society (Double lock centre, Single lock centre), private agriculture input dealer, *Sinchai Samittee*, nature clubs and other cultural organizations, rural health centres, Aganwadi, schools and colleges.	

Source: Rural Agriculture Development Officer, Area – Ratanpur, Block Name – Kota, Gram – Ratanpur, District – Bilaspur (Chhattisgarh), 2019–20

4.3 Wetland Complex: Unique Attributes

Over the centuries, Kalchuris developed about 1,400 interconnected ponds utilizing brilliant water-science principles that factored in the geo-morphological characteristics of the area i.e. slope direction, soil depth, fertility levels, and natural drainage networks. The slope varies largely from nearly level and very gently sloping to moderately steep to steep. All the above parameters made the cascade pond system sustainable for centuries. Water flows through natural drainage networks from one pond to the next filling each one in turn and forming the pond chain system. The complex is unique in terms of pond density and interconnectedness. However, in the last hundred years, particularly after independence, a number of ponds disappeared one by one due to siltation, encroachment, low water-retention capacity, decreased average depth of water, increased density of tube wells in the wetland complex area, and loss of traditional, informal institutional arrangements.

Recently, the Directorate of Fisheries, GoCG, has created a database of the ponds and tanks of Ratanpur, using a remote sensing technique to assess the seasonality, water-retention months, the growth of aquatic weeds, and the possibilities of cultivating fish and aquatic crops. Remote sensing data revealed that there were 174 ponds in Ratanpur Nagar Palika (local governing body) during 2014. Remote sensing data shows that the number of ponds and their area are higher than that recorded for the same parameters in the official records of Ratanpur Nagar Palika. The distribution of ponds according to the number and size are given in Table 4.2 and Figure 4.1, and it can be seen that about 92.70 percent of the total number of ponds of the complex have 66.60 percent water-spread over an area of under 0.5 ha to 4 ha.

Table 4.2 Distribution of Ponds in the Wetland Complex (*Number and Water-Spread Area of Ponds*)

Pond Area (ha)	0–0.5 ha		0.5–2ha		2–4 ha		4–6 ha		6–8 ha		8–10 ha		10–20 ha		above 20 ha		Total	
	No.	Area	No.	Area	No.	Area	No.	Area	No.	Area	No.	Area	No.	Area	No.	Area	No.	Area
Perennial (P)	3	12.091	60	74.174	32	84.66	5	22.895	2	13.36	1	8.497	1	10.976	2	43.937	106	259.703
Seasonal (S)	30	10.885	38	27.882	0	0	0	0	0	0	0	0	0	0	0	0	68	38.767
Aquatic weeds	1	0.367	3	2.556	0	0	0	0	0	0	0	0	0	0	0	0	4	2.923
Under Fisheries cultivation	17	6.755	98	102.056	32	84.66	5	22.895	2	13.36	1	8.497	1	10.976	2	43.937	158	293.134

Sources: Remote sensing data of Ratanpur water bodies, State Department of Fisheries, GoCG, 2014 (Compiled by the author)

Note: According to remote sensing data, there are 174 water bodies covering an area of 298.47 ha. This number is higher than that in physical records of Ratanpur Nagar Palika (see Table 4.4).

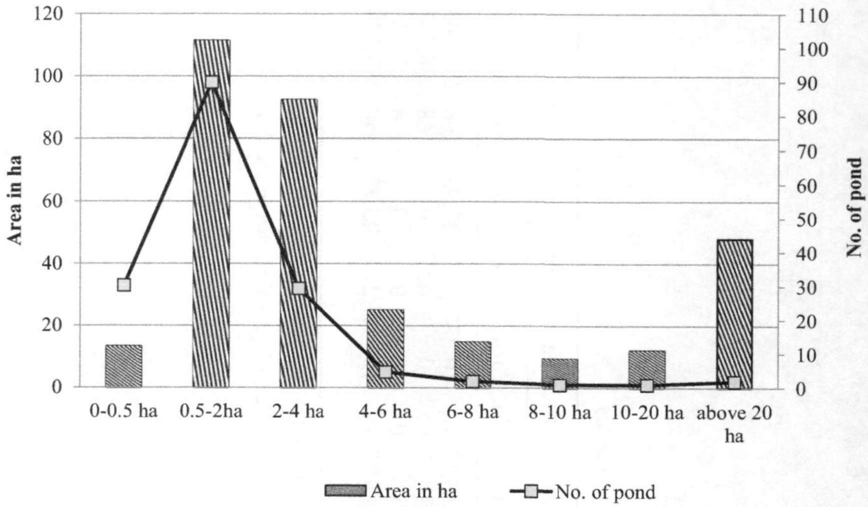

Figure 4.1 Area and Number of Ponds in Wetland Complex

Source: Drawn from remote sensing data of Ratanpur water bodies, Dept. of Fisheries, GoCG, 2014

Against this, 7.30 percent of total ponds cover 33.40 percent water area under 4 ha and above 20 ha categories. These water bodies have also been marked with Unique Identity Numbers using the remote sensing technique (Map 4.2).

As mentioned, the Directorate of Fisheries, GoCG, has created a database of ponds and tanks in Ratanpur using a remote sensing technique. Unique Identity Numbers have been assigned to each water body (Map 4.2). The location of wetlands and contours are given in the Map 4.2.

The topographical map of Ratanpur has been drawn on the GIS platform using ArcGIS version 10.3. Water bodies have been delineated as per the contour lines of Ratanpur. The elevation of the area declines from west to east indicating a slope from 410 m to 280 m. The largest number of water bodies is located at or below 310 m elevation, which have formed in a natural depression. As a result, the flow of rainwater is towards the lower elevation. The depression supports more water retention as it is associated with low turbulence flow. A total of 203 water bodies are present in and around the historical city of Ratanpur, ranging from small to large, and they can store a huge quantity of water with a uniquely interconnected system to equalize the flow. While constructing water bodies, the surface morphology of the town was reviewed and designed using the concept of rainwater harvesting and conservation.

Iso-contour lines follow the contours of the land, with elevation ranges between 346 m and 447 m, smartly moving water to the water-storage

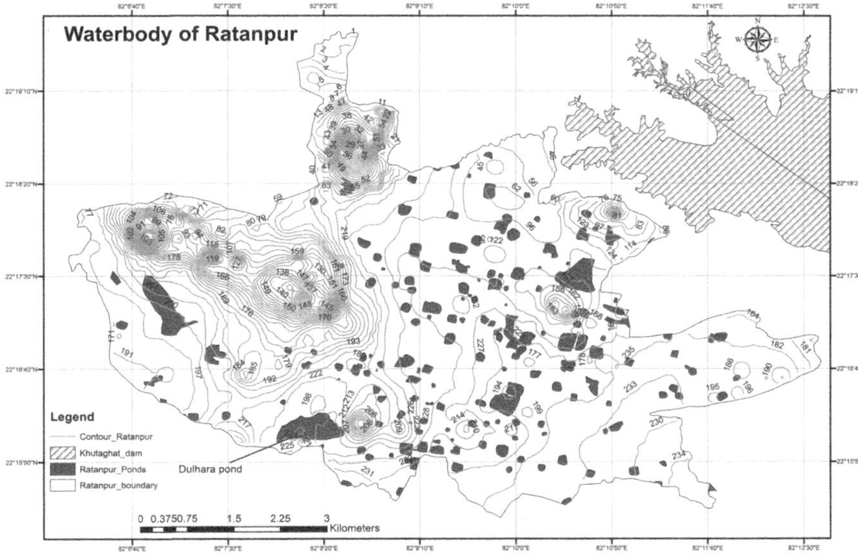

Map 4.2 Water Bodies and Contour Mapping of Wetland Complex Within the Area of Ratanpur Nagar Palika

Source: Govt. of Chhattisgarh, State Department of Fisheries, Village wise water bodies map, District – Bilaspur, Block-Kota, Village (Ratanpur NP) Reproduced with permission of Director, State Department of Fisheries, GoCG)

Note: Contour mapping was developed using the basic remote sensing map (State Department of Fisheries, GoCG, 2014) of wetlands of Ratanpur.[4]

structures and maintaining surplus flow. Further, the flow of water joins natural streams to distribute surplus water and subsurface water equally among the different water bodies of the area, which ultimately ends with a natural drain (river system).

4.4 Dulhara *Sarovar* (Tank)

Dulhara pond is a historical water storage tank constructed during the rule of the Kalachuri to provide water for irrigation and the domestic needs of the people of Ratanpur and nearby villages. Dulhara pond is surrounded by hills on all three sides, Lakhani Devi mountain (Isha mountain) to the north and Siddha Baba mountain to the east.

The name "Dulhara" also has its own special meaning. During the medieval period, it was a beloved lake of sages who stayed for long periods to attain their spiritual goals. Many of them often observed overlapping waves coming from the north and the south banks of the wetland, believed to be the waters of the rivers Ganga and Yamuna merging. Due to this belief, this wetland has attracted socio-economic and religious activities for centuries.

The name is derived from the Chhattisgarhi language, literally meaning the progression of two (*Du)* waves (*Lehara)*. With great reverence, saints over generations started calling this rhythm "Dulhara", and even today the lake is popularly known by this name (translated from Sukhdev Prasad Kashyap, 2004, pp. 89 and 90).

Rainwater flows into the tank from the surrounding hills and from Muse Khan's tomb (Sukhdev Prasad Kashyap Kashyap, 2004, p. 90). A large part of the tank has been encroached upon for farming and poor households living on its periphery, observed by this author during a field visit in the summer of 2019. This has changed the ecological landscape of the tank.

A brief profile of the Dulhara tank is given in Table 4.3. Dulhara is the largest water body, designated as an irrigation tank, in the area of Ratanpur town (see Map 4.2). Since independence, the tank has been controlled and managed by the DWR, Govt. of Chhattisgarh. The tank provides mainly irrigation to kharif and rabi crops for the farmers of four villages. Dulhara also receives water from the overflow of the Chapi Tank through a feeder canal.[5] Cases of canal damage have been reported in the last few years. Farmers located between the feeder canal outlet and the inlet into Dulhara damaged

Table 4.3 Profile of Dulhara Irrigation Tank

S.No.	*Particulars*	*Details*
1.	Year of transfer to DWR Govt. of Madhya Pradesh	1960
2.	Water-spread area (ha) in 1950	30.756
3.	Depth with full tank capacity, live-storage capacity, and dead-storage level	5.50 m/1.075mcm/0.32 mcm
4.	Total irrigated area under kharif and rabi (ha)	300
5.	Total command area (ha)	300
6.	No. of farmers receiving irrigation (as per agreement)	948
7.	Duration of water availability (June to May)	0.32 mcm
8.	Feeding canal connecting to Dulhara for water supply	3.09 km Canal
9.	Other activities in Dulhara	
	a. Fisheries	Around the years
	b. Lotus growing	
	c. Social, cultural, and religious activities	
10.	Villages receiving irrigation (as per agreement)	Ranigaon, Ratanpur, Janjiradih, Madnupar
11.	Irrigation fees for kharif + rabi (Rs)	Rs. 230 (kharif) Rs. 600 (rabi)
12.	Collection of irrigation fee	DWR

Source: Office of Executive Engineer, Water Resources Division, Kota, Dist. Bilaspur, DWR, Govt. of Chhattisgarh, 2019–20

the canal at several points to divert water to their fields. Due to the uncertain supply of water, only about 150 farmers are able to receive water for irrigation. Besides irrigation, the tank has also been used for fisheries and aquatic-crop cultivation. Lotus products from this tank fetch a premium price, particularly for rhizome due to its whiteness, nutritional value, and softness. Similarly, vegetables grown in the command area of the tank are in high demand. The Fisheries Department earns a good amount of money in rent and from royalties on rhizome production.

Dulhara is practically encircled by floating macrophytes. In fact, macrophytes provide cover for fish, substrate for aquatic invertebrates, and oxygen, and act as food for fish and other wildlife. The appearance of macrophytes in a waterbody is indicative of ecological succession. These macrophytes play both functional and structural roles for the ecosystem in addition to adding to the aesthetics to the surroundings. They regulate, control, and influence the micro-ecosystem of the surrounding areas. However, the exotic and floating water hyacinths play a negative role to the ecosystem. Lake management authorities should be vigilant about their appearance. A late response to their appearance may cause severe ecological damage and incur huge costs. Dulhara is suffering from pollution due to chemicals and human encroachment, but a large number of trees and the ecosystem generally attracts migratory birds.[6]

4.5 Activities and Ecological Status

According to the census of 2019, the Ratanpur Nagar Palika wetland complex has only 156 ponds. Out of these ponds, Dulhara, Bikma, Ratneshwar, and Bairagban have water-spread areas above 4 ha. The remaining ponds have water-spread areas below 4 ha. Of the 156, only 118 are suitable for fish, lotus, and water chestnut cultivation, 21 ponds are seasonal or have dried up and have no water-retention capacity, 15 ponds have been encroached (either by homeless people or farmers farming on the periphery of wetland) and two have been lost to highway construction (Table 4.4). Today, the average depth of the ponds varies between five and seven feet, against the original depth of 20 feet, and only in 20 ponds water flow from one pond to the next. The average distance between two ponds was about 50 meters. Fifteen to 20 years ago, ponds met the demands for irrigation and domestic use.[7] Today, the ponds' water is not used for irrigation, except Dulhara. Water for irrigation in Bikma, Ratneshwar, and Bairagban pond/tank is rarely available due to poor water-retention capacity and siltation.

Changes in the drainage network in the catchment area due to encroachment (either by homeless people or farmers farming on the periphery of wetland), waste disposal, and siltation has distorted the flow of water from one pond to another. For example, Kasonti pond near Lakhni Devi Temple, Badhwa pond near Gayatri Temple and Bikma pond are highly infested with weeds and are polluted. Hathni pond has almost disappeared. Krishnoti pond

Table 4.4 Status of Complex Ponds and Their Use for Fish and Aquatic Crops

S. N.	Pond Area (ha)	0–0.5 ha		0.5–2.0 ha		2.0–4.0 ha		4.0 ha Above		Total		% of No. of ponds	% of Area
		No.	Area	No.	Area	No.	Area	No.	Area	No.	Area		
A	WSA (ha.) available for fish	32	8.639	73	77.275	25	71.285	3	38.707	118	195.906	75.641	92.272
	Pond under sole fisheries	24	6.257	56	58.840	18	50.808	1	11.342	99	127.247	63.462	59.933
	Fish + water chestnut	3	1.095	6	6.530	3	8.354	1	4.120	13	20.099	8.333	9.467
	Fish + lotus	0	0.00	2	2.752	3	8.777	1	23.245	6	34.774	3.846	16.379
	Total	27	7.352	64	68.122	24	67.939	3	38.707	118	182.12	75.641	85.779
B.	Unused for fish + aquatic crops												
	Seasonal ponds	5	1.705	9	9.153	1	3.346	0	0.00	15	14.204	9.615	6.690
	Dry ponds	1	0.362	1	0.919	0	0.00	0	0.00	2	1.281	1.282	0.603
	No retention of water	1	0.073	3	2.624	0	0.00	0	0.00	4	2.697	2.564	1.270
	Total	7	2.14	13	12.696	1	3.346	0	0	21	18.182	13.461	8.563
C.	disabled ponds												
	Encroached	8	2.541	6	4.372	1	2.338	0	0.00	15	9.251	9.615	4.357
	Lost under highway construction	0	0.00	1	0.524	1	2.237	0	0.00	2	2.761	1.282	1.300
	Total	8	2.541	7	4.896	2	4.575	0	0.0	17	12.012	10.897	5.657
Total (A+B+C)		42	12.033	84	85.714	27	75.86	3	38.707	156	212.314	100.00	100.00

Source: Compiled by the author from Ratanpur Nagar Palika and Fisheries Inspector records (Directorate of Fisheries, GoCG), Bloch Kota, Distt. Bilaspur as on April 2020

Note: Total area covered by 156 ponds is 212.314 ha; Total water-spread area available for fish, water chestnut and lotus is 195.906 ha.

is polluted, though even in the very hot month of May, thousands of lotus flowers can be seen there. Bairagban pond near Mahamaya Temple, Krishna bandh, Sukhari dabri, Pandri talab, Kadmani talab, Rani talab, Aathbias talab, and the kund of Budha Mahadev, are in bad shape. Kuwar pond and Krishna-arjuni pond have been encroached upon by farmers slowly, almost unnoticed (this observation is based on visits of the author to these ponds and his discussions with different stakeholders). Each pond is associated with some *puranic* (ancient) story which can be traced out in ancient literature (see endnote 3).

The flow of devotees during two *Navaratri* seasons (Navaratri is a Hindu festival that spans over nine nights) who come to perform rituals within and around temple premises, has polluted several ponds, particularly temple ponds. Poor maintenance has led to eutrophication due to the disposal of ritual offerings that are rich in organic matter. This has posed a serious threat to the existence of many temple-pond ecosystems. Recent exploratory research studies have estimated the level of pollution using biophysical and chemical parameters in temples and other ponds of the wetland complex (Chhattisgarh State Planning Commission, 2018; Nayar, 2019). The water-pollution level has increased in Dulahra pond due to religious activities and has caused adverse effects on the entire aquatic ecosystem. A far greater impact of pollution is seen during the festival season, when the immersion of idols in ponds destroys the whole ecological balance. Heavy metal pollution caused by idol immersion can damage the ecosystem as it kills off aquatic animals, damages plants, blocks the natural flow of water, and causes stagnation.

A quick preliminary study of a few ponds of Ratanpur revealed that immersion of idols releases 8.8 tonnes of paints and oil and 20 tonnes of other colouring agents into the water. These colours contain large amounts of heavy metals like lead, mercury, cadmium, manganese, and chromium. The study also found that, during festival time, oil and grease in the pond increases by 0.86 mg/litre and the concentration of heavy metals increased by 0.104 mg/litre (Chhattisgarh State Planning Commission, 2018; Nayar, 2019). The level of pollutants in Bikama and Ratneshwer ponds was significantly lower in comparison to other nearby ponds investigated (Nayar, 2020).

Numerous ponds are also infested with the uncontrolled growth of exotic species, particularly water hyacinths. In recent years, there has been little effort to remove them and, as mentioned earlier, there is a misplaced understanding that these plants can remove pollutants and help improve water quality. However, the decay of these weeds releases the nutrients and pollutants back into water and the huge quantities of decomposed organic matter that accumulate in the water body eliminate dissolved oxygen and cause mass fish deaths (Gopal and Marothia, 2017). A recent visit by the author to various ponds revealed a sad story of pollution and encroachment. Members of the nature club expressed their concern over air pollution from NTPC and the increased frequency of coal transportation. According to them, most of the trees, like *Palash* (Butea gum and a sacred tree) have died due to pollution.

This has resulted in complete loss of lac cultivation and the livelihoods of many households. Fine particles of coal floating through the air have polluted some ponds, like Chokrabandha, and some water bodies in Medrapara (observations based on the author's interactions with members of the nature club and senior citizens in the area).

4.6 Levels of Institutional Hierarchy for Managing the Wetland Complex

Over generations, Kalachuri kings developed and used a decentralized governance structure to manage the area's natural and financial resources, from the village level to the corridors of power. Under a decentralized governance structure, until their fall, kings sustainably managed the wetland complexes, temple ponds, and other sacred commons. After the fall of the Kalachuri rulers, the Maratha dynasty ruled Ratanpur and Chhattisgarh. The Marathas passed the control of Ratanpur to the British, who constructed Khutagath dam to enhance irrigation and drinking water in Ratanpur and its neighbouring villages. During Maratha and British rule, wetland complexes sustained their original networks to a great extent, despite deviations from the decentralized natural resource-management system of the Kalchuris.

After independence, particularly after 1952, most of the ponds and tanks were transferred to the Panchayats, municipal bodies, and the Irrigation Department, depending on the size, water-spread area, and location of these water bodies. All ponds of the wetland complex were transferred to RNP. Dulhara Tank was transferred to the DWR, GoCG.

The issues related to managing the wetland complex are multifaceted, due to sharing the water for so many uses, like fisheries, aquatic agriculture, domestic use, as well as socio-cultural-religious activities. Also, multiple agencies are involved in governing the water resources. The RNP has ownership and control over all the ponds of the Ratanpur complex and is responsible for maintaining and protecting the ponds and the whole complex, including a considerable number of temple ponds. Only three ponds are under the ownership of Mahamaya Temple Trust (MTT). Three ponds – namely, Jagannath temple pond (Khiyaa pond), Tulja bhawani pond, and Lakhanidevi pond, near Dulhara tank – were annexed by the MTT during the early 1990s (the MTT was formed in 1982), and these are maintained (basically, for desilting) by the trust. The two ponds in the Mahamaya Temple complex remain almost pollution-free as MTT has in recent years adopted methods to recycle ritual offerings and has not permitted devotees to dispose of flowers and other offerings in the ponds. Such an initiative is yet to be seen in other temple ponds as they are owned by the RNP.

As per the Chhattisgarh State Fisheries Cooperative Leasing Policy, RNP leased out all ponds below 100 ha to FCSs under a three-tier cooperative arrangement (Village Panchayat, Janpad Panchayat/Nagar Palika, and District Pancha). Only Dulhara irrigation tank has been leased out by the Chhattisgarh State Fish Cooperative Federation to an FCS. The DWR, GoCG, controlled and administered Dulhara irrigation tank under the state property regime. The administrative and functional jurisdiction of Village Panchayat, Janpad Panchayat, and District Panchayat are restricted according to the water-spread area of ponds/tanks. Lease of ponds or tanks with a water-spread area of below 10 ha, 10 ha. to 100 ha., and 100 ha. to 200 ha. is assigned by Village Panchayat, Janpad, and District Panchayats respectively. The SDF assigned leases to all ponds/tanks above 200 ha.

The lease amount for seasonal (where water is available until February) and perennial ponds is currently fixed at Rs. 1,500 and 3,000 per ha. per year respectively, with provision of ten percent increase after every two years. In the wetland complex, fish, lotus, and water chestnut are grown in 118 ponds and leased out at Rs. 3,000 per ha. The lease amount can be paid in three instalments of 35 percent, 30 percent, and 35 percent respectively during the financial year. The lease amount is used for promoting the activity of fisheries by Panchayats and the State Department of Fisheries (SDF)

Ten years ago, priority was given to irrigation at the critical crop-growth stage of paddy. However, after a substantial reduction in the water-retention capacity of the ponds, irrigation was only possible from the Dulhara tank. From Bikma, Ratneshwar, and Ved ponds, irrigation was provided if there was an overflow of water due to good rains. However, for the last five to seven years, irrigation has not been available from these ponds. Under the changed institutional arrangements, priority in leasing out pounds is given to FCSs followed by fishermen. Water for irrigation can be provided after ensuring the minimum water requirement for the growth of fish is reached. Occasional conflicts have been observed between fishing cooperative societies and *Sinchai Samittee* (village-level irrigation committees) over water sharing. Unfortunately, the role of RNP was not effective in resolving the conflict.

Sometimes, due to excessive and indiscriminate use of pesticides and insecticides, fish mortality has been observed, which further creates social tension between fishermen and farmers. RNP leased out ponds for fish and aquatic-crop cultivation with the condition that feed and manure cannot be used by FCSs. Ineffective authority of Ratanpur Nagar Palika resulted in poor maintenance, heavy siltation, and encroachment (by locals, homeless, and farmers nearby wetlands), and, as a result, today the wetland complex is experiencing an existential crisis. Members and trustees of MTT are willing to pay an even higher lease rent over the prevailing rate to RNP in order to lease other temple ponds. However, the priests of other temple ponds are not in a position to lease ponds attached to their temples due to poor financial and human resources available to them. Currently, FCSs under an informal contract have subleased ponds to flower sellers for lotus cultivation. A large number of flower sellers would prefer to lease ponds from the RNP, even if this meant paying little more than current lease rents. Such an institutional arrangement will give them an opportunity to cultivate seasonal flowers on the bund of a pond and to undertake fisheries activities.

Almost 50 water bodies are suffering from encroachments around their boundaries; the extent varying from pond to pond. Between 500 to 700 households are currently living in the encroached area and also cultivating vegetables on the bunds, and using water for various domestic activities. These families are responsible to a large extent for pollution and can create potential externalities for fishermen and other users and non-users. Frequent conflicts have been observed between the fishermen and the households of the area. The local administration and the State Ministry of Urban Development should take a considered decision to relocate these households to a place with adequate facilities, including sanitation and water availability. Such an administrative decision will not only save the ecological landscape of the water bodies, but also provide dignified living circumstances to the displaced households.

4.7 Analysing Wetland Complex Using Synthesis of Institutional Frameworks

A synthesis of institutional analysis frameworks developed in Chapter 2 is used here to analyse the ponds and tanks of the wetland complex. The details of resource and community attributes, institutional arrangements, and patterns of interaction for managing fisheries under a common property regime in the ponds of the wetland complex and for the Dulhara irrigation tank under the state property regime are presented in Table 4.5 and Table 4.6. As mentioned in Chapter 2, we have used four attributes to analyse the performance of FCSs under a common property regime and Dulhara tank under the state property regime (see Table 4.5 and 4.6). These are the physical and technical attributes of a resource, the characteristics of the user's community, external and internal institutional arrangements, and the patterns of interaction and outcomes of an institutional framework. Due to the relatively strong interrelationship between the four attributes and rule conformance, FCSs performed very well in terms of patterns of interaction among fishermen and in terms of outcomes (Table 4.5).

The physical and technical characteristics of the Dulhara tank, farmers, fishermen, and lotus growers are presented in Table 4.6. The physical and technical attributes and the socio-economic characteristics of farmers and fishermen jointly affect the overall performance of the tank-irrigation system. For example, as shown in Table 4.6, the Dulhara irrigation supply system suffers from continuous degradation due to poor maintenance above and below the water outlets, damage to feeder channels in many places, and encroachment of the catchment area (mostly by farmers). The poor physical and technical health of the tank has resulted in the reduction of the command area and impeded water flow from the feeder channel to the tank to farmers' fields. Poor irrigation flow has created conflict among farmers located at the head, middle, and tail-end of a water course during periods of water scarcity, particularly at critical crop-growth stages when all the farmers located at the

Table 4.5 Resource and Community Attributes, Institutional Arrangements, and Patterns of Interaction for Managing Fisheries in Ponds of the Wetland Complex Under Common Property Regime

Attributes	*Fisheries cooperative societies (FSCs)*
A. Physical and technical resource attributes	
1. Original property rights regime	All ponds of wetland complex constructed, controlled, and managed by Kalachuri rulers; after their fall, Maratha and British rulers had ownership of all ponds.
2. Current property rights regime	Ratanpur Nagar Palika
3. Management system	*Usufructuary* rights to FCSs. FCSs managing ponds for fish culture under common property regime
4. No of ponds of the wetland complex under fish +aquatic crops	118 ponds with 195.906 ha water-spread area
5. Size and boundary	Physical size and boundary of ponds are demarcated as per revenue and land record of RNP
6. Variability in water-spread area of ponds	Out of 118 ponds, Dulhara, Bikma, Ratneshwar, and Bairagban have water-spread area above 4 ha. Remaining ponds water-spread area is below 4 ha (see Table 4.4)
7. Source of water and water supply system	Rainfed, all ponds are interconnected and water flows from one pond to another in a sequential manner
8. Current status of ponds	Encroachment distorted flow of water from one pond to another, waste disposal and siltation is also a serious problem
9. Current status of catchment area of ponds	Due to encroachment, drainage network has been distorted over the years
10. Current status of ponds under fish culture	At the periphery of pond's bunds, illegal encroachment is common, most pond beds are silted, infected with weeds and pollution
11. Uses of pond's water	Fish, aquatic crops, domestic, socio-cultural, and religious activities
12. Alternative water sources available for fishery	Nil

(*Continued*)

Table 4.5 (Continued)

	Attributes	*Fisheries cooperative societies (FSCs)*
B.	Attributes of community	
	1. Social structure of users	Homogeneous fisheries community
	2. Number of FCSs and SHGs	Six FCSs and one women's SHG
	3. Number of households engaged in fish and aquatic-crop production	158 households
	4. Main source of income of fishermen community	Fisheries, lotus, and chestnut cultivation
	5. Variability in income and living standards of fishermen	Almost in same loop
C.	Decision-making arrangements	
	1. Legal and administrative relationship with the state	Registered FCS under the Cooperative Act
	2. Functional jurisdiction of FCSs	Restricted to leased in ponds
	3. Organizational setup of FCS	
	a. President, Vice-President, Secretary, Treasurer	1 (each)
	b. Executive members	3
	4. Arrangement for inputs, netting, price fixation, marketing, and distribution	Collective decision
	5. Ability to raise funds	Through selling fish among members and outsiders, they can take loans from banks, subsidies from the SDF under different schemes, if needed. Members contribute 60 percent of the net income for creating assets to the fisheries cooperative
	6. Ability to influence RNP	Organize to ensure adequate water in ponds for fish cultivation
	7. External organizational and technical assistance	The SDF provides technical support to increase fish production, supply of quality seeds, nets, boats, and arranging leases
D.	Patterns of interaction	
	1. Membership, legitimacy, and outsiders	All fishermen households are members, membership by household, elections every five years under the Cooperative Act. Outsiders have no rights to participate in the fisheries' cooperative meetings, and hence have no right to products

2. Conflict between fishermen and other water users	No, RNP leased out ponds with the condition that other services (socio-cultural and religious) continue as long as these activities do not create limitations on fisheries' activities
3. Conflict between FCSs over water flow from one pond to another at the demarcated area of FCSs	Since all ponds are interconnected, FCSs solve conflicts over water flow issues from ponds under the jurisdiction of one FCS to another. Fair understanding between/among FCSs about the nature of water flow and interconnected nature of ponds
4. Reciprocal interaction:	Conducive to collective behaviour
5. Protection	Members watch against poaching and killing of fish. Offenders are fined by the village Panchayat. If the offender is from among the members, the executive body heavily punishes, members exclude him from the cooperative. If farmers require water for irrigation, they have to seek permission from the fisheries cooperative. It depends on FCS members to accept the request of the farmers
6. Use regulations	Members can buy fish for their consumption; they are paid only wages, must participate in all operations of fish culture, attend meetings regularly and abide by the by-laws. Members who do not participate in fishery operations are not paid wages

Table 4.6 Tank and Community Attributes, Institutional Arrangements, and Patterns of Interactions Under State Property Regime: A Case of Dulhara Tank-Irrigation Management

	Attributes	*Description*
A.	Physical and technical attributes of tank	
	1. Original property rights regime	Constructed, controlled, and managed by Kalachuri rulers, after their fall, Maratha and British rulers had ownership over all ponds of the complex
	2. Current property rights regime	State property regime, managed by DWR, GoCG
	3. Water source for tank	Largely rainfed and linked with canal system
	4. Size and boundary	Water-spread area and physical boundary is well demarcated as a hydrologic unit.
	5. Current status of catchment area	Catchment area drainage network has been distorted over the years; waste disposal and encroachment
	6. Current status of tank	Periphery of tank's bunds are illegally encroached, tank bed is silted, infected with weeds and pollution
	7. Current status of feeder channel	Damaged at many points, affecting continuous flow of water to tank and from channel to tank
	6. Current status of irrigation supply structure	Most of the water outlets along feeder and field canals are distorted by farmers to get the maximum possible water to their fields
	8. Major uses of tank water	Irrigation, fisheries, lotus cultivation, socio-cultural, spiritual, and religious activities
	9. Alternative water sources available to farmers	Tube wells
	10. Alternative water sources available to fishermen	Nil
B.	Attributes of community	
	1. Social structure of water users	Heterogeneous (irrigators, fishermen, local residents (male and female), pilgrims, religious practitioner)
	2. Number of irrigators in command area	948

	3. Number of villages receiving irrigation	Four
	4. Number of FCSs and number of fishermen engaged in fish and lotus cultivation	One FCS with 21 households
	5. Irrigators' main income and livelihood sources	Paddy and vegetables
	6. Main source of income of fishermen community	Fisheries and lotus cultivation
	7. Legal status of irrigators	Registered under irrigation agreement with DWR
	8. Legal status of fishermen	Member of registered FCS under the Cooperative Act
	9. Variability in income and living standards of farmer community	Almost equal – no visible inequality
	10. Variability in income and living standards of fishermen community	Almost in equal – no visible inequality
	11. Variability in farm size and numbers	High concentration of marginal and small farmers with lower share in total land
	12. Dependency of irrigators and fishermen	A large number of farmers depend on tank water for irrigation, but a good number of farmers have their own tube wells; fishermen, however, totally depend on adequate availability of water for potential growth of fish and lotus products
	13. Relationship among farmers located at head, middle and tail-end of water channel	Due to flood-irrigation system, conflicts occasionally observed, particularly during water scarcity and concurrent requirement of water at critical crop-growth stages; inter and intra village conflicts also observed among farmers of
	14. Conflict between farmers and fishermen over water use	Occasional, but settled with the intervention of SDF
C.	Decision-making arrangements	
	1. Organizational setup	Due to the defunct WUA, Executive Engineer, Water Resources Division, Kota, Dist. Bilaspur, CG, administered the tank irrigation
	2. Method of irrigation	Field-to-field, should be rotational

(Continued)

Table 4.6 (Continued)

	Attributes	*Description*
	3. Recovery of irrigation fee	Poor recovery of irrigation fees
	4. Maintenance of tank	Routine maintenance of outlet, feeder, and minor branch canal by the Irrigation Department. Due to lack of participation of farmers the tank and irrigation structures are in bad shape
	5. Ability to influence other Panchayats/ government activities	Due to non-functional WUA, farmers are not in a position to ensure adequate water supply and maintain tanks. Since RNP and other three village Panchayats have no control over this tank, farmers are unable to influence RNP/village Panchayats for the proper maintenance of the tanks
	6. External organizational and technical assistance	Non-existence of WUA resulted in absence of coordination with DCs and PCs. Hence, poor management and continuous degradation of tanks and irrigation structures
D.	Patterns of interaction	
	1. Membership, legitimacy, and outsiders	Due to non-functional WUA, the basic institutional arrangement of PIM is not effective, collective irrigation management collapsed
	2. Reciprocal interaction	Not contributing to induce collective behaviour
	3. Protection	Due to the absence of institutional arrangements on maintenance and irrigation distribution systems, the tank area is not protected by irrigators. Irrigators think it is job of officers of the water resource department
	4. Use regulations	Water distribution rules are non-operative. Hence, conflict over water use among farmers at the head, middle, and tail ends has become common. Also conflicts among farmers of four villages is common
	5. Consequence	Selfishness among farmers to extract maximum water for their fields, poor excludability due to unregulated irrigation supply, water insecurity and uncertainty due to ineffective rules and authority systems

head, middle, and tail-end require adequate water at the same time. Also, a minimum water level is required for the growth of fish in the fisheries. Chhattisgarh State Fish Cooperative Federation (CSFCS) leased out the tank for ten years (Sept 2017 to June 2027) to Jai Bhairav Baba Fish Cooperative Society Ratanpur, for Rs. 5,379 for fish culture and a royalty of Rs. 21,106 on the sale of lotus rhizome per year.

To rejuvenate the canal and tank (connected with the distributary), the irrigation system, and to improve the tank's irrigation efficiency, the Government of Chhattisgarh (GoCG), with the Farmers Participation in the Management of the Irrigation System in Chhattisgarh Act (CG FIMS Act 2000), transferred the management of the irrigation system to farmers' organizations (Fos) in 2002 under a three-tier of management system. For include WUAs at the primary level, distribution committees (DCs) at distributor level, and project committees (PCs) at the project level. Institutional structures at each level are well-defined.[8] Water users' area, the number of villages and farmers, the procedure for electing members of sub-committees, the financial management, and the functions and responsibilities of WUAs are clearly defined in the Farmers Participation in the Management of the Irrigation System in Chhattisgarh Act (CG FIMS Act 2000) for PIM. Prior to the implementation of this act, the tank had total control of the state through its Department of Water Resources; from 2000–2006, all farmers under the command area of the tank were part of WUA (Marothia, 2002b, 2005a).

The area of a WUA Is subdivided into territorial constituencies with a minimum of four and a maximum of ten, depending upon the size of the water users' area. The delineation of water users' area and its sub-division (territorial constituencies) is based on hydrologic parameters and must be hydrologically viable. Every WUA is known by its local, district name. At the WUA level, every command area in the irrigation system is delineated on a hydraulic basis as water users' area for each WUA. The water users' area is then divided into territorial constituencies that are not generally less than four, but not more than ten, according to the size of the command area under the control of the WUAs (Marothia, 2002a, 2005a). According to this norm, the water users' area of Dulhara has one WUA and four Territorial Constituencies representing four villages covered under the command area of the Dulhara tank. After the WUA irrigation became defunct, it is now controlled by the DWR, with ineffective monitoring and a lack of field staff to resolve the issues around maintenance. As a result, the tank and water-carrying structures are in a degraded condition.

Though all irrigators in the command area of the Dulhara tank were part of the defunct WUA and had full information about the decision-making arrangements, the operational rules, and the protection of water structures, in the absence of WUAs, most of them engaged in the "free rider" trap to exploit as much water as possible from the total flow. In the process, they created multi-party externalities – getting into each other's hair. Further, poor maintenance, lack of field channels, poor water control, inadequate labour and non-payment, or poor recovery of irrigation fees, were the most serious problems that the Dulhara tank was confronted with. The impact of non-implementation of operational rules and internal and external institutional arrangements resulted in a distorted pattern of interaction in terms of negative reciprocity, and protection and use regulations that are presented in Table 4.6.

Despite the degraded condition of the Dulhara tank, the non-functional WUAs, the failure of institutional arrangements, and the poor monitoring of water resources beyond the outlets, it being a complex irrigation system, the tank has played a significant role in providing life-saving irrigation, high-quality lotus products and fisheries. Also, Dulhara provides valuable ecosystem services in terms of groundwater recharge and socio-cultural and religious activities, and is one of the favourite sites of local and migratory birds. To this end, ecosystem services provided by the wetland complex are discussed in Table 4.7.

4.8 Ecosystem Services Derived from the Ratanpur Wetland Complex

All kinds of wetlands, including the Ratanpur wetland complex, provide benefits to users and non-users. These benefits arise either directly from the use of water, plants, animals, soils, and other direct components of the wetland ecosystem, or indirectly from its functions. These benefits are now known as ecosystem services and are categorized into provisioning, regulating, supporting, and cultural services (see Chapter 2 and Chapter 3 for sub-attributes of ecosystem services). The Ratanpur wetlands provide a wide range of provisioning ecosystem services. Besides providing provisioning and social and cultural services, the wetlands also support aquatic and avifaunal biodiversity; regulate the water cycle, micro-climate, hydrology, and water quality; and recharge groundwater; while promoting social, cultural, religious, and recreational activities, together with enhancing the aesthetics of the area. Often, they support the subsistence and livelihoods of poor and wetland-dependent local communities.

A few avifaunal studies have been conducted in Ratanpur (see Chapter 3 for a summary). The findings of these studies cover the IUCN status of the wetland's domestic and migratory birds: Least concern, Nearly threatened, Winter visitor, Summer visitor, Monsoon visitor, Very common, Common, Not rare, and Rare, and globally threatened (Rahalkar and Tiwari, 2014; Porte and Gupta, 2018; Dey, 2017; Sahu, 2017).

4.9 Financial and Employment Gains from Aquatic and Other Activities

Financial gains, income, and employment opportunities for different stakeholders from different activities undertaken in the ponds of the wetland complex are given in Table 4.7 and Table 4.8 respectively. These attributes are

Table 4.7 Gains From Provisioning Services of Wetland Complex

Activities / Gains	*Yield per ha*	*Price per kg*	*Total Area (ha)*	*Total Yield (kg) of Total Area*	*Income Rs/ha*	*Cost Rs/ha*	*Net Return Rs/ha*	*BC Ratio*	*Total Income (Rs) from Total Area*	*Income to RNP@ 3000/ha*
	(a)	*(b)*	*(c)*	*(ac)*	*(ab)*	*(d)*	*(ab-d)*	*(ab/d)*	*(ac.b)*	*(3000.c)*
A. Activities in Ponds under RNP										
1. Fish (Sole Activity) (kg)	3400	115	127.247	432639.80	391000.00	204000.00	187000.00	1.92	49753577.00	381741.00
2. Fish+ water chestnut										
(i) Fish (kg)	2000	115	20.099	40198.00	230000.00	46000.00	184000.00	5.00	4622770.00	60297.00
(ii) Water chestnut (kg)	25000	20	20.099	502475.00	500000.00	175000.00	325000.00	2.86	10049500.00	
3. Fish + lotus										
(a) Fish (kg)	2000	115	34.774	69548.00	230000.00	48000.00	182000.00	4.79	7998020.00	104322.00
(b) Lotus products					628600.00	101906.00	526694.00	6.17	21858937.00	
(i) Lotus flower (no.)	13500	10*	34.774	469449.00	135000.00				4694490.00	
(ii) Rhizome (kg)	6000	80	34.774	208644.00	480000.00				16691520.00	
(iii) Pods (no.)	700	10*	34.774	24341.80	7000.00				243418.00	
(iv) Seeds (kg)	12	550	34.774	417.29	6600.00				229508.40	
Total (A)					1979600.00				94282803.40	546360.00
B. Activities in irrigation tank under DWR										
Fish + lotus										
(a) Fish (kg)	400	115	30.756	12302.40	46000.00	6000.00	40000.00	7.67	1414776.00	Yearly lease Rs. 537900/- Royalty on sale of Rhizome 21106/-
(b) Lotus products					362000.00	90000.00	272000.00	4.02	11133672.00	
(i) Lotus flower (no.)	20000	10*	30.756	615120.00	200000.00				6151200.00	
(ii) Rhizome (kg)	2000	70	30.756	615120.00	140000.00				4305840.00	
(iii) Pods (no.)	1000	10*	30.756	30756.00	10000.00				307560.00	
(iv) Seeds (kg)	20	600	30.756	615.12	12000.00				369072.00	
Total (B)	-	-	-	-	408000.00				12548448.00	

(Continued)

Table 4.7 (Continued)

Gains / *Activities*	*Yield per ha*	*Price per kg*	*Total Area (ha)*	*Total Yield (kg) of Total Area*	*Income Rs/ha*	*Cost Rs/ha*	*Net Return Rs/ha*	*BC Ratio*	*Total Income (Rs) from Total Area*	*Income to RNP@ 3000/ha*
	(a)	*(b)*	*(c)*	*(ac)*	*(ab)*	*(d)*	*(ab-d)*	*(ab/d)*	*(ac.b)*	*(3000.c)*
(A+B)										
(a) Fish	7800	115	212.876	554688.20	897000				63789143.00	
(b) Lotus products										
(i) Lotus flower (no.)	33500	10	65.53	1084569.00	335000.00				10845690.00	
(ii) Rhizome (kg)	8000	70	65.53	270156.00	620000.00				20997360.00	
(iii) Pods (no.)	1700	10	65.53	55097.80	17000.00				550978.00	
(iv) Seeds (kg)	32	600	65.53	1032.41	18600.00				598580.40	
(c) Water chestnut	25000	20	20.099	502475.00	500000.00				10049500	
Total Income (Rs) (A+B)	-	-	-		**2387600.00**				**106831251.40**	

Source: Author's analysis based on field work

Note: Cost of production of lotus products includes lotus flower, rhizome, pods, and seeds
* indicates price per flower and pods

Table 4.8 Income and Employment From Activities Undertaken in Wetland Complex and Temple Ponds

Activities	*Total Income (Rs)*	*Employment (Man-days)*	*Total Employment From Total Area Under Activities (Man-days)*
A. Activities in ponds under RNP			
(i) Fisheries	62374367.00	137 man-days for 365 days	24950.44
(ii) Water chestnut	10049500.00	360 man-days for 90 days	7235.64
(iii) Lotus products*	21858936.40	670 man-days for 365 days	23298.58
Total (A)#	94282803.40		55484.66
B. Activities in irrigation tank (Dulhara) under DWR			
(i) Fish	1414776.00	21 man-days for 365 days	645.876
(ii) Lotus products*	11133672.00	680 man-days for 365 days (for doing alone 225 man-days for three seasons)	20914.08
(iii) Paddy and vegetable production	250000.00	460 maydays	834900.00
Total (B)#	12798448.00		856459.956
C. Flow of devotees (Mahamaya Temple Trust)			
1. No of pilgrims**			
a) Winter (10,00,000)			
(i) Puja Thali @ Rs. 20/-	20000000.00	150 shops in the MTT premises and 300 families individually gain employment for at least 180 man-days including festival seasons (man-days include winter +summer seasons)	54000.00
(ii) offerings @ Rs. 30/-	30000000.00		
b) Summer (7,00,000)			
(i) Puja Thali/Plate @ Rs. 20/-	14000000.00		
(ii) offerings @ Rs. 30/-	21000000.00		
Total (C)	85000000.00		54000.00
D. Park and water recreation	800000.00	10 man-days year around	3650.00
Total (D)	800000.00		3650.00
Total (A+B+C+D)	192881251.40		969594.62

* Includes flower, pods, seeds, and rhizome

** Flow of pilgrims and rate of *Puja Thali* and offerings average at minimum level

\# Total income from ponds + tank = Rs. 107,081,251.40

part of the outcome and evaluative criteria of the synthesized institutional framework. It can be seen from Table 4.7 that fish and lotus provide higher gains per rupee invested in ponds managed by FCSs under control of RNP and DWR. Per hectare and total income from ponds and tanks of the wetland complex contribute to aquatic farmers. The complex also generates income and employment from other activities (Table 4.8). These facts demonstrate that, for underprivileged people, the wetland complex is a lifeline that has been sustaining provisional and other services for several generations.

4.10 WTP for Restoration of the Wetland Complex

To assess the WTP of various parties for conserving and managing temple ponds and wetland complexes, the opinions of direct users and non-users have been recorded for the period of March to April and September to October 2018. The range of WTP is reflected in user and non-user groups (Table 4.9). The WTP of Fishermen and paddy and vegetable growers is a little higher than the currently prevailing lease rents if water supply is regulated to ensure the minimum requirement for the crop-growth period. Pilgrims are willing to pay for the conservation of temple ponds and the surrounding biodiversity. However, more than 70.3 percent think that the temple pond and wetland complex should be managed with the financial resources of the Ratanpur Nagar Palika. Visitors to the recreation park are willing to pay a little more if the park can be maintained and the flow of visitors can be regulated.

Among the non-user group, a very high percentage of local residents are willing to contribute to conserving the wetland complex and temple-pond ecosystems. Non-residents, on the other hand, have expressed their WTP to maintain the temple ecosystems but the response was not very encouraging for maintaining the rest of the pond's ecosystem. It is interesting to note that MTT is willing to pay a higher amount than the existing lease rents fixed by the Ratanpur Nagar Palika to lease out the temple ponds. However, the priests of the other temple ponds are not willing to pay, nor are they willing to lease the ponds due to financial and human-resource constraints. Flower sellers are willing to pay the prevailing lease rates to lease the ponds. FCSs currently informally sublease ponds to flower sellers for lotus cultivation and the cultivation of other types of flowers on pond bunds.

4.11 How to Reverse Ecological Degradation

The poor ecological health of the wetland complex is solely due to institutional failure at different levels of the management hierarchy. The ineffective role of RNP in carrying out de-siltation, its inability to control encroachment, and its failure to resolve conflicts between fishermen and farmers' communities over water-sharing has resulted in degradation of the wetland ecosystem. To restore the wetland complex, effective and implementable institutional

Table 4.9 Responses of Users and Non-Users on WTP for Restoration of Wetland Complex

S.No.	*Users and Non-users*	*Total No. of Respondents*	*% of Total Respondents WTP*	*Range of WTP (Rs.)*
	Users			
1.	FCSs for regular desilting of ponds (WTP over current leasing rent per ha)	158	100	100–150
2.	Water chestnut and fish cultivation (lease rate/year)	13	100	40–50
3.	Fish with lotus (lease rate/year)	6	100	50–70
4.	Farmers for adequate and timely supply of water to avoid conflict with fishermen community (Rs./ha)	30	70	10–15
5.	a. Pilgrims willing to pay for conservation of temple ponds and surrounding biodiversity	500	60	15–20
	b. Pilgrims willing to pay for conservation of wetland complex and biodiversity	500	40	5–10
	c. Pilgrims who think respective temple/trust should manage temple ponds and biodiversity	500	75	NIL
	d. Pilgrims who think Ratanpur Nagar Palika or government should manage wetland complex and temple ponds and biodiversity	500	70	NIL
6.	Daily visitors in recreation park	100	70	5–7
7.	MTT willing to lease in and manage all temple ponds	11*	100	3000+
8.	Priests of temple ponds (other than MTT owned ponds)	15	100	Financial and human resources Constraints
9.	Flower sellers WTP to lease in ponds	50	100	3000
10.	Male groups WTP to conserve temple and complex ponds	70	90	100–120

(*Continued*)

Table 4.9 (Continued)

S.No.	*Users and Non-users*	*Total No. of Respondents*	*% of Total Respondents WTP*	*Range of WTP (Rs.)*
11.	Female groups WTP to conserve temple and complex ponds	100	80	150–175
	Non-users[**]			
	i. Residents (local)	200	80	10
	a) For temple ponds ecosystem	150	90	15–20
	b) For rest of the ponds' ecosystem	150	80	10–20
	ii. Writers, researchers, bird watchers			
	a) For temple ponds ecosystem	25	90	30–40
	b) For rest of the ponds' ecosystem	25	90	35–45
	iii. Environmentally conscious citizens			
	a) For temple ponds ecosystem	150	80	25–30
	b) For rest of the ponds' ecosystem	150	95	35–40
	iv. Non-residents (outside Ratanpur)			
	a) For temple ponds ecosystem	100	90	20–25
	b) For rest of the ponds' ecosystem	100	40	5–10

Source: Author's analysis based on field work

[*] Trustee members representing MTT [**] Respondents not involved in water using activities but willing to pay for non-use values or ecosystem services (option, bequest, and existence values)

arrangements, enforceable by the authority of Ratanpur Nagar Palika, need to be designed to restore the original water-spread area of the wetland complex. Ratanpur Nagar Palika should remove encroachments. This action will restore the original land use of the wetland complex to that indicated in the historical revenue records. Significant action is needed to carry out the de-siltation process to restore the potential water-storage capacity of the ponds. This will create more water availability for fish and irrigation in big ponds, and thereby reduce the tension between farmers and fishermen.

The Water Resources Department should ensure that the feeder canal is not damaged by farmers illegally extracting water. This will increase the irrigation area for Ratanpur farmers, and the department can generate revenue through irrigation fees. GoCG should make sincere efforts to revive WUAs with an apolitical structure (Marothia, 2002b, 2005c). The immersion of idols in water bodies should be immediately stopped. Ratanpur Nagar Palika can also promote aquatic-crop cultivation by leasing out tanks. Under the CSR fund, NTPC coal companies and distilleries may contribute to conserving the wetland complex.

MTT, being a major economic beneficiary, should liberally invest in the ecological restoration of other temple ponds which are currently under the ownership of Ratanpur Nagar Palika. It is worth mentioning that MTT has been investing in creating drinking water facilities and de-silting ponds under its ownership.

In recent years, many sacred trees have been uprooted or felled due to the end of their vegetative life cycle. New plantations of trees are urgently required to maintain the uninterrupted visit of local, domestic, and migratory birds. A sizable part of the area's plant and animal biodiversity has disappeared due to encroachment and pollution. Hopefully, some renewable biodiversity can be revived. Fortunately, Ratanpur town has many environmentally conscious citizens and a vibrant nature club. Their services can be utilized to restore the wetland complex ecosystem in a sustainable manner. Institutions like INTACH can play a bigger role in collaborating with the different agencies of the government to restore the historical status of Ratanpur town. Such efforts can facilitate the process of getting Ratanpur designated as a Ramsar wetland city.

The findings of this chapter adequately establish that the wetlands of Ratanpur make an immense contribution to use value and non-use value. The total economic value includes direct-use value, indirect value, option value, bequest value, and existence value. Sub-attributes of these values are listed in Table 4.7 and Table 4.8. Hopefully, the lessons from this chapter may induce interest in the concerned government departments, policy makers, academicians, civil societies, and the environmentally conscious citizens of Ratanpur and Chhattisgarh for the conservation and sustainable management of this unique wetland complex. These issues are further discussed in Chapter 9 of this book. A few researchers have also highlighted these issues.[9] Examples of wetland complex restoration with redesigned technical and institutional arrangements are available in Asian and African countries. Lessons from case studies in these countries and Ratanpur may improve researchers' and managers' understanding of wetland chain systems.[10]

Notes

1 Even before the Kalachuri Kings started ruling Ratanpur, it was an important city, existing through all four *yuga* (Ages). Raja Ratnadeva I made Manipur, a village renamed Ratnapur, its capital. It was known as Manipur in *Satya Yuga* and *Dwapar Yuga*, as Heerapur in *Treta Yuga*, as Ratnapur in early *Kaliyuga*, and as Ratanpur in modern times. All these *pauranic* accounts about ancient Ratanpur reflect that it was a prominent city and might have been in close interaction with other kingdoms in the region (Jha, 1997). Descriptions of Ratanpur are first recorded in the *Ramayana/Mahabharata* age and later in *Jaimini* and *Rewa khand of Skand Puran*. According to one mention, it is believed that during *Treta yug*, Kapardi – the husband of Surpanakha, who was the sister of Ravana – stayed here. Once he kidnapped the Goddess Parvati from Kailas while accompanying Ravana on his way to meet his *guru*. After a fierce battle with Lord Shiva, he was killed and Surpanakha cried over his death. She cried so much that her tears mixed with kaajal (collyrium) and formed a pond called *Kajara talab*, which still exists in Ratanpur (Gupt, 1973; Sukhdev Prasad Kashyap, 2004). According to mythology, the shoulder of Lord Shiva's consort Sati fell at Ratanpur, making it one of the 52 *Shakti peeths* worshipped in the country. Ratanpur city is mentioned in *Ashwamedha Parva* of the *Mahabharata*where the Sacred Horse of *Pandawas*-was visited by prince Moradhwaj of Ratanpur. Then Krishna and Arjun came to Ratanpur to take power from the king of Ratanpur. The pond *Krishna-Arjuni* is still there, with its long story (description drawn from Cunningham (1872, 1881); Jha (1982a, 1982b, 1982c, 1997); Kielhorn (1888); Kulke and Rothermund (1998); Lal (1916); Mirashi (1955, 1957); Misra (2003); Roy (1980); Sampath (2001); Sukhdev Prasad Kashyap (2004); Verma (2015); Sharma (2017)).

2 The Kalachuri Kings commanded the *Dakshin Koshala*, the area covering a major part of today's Chhattisgarh State, from Ratanpur. Thus, Ratanpur was a historic capital of Chhattisgarh and as such it has an important place in history and culture of Chhattisgarh State. Ratanpur kingdom was ruled by four dynasties. It was under the rule of a Jain dynasty from the 2nd century BC till the third quarter of the 4th century AD. After the fall of Jain rule, Buddhists ruled the kingdom of Mahakaushal, including Ratanpur, until the 5th-6th century. Kalachuri of Tripuri entered into Ratanpur in 1048 AD. Kalachuries, most famous among all four dynasties, were the hayhaiwansi Kalachuries of central India, who ruled Ratanpur and a large part of Chhattisgarh from the 7th to the 18th centuries. Around 1050 AD, King Ratan Dev I shifted the capital of Kalachuri kingdom from Tumman to Ratanpur (Ratanpur stone inscription of Prithvideva II: Kalachuri, year 915 (1163 AD), *Corpus Inscriptionium Indicarium Vol IV Part 2 Inscriptions of the Kalchuri-Chedi Era*, (Mirashi, 1905)). This inscription was first noticed as early as 1825, mentioned by Sir Richard Jenkins, who published a short account of it in the *Asiatic Researches, Vol. XV*, pp. 504. There are two interesting suppositions related to King Ratan Dev I's decision to shift the capital city. One focuses on the strategic geographical location of Ratanpur due to its defensive and riverine site. A village, Koni, was chosen as a suitable place for the cantonment of the army, and the forest of Ratanpur provided an ideal training ground for jungle warfare; also, the bridge served as a perfect means of regularly supplying the material needs of soldiers. While the other writers suggest that, once during a night

stay under a tree near the Mahamaya temple, King Ratandev I saw the Goddess Mahamaya and, later in his dream, the Goddess told him to move the capital city. The divine and magnificent temple of Shri Adishakti Maa Mahamaya Devi was built by King Ratnadeva I in the 11th century. The artistic statues of Mahakali, Mahasaraswati, and Mahalakshmi are enshrined in the temple. It is believed that Tantra must have been the center of spiritual practice in this temple. As the right shoulder of the Goddess Sati fell in Ratanpur itself, Lord Shiva gave the place the name of Kaumari Shakti Peetha, for which reason virgin girls seek blessings for a happy life on the occasion the Navratri festival, when thousands of *Manokamna Jyoti Kalash* are lit by devotees. In an inscription of Jajalladeva, dated in 1114 AD (a descendant of Ratnaraja), Ratanpur is described as the city of Kubera (Lord of Wealth) and the city, with many-coloured and wonderful jewels, looks like an ocean decorated with many temples, making it appear like heaven. A later inscription of King Vahara in 1495 AD compares this city with the city of Lord Indra. The Marathas attacked Chhattisgarh in 1741 and destroyed Haihaya's power. In 1745 AD, after conquering the region, they deposed Raghunath Singh ji, the last surviving member of the Ratanpur royal dynasty. In 1758, the Marathas finally annexed Chhattisgarh and it came directly under Maratha rule, with Bimbaji Bhonsle appointed the ruler. The Maratha officials served the interests of the British. As a result, the region became extremely poor. J. D. Beglar visited Ratanpur in 1873–74 and reported its antiquities. He describes it as an old capital reduced to a decayed village. During the Second World War, the British army and the Royal Air Force developed two stations near Ratanpur and Mohanbhata – a nearby village. They acquired thousands of acres of land nearby and a proper aircraft refueling and repairing centre was installed there. This station was a backup centre for the north-east air force command. It was wound up around 1945, but the airstrip and some water tanks are still there near Kaka pahad and Mohan bhata. (Description drawn from Cunningham (1872, 1881); Jha (1982a, 1982b, 1982c); Kielhorn (1888); Kulke and Rothermund (1998); Lal (1916); Mirashi (1955, 1957); Misra (2003); Roy (1980); Sampath (2001); Verma (2015); Gupt (1973); Sharma (2008)).

3 Mythological stories attached to several ponds have travelled through generations and are still being told and documented in many anthropological and historical studies. For example, the equestrian statue of Shiva and Bhavani is enshrined at Shri Khandoba Temple. This temple was built by the queen of the Maratha King Bimbaji Bhansale in memory of his nephew, Khando ji. The ancient and huge lake of Dulhara Talab is located behind Khandoba temple. There is a grand, nine-foot-high idol of Bhairav built by Baba Gyanagiri. There is a pond adjacent to the temple. A historic temple of Shri Mahalakshmi Devi is situated on Ekbira Hill on the Ratanpur-Kota main road. This temple was built by Gangadhar, the Prime Minister of King Ratna Dev III. Its local name is Lakhni Devi Temple. Mangal Jawara is sown here on the occasion of Navratri and religious rituals are performed. Kasonti Talab is still exciting near Lakhni Devi Temple, at Motipur, near the last section of the elephant fort, where 20 queens of King Lakshman Sai self-immolated on the banks of the Beas Duwariya Pond. The Beas Duwariya Temple was built in memory of the queens. Some distance behind Sri Mahamaya Devi Temple, there is a pond in Bairagavan surrounded by Amrakunj and on its side there is a temple of Lord Narmadeshwar Mahadev, while on the other side there is a grand monument of King Raj Singh, which is known as Beas Duwaria Temple. This temple is idol-less and has 20 gates. The ancient temple of Khichri Kedarnath is situated near Bairagavan. Kantideval – a grand temple of Shri Neelkantheshwar Mahadev – is located in the Sri Mahamaya Devi Temple complex. This temple, built in the 15th century AD, has since been rebuilt by the Central Archaeological Department and is included in the National Heritage List. It has four entrances

and engraved images of deities on the walls. Near this temple is a temporary museum of statues found in Ratanpur and the surrounding areas, and a permanent museum is under construction. Sri Ratneshwar Mahadev Temple is situated on the banks of Ratneshwar Pond in Karaipara. It was founded by King Ratneshwar. A 400-year-old Kabir Ashram on the banks of Ved-Ratneshwar pond is still functional. It was created by Shri Sudarshan Naam Saheb of the Kabir Panth. Bhubaneswar Mahadev ancient temple is situated on the banks of Krishnarajuni pond on Ratanpur-Chapora road. The idol of Lord Bhaskar (Suryadev) is installed in this temple and it is also called Suryeshwar temple. This temple and Shivling is built on a geometric base. An inscription is engraved on the entrance to the sanctum sanctorum (Gupt, 1973, Sukhdev Prasad Kashyap, 2004).

4 Thanks to Adikant Pradhan, IGAU, for technical input to develop contour map of the wetland complex.

5 Executive Engineer, Water Resources Division, Kota, Dist. Bilaspur, Department of Water Resources, GoCG, provided detailed information of Dulhara tank. Thanks to him.

6 I am thankful to M.A. Hassan for this observation.

7 Personal communication with V.K. Shukla, Director Fisheries, Sukhdev Prasad Kashyap, author of a book on Ratanpur (in Hindi), and other senior citizens of the town.

8 For critical analysis, see the Farmers Participation in the Management of the Irrigation System in Chhattisgarh Act (CG FIMS Act 2000). See also Marothia (2002b, 2005a).

9 For further details, see also *A Status Paper on Wetlands of Chhattisgarh* (Gopal and Marothia, 2017, Marothia, 2021; Marothia et al., 2017; Wanganeo, 2017), Factsheet on Ratanpur wetlands complex (Sharma, 2017), and the preliminary report on inventory prepared by Chhattisgarh Council of Science and Technology (CCOST, 2018) (see Marothia et al., 2017).

10 See studies reported in Euisoon et al. (1997); Frenken and Mahapatra (2002); Galbraith and Burns (2007); Kingsford et al. (2021a).

5 Restoration Journey of Lake Vivekananda

Ecological Gain or Loss

Following a brief description of site characteristics and the history of LV, an attempt is made in section two to assess the nature and extent of water pollution and degradation status of LV. In section three, concepts of entitlement systems, property rights structures and the role of authority at different decision-making levels are applied to understand policy attributes associated with interdependent activities of LV. Results of the valuation of LV and WTP for improvements to the lake ecosystem are presented in section four and its subsections. Section four also provides an analysis of the perceptions of use value and non-use values accruing from LV, as assigned by visitors for recreation, the local population living on the periphery, environmentally conscious citizens, and scientists. Section five takes into account the purpose, process, and social-ecological impacts of reversing the degradation of the LV ecosystem. Major restoration initiatives were undertaken between 1977 and 2020. Critical socio-ecological analysis of the major activities reveals that urban amenities were created at the cost of ecological chaos, as described in subsection five. Issues and lessons learned from the restoration efforts are discussed in the last section.

5.1 Characteristics and History of Lake Vivekananda

LV, popularly known as *Budhatalab*, is a distinct historical wetland site in Raipur. Raipur, the capital city of Chhattisgarh, has many historical wetlands, but a lot of them are ecologically beleaguered. Raipur city has lost a sizable number of medieval historical wetlands in the post-independence era. In 1954, the city had 227 water bodies with a total water-spread area of 490.84 ha within the boundary of Raipur Municipal Corporation (RMC). A survey conducted in 2013–14 revealed that 173 water bodies still remained with a 328.12 ha water-spread area (Dey, 2015).

LV is located in the southern part of Raipur city (21°23'N North, 81°63'N East and 260 m above sea level), which is in Raipur district in the State of Chhattisgarh (see Map 5.1). LV is one of the oldest water bodies and landmarks in the city of Raipur. LV was constructed in 1402 AD by the King Brahma Dev. Subsequently, King Bhuvneshwar Dev developed *Ghats* around

DOI: 10.4324/9781003456742-5

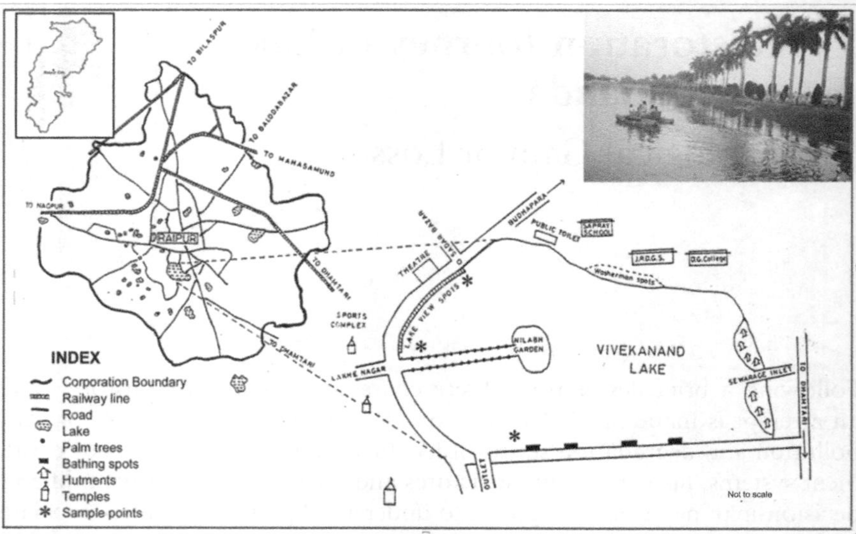

Map 5.1 Location Map of Lake Vivekananda

Source: Marothia (1995a, 1997c)

Note: Dhamtari narrow gauge railway line dismantled and converted to road.

the lake. During that time, the water-spread area of the lake is believed to have been between 2.50–2.70 sq km. King Rai Singh Dev further developed and named the lake after his presiding deity "*Budha Dev*". He established a city on the banks of the lake which is today known as Raipur.[1]

LV was hydrologically connected with seven ponds through a natural overflow system. The complex of seven ponds covers Kankali talab, Maharajbandh and Panchbhai talab, Tikrapara talab, and the Khokho group of talabs, which consists of three talabs, namely Khokho talab, Pehladwa talab and Bandhwa talab.[2] Due to encroachment of the catchment area and boundaries of ponds, urban settlement, and acute pollution, the overflow of water from one pond to the next vanished decades ago. LV is encircled with a road on which Budheshwar Temple, Durga Temple, Batukeshwar Temple, Dattreshwar Temple, and Fort Wale Baba are located. Around the same road, a high density of residential and commercial settlements can be seen. Four hundred years old, *Dudhadhari Math* is also located at the edge of LV.

LV is connected with memories of Swami Vivekananda. Swami Vivekananda stayed in Raipur between 1877–79 (at the age of 14–16), in Rai Bahadur Bhootnath De's house situated between Kotwali Chowk and Kali Bari Chowk.[3] During this period, he used to spend a lot of time alongside the lake. Some people also believe that he used to swim across the lake, and though this is not confirmed, it is known that he was a good swimmer. To make his memories indelible, the lake was named Vivekananda *Sarover* (lake) in 1984 and, in April 2005, a 37-feet tall statue of Swami Vivekananda was installed at the centre of the atoll in the lake. The statue was designed by *Padmashri* – John Martin Nelson. It was unveiled by the former Prime Minister of India, Atal Bihari Vajpayee. A water wall was also built behind the statue. At that point, a garden was also developed with boating facilities.

After independence, LV was brought under the state property regime (local urban body). The total water-spread of the lake during the year 1954 was around 64.75 ha, which has reduced today to 35.208 ha (Raipur Municipal Corporation, 2020). According to 1954 land records of Budhatalab, the Khasra numbers 1937 and 1938 are related to the water body and the island/atoll (tapu) respectively. During 1972, an approach road of about 300 metres was constructed to connect the atoll in the middle of the lake with the main entrance to the lake (now, it is a rear gate, after development of the lake area under the Smart City Development Plan). In 1993, the area of the lake was assessed to be 34 ha (Marothia, 1995a, 1997c). However, in a survey conducted during 2013–14, it was found to be 27.255 ha (Dey, 2015). In response to a Public Interest Litigation, the local urban authority (RMC) has informed the Chhattisgarh High Court that the area of Budhatalab is currently 35.22 ha. As per a recent survey done by an authorized surveyor, the total areas of the Budhatalab and the garden were measured to be 28.619 ha and 1.48 ha respectively (Er. P.R. Chouhan, 1st July, 2020).

The water depth during monsoon and non-monsoon seasons remains around 3.5 metres. The RMC is responsible for the management of the lake. In 1997, the RMC looked into the pollution problems of the lake. The community living in the surrounding area of the lake have traditionally been using the water of the lake for bathing and washing, tending cattle, and disposing of domestic waste, including sewerage. The lake has also been receiving sewage waste and other domestic discharge from the up-lake colonies, schools, colleges, and the hutment population occupying the north-east corner of the lake. Most of these settlements were developed on the water lungs of LV. According to senior citizens living on the area of LV, the outdoor stadium was built on land that was soaking wet from the overflow of surplus water from LV.

Prior to 1977, LV was completely covered with water hyacinths and other aquatic weeds, and was full of mosquitoes. The RMC has been leasing LV to the FCS for aquaculture development since 1985 and earning approximately Rs. 200,000 per year. The FCS was only given usufructuary rights for aquaculture development with the understanding that other users of the LV cannot be excluded from using the lake's waters (Table 5.1). There were also viewing spots on the north-west boundary of the lake (Map 5.1).

Water from LV is used for multiple purposes by various users (Table 5.1). In addition to the activities listed in Table 5.1, LV was used until 2011–12 for the immersion of flowers from temples and clay statues (idols) of the god Ganesh and goddess Durga during Ganesh and Durga puja festivals. These clay statues contain colouring agents, chemicals, and metallic substances.

Table 5.1 Multiple Use of Lake Vivekananda

S.No.	*Activities*	*Users (No./day)*
1.	Fisheries (fisherman)	140
2.	Recreation (excluding rainy days)	
	a. Boating	50
	b. Park visits	150
	c. Swimming	25
3.	Bathing	1500
4.	Washing (washerman)	25
5.	Cattle tending (buffaloes)	30
6.	Dumping of sewerage and domestic wastes	
	a. Discharge from hutment households located at the mouth of lake	1300
	b. Discharge from up-lake households (colonies)	40000

Source: Based on a survey conducted in 1993 by the author. Also appeared in Marothia (1995a, 1997c, 2004d)

Due to the continuous flow of domestic wastewater into the LV, the growth and area of water hyacinths has been increasing, choking the waterways, and hampering recreational boating, creating suitable conditions for breeding mosquitoes, and reducing the bathing and swimming areas. Through biological control, the RMC eradicated the water hyacinth in 1990 (Marothia, 1995a, 1997c, 2004d).

The RMC also introduced two- and four-seater paddle boats for recreation in 1986–1987. The Nilabh Park remains open every day for visitors from 4 P.M. to 8 P.M. There is no entry fee for the recreation park, however, there are fees for hiring the boats (Rs. 25 for two-seater and Rs. 30 for four-seater for 30 minutes). Sometimes, the unrestricted access to the park results in congestion and visitors have perceived a deterioration in the quality of recreation provided. Realizing the deteriorating environment of the lake, the Environment Planning and Coordination Organization of the Environmental Department of the Government of Madhya Pradesh sponsored a project for the environmental improvement and beautification of LV in 1992. An amount of Rs. 1.25 million has been sanctioned and Rs. 950,000 has been invested for improving bathing spots, lighting facilities, and for some plantation. Provision for diverting sewage was also made in the project but this was not implemented.

5.2 Nature and Extent of Pollution and Degradation

Prior to the formation of Chhattisgarh State, Madhya Pradesh Pollution Control Board (MPPCB) was responsible for monitoring the environmental quality of the lake, as well as enforcing pollution control measures. Its regional office, located at Raipur, has been monitoring, on a monthly basis, the water-pollution levels from three sample spots (see Map 5.1). Since the formation of the state, Chhattisgarh Pollution Control Board (CGPCB) has been responsible for monitoring the physio-chemical parameters of the water bodies located in different parts of the state, including LV. Information about different pollution parameters between 1990 and 1993 was obtained from MPPCB and for the period between 1997 and 2001 from CGPCB.

The average figures of various physio-chemical parameters, together with their range, have been calculated and are presented in Table 5.2. The Indian standard for the range of acceptable values was used to assess the damages associated with each of these parameters. It is important to note here that the permissible values vary according to the socio-ecological and technical conditions of the water bodies concerned.

The sewage waste and other domestic discharge from the up-lake colonies, schools, colleges, and the hutment population occupying the north-east corner of the lake are quantified as total solids (TS), suspended solids (SS), and dissolved solids (DS). Coliforms are the principal bacteria associated with these solids.

The average and the maximum values of TS exceeded the permissible limits between 1997 and 2000. The average, minimum, and maximum calculated values of SS also exceeded the permissible limits by many times. The large quantity of SS has caused a high degree of turbidity in the lake water. This has intensified pollution, at least in the summer months. The maximum calculated value of DS was also much higher than the permissible limits. At the same time, except in 1997, the calculated average values were slightly lower than acceptable limits. Also, the minimum values in all four years were much lower than allowable limits for DS in water bodies.

Information regarding coliform most probable number (MPN) values for the lake was only recorded during 2000–2001. MPN values were almost 48 times higher than the permissible limits indicating that the lake water needs to be treated immediately. Looking at the magnitude of the values of other parameters, higher MPN values could be expected in subsequent years (2000 onwards).

Most of the time, the average and maximum values of dissolved oxygen (DO) remained above the permissible limits. However, the minimum value of DO fell slightly below the allowable limits in 1998. Variations in the level of DO depend upon a multiplicity of factors including the size and decomposition of the discharge, movement in the water bodies, and water temperature.

Biological oxygen demand (BOD) is the quantity of oxygen required to decompose the organic matter contained in domestic waste and sewage. The calculated average, minimum, and maximum values of BOD reached a very high level during 1997 to 2001, indicating that the process of decomposition had exhausted DO in these years. The lower levels of DO are reflected in the calculated range values. Low-DO decomposition in the lake still continues, but it takes place anaerobically. Anaerobic conditions are common in the lake during the summer months.

Table 5.2 Physio-Chemical Characteristics of Lake Vivekananda (1990–2001) (Annual Average of parameters)

Parameters	*Unit*	*Allowable limits**	*1990*	*1991*	*1992*	*1993*	*1997*	*1998*	*1999*	*2000*	*2001*
Temperature	°C	-	26.8 (25.0–27.0)	29.1 (25.0–31.0)	29.0 (26.0–33.0)	31.4 (24.0–38.0)	26.96 (21.0–32.0)	27.14 (24.0–29.0)	26.08 (23.0–29.0)	28.25 (21.32)	26.83 (25.29)
pH	pH unit	6.5–8.5	7.8 (7.4–8.3)	8.0 (7.4–8.7)	7.8 (7.1–8.9)	7.7 (7.1–8.1)	8.16 (7.47–8.96)	8.48 (6.70–9.28)	8.06 (7.21–8.81)	7.61 (7.40–8.70)	8.52 (8.19–8.98)
Total Solid	mg./lit	500	584.1 (226.0–904.0)	453.4 (370.0–688.0)	480.5 (186.0–842.0)	596.8 (422.0–656.0)	729.11 (402.0–1690.0)	826.78 (462.0–1428.0)	915.72 (438.0–1230.0)	713.25 (440.0–1008.0)	668.17 (480.0–986.0)
Suspended Solid	mg./lit	10	76.0 (32.0–144.0)	46.7 (26.0–92.0)	95.9 (32.0–308.0)	144.0 (52.0–196.0)	246.39 (24.0–642.0)	252.72 (84.0–518.0)	269.78 (120.0–434.0)	163.92 (118.0–204.0)	192.56 (120.0–306.0)
Chloride	mg./lit	250	78.5 (67.0–101–2)	80.8 (67.7–100.1)	93.9 (23.9–124.0)	90.0 (64.5–110.2)	106.22 (70.50–155.95)	116.65 (77.50–154.0)	119.22 (74.97–154.0)	88.97 (60.0–115.0)	72.05 (58.0–90.0)
Dissolved Solid	mg./lit	500	517.7 (132.0–785.0)	382.4 (344.0–662.0)	391.0 (114.0–758.0)	453.9 (346.0–582.0)	482.72 (221.0–846.0)	574.33 (282.0–930.0)	645.69 (316.0–884.0)	549.33 (280.0–804.0)	475.61 (310.0–774.0)
Total Alkalinity	mg./lit		- -	- -	- -	- -	184.69 (138.0–232.0)	230.33 (165.0–345.0)	217.72 (155.0–340.0)	196.69 (170.0–250.0)	219.78 (150.0–295.0)
Total Hardness	mg./lit		- -	- -	- -	- -	126.83 (64.0–198.0)	177.94 (75.0–290.0)	183.33 (96.0–315.0)	168.14 (125.0–195.0)	197.39 (140.0–265.0)
Calcium Hardness	mg./lit	200	68.0 (148.0–86.0)	63.4 (30.0–102.0)	58.2 (40.0–80.0)	66.7 (35.0–110.0)	70.86 (20.0–116.0)	115.17 (60.0–175.0)	123.67 (64.0–220.0)	115.53 (90.0–165.0)	136.28 (90.0–195.0)
Magnesium Hardness	mg./lit	200	68.0 (148.0–86.0)	63.4 (30.0–102.0)	58.2 (40.0–80.0)	66.7 (35.0–110.0)	55.94 (15.0–92.0)	62.69 (15.0–116.0)	59.33 (25.0–120.0)	52.61 (20.0–100.0)	61.11 (40.0–85.0)
Dissolved Oxygen (DO)	mg./lit	6	7.5 (6.0–9.8)	7.7 (3.3–11.6)	6.5 (2.7–9.4)	5.4 (3.4–10.0)	7.67 (6.0–8.50)	7.52 (5.50–8.90)	7.69 (6.10–9.60)	7.81 (6.30–10.50)	8.41 (7.90–9.60)
Biochemical oxygen demand (BOD)	mg./lit	3	NA	5.7 (2.0–8.0)	17.0 (2.0–33.0)	11.7 (5.0–26.0)	30.32 (20.0–62.0)	38.19 (24.0–80.0)	36.27 (20.0–80.0)	33.06 (16.0–62.0)	58.78 (34.0–95.0)

Chemical oxygen demand (COD)	mg./lit	10	36.5 (10.5–66.7)	59.4 (14.8–80.0)	75.1 (24.0–130.0)	85.1 (41.0–127.0)	113.78 (40.0–240.0)	182.76 (80.0–320.0)	201.02 (80.0–320.0)	157.18 (38.40–250.0)	292.78 (184.0–460.0)
Calcium	mg./lit	75	33.9 (23.2–118.2)	25.4 (12.0–40.0)	23.3 (16.0–33.0)	28.8 (21.0–44.0)	21.54 (11.50–36.80)	46.91 (24.0–68.0)	47.27 (25.6–88.0)	46.21 (36.0–66.0)	54.51 (36.0–78.0)
Magnesium	mg./lit	125	23.6 (15.1–28.8)	15.4 (13.2–18.5)	14.8 (9.7–31.5)	13.8 (7.8–31.9)	7.29 (3.0–22.4)	15.15 (4.87–29.26)	14.67 (6.09–29.26)	12.82 (4.87–24.39)	14.66 (9.75–19.51)
Coliform	MPN/ 100ml	50	75.0 (75.0–75.0)	NA	NA	NA	NA	NA	NA	2400.00	2400.00
Sulphate	PPM	250	36.2 (19.2–70.4)	NA	NA	18.7 (16.0–30.8)	-	-	-	-	-

Source: Regional Office, MP Pollution Control Board, Raipur (data from 1990–93 and 1997–2001) compiled by the author

*Note:** As per IS:2296:1982 NA = Not available. Figures in parenthesis are range values. MPN= most probable number. PPM= parts per million.

The calculated average as well as the range values of chemical oxygen demand (COD) for Lake Vivekananda for all four years was higher than the permissible levels. The BOD and COD levels in the lake were much higher than DO levels, as a result of which the assimilative capacity of the lake has been greatly impacted and has reached a crisis point. The calculated values of the other parameters given in Table 5.2 are largely within acceptable limits. However, with the high level of waste disposal, these values have crossed the permissible limits in subsequent years (see Dey, 2015). Between 1977 and 2000, water-recreation facilities were introduced.

Based on values of the pollution-measuring parameters, water pollution has been identified as a serious problem in LV (Marothia, 1995a, 1997c, 2004d). It has caused a number of water-borne diseases, thereby affecting the health of a large number of the local population (Table 5.3).

A study was conducted during 2013–14 to assess the values of physio-chemical parameters of LV's water quality by collecting water samples for testing during the summer, rainy, and winter seasons (Dey, 2015, p. 15). Analysis of pH, total alkalinity (mg/L), turbidity, TDS, total hardness, calcium hardness, magnesium hardness, chloride, BOD, and COD parameters indicated that LV, during the period, had almost reached beyond its carrying and purification capacity. Since this study, no information on pollution parameters has been available, even with CGPCB. However, the populations living on the area have raised their concernsd from time to time about foul odor and water contamination.

5.3 Interdependent Activities and Their Impact

The degradation of LV's ecosystem is directly related to conflicts between users, the absence of property rights structures resulting in conflicting rights, or falsely claimed rights over the boundaries of the lake for the construction of huts. In addition, externalities, institutional failures and weakness, and poor functioning of the RMC and a lack of coordination between the RMC

Table 5.3 Users Affected by Water-Borne Diseases

S.No.	*Diseases*	*Percentage*
1.	Water-borne & water washed	25
	a. Diarrhoeal diseases and dysentery	05
	b. Enteric fevers (paratyphoid, typhoid)	05
	c. Ascariasis (roundworm)	10
	d. Skin & eye infections	10
2.	Water-related insect vectors	
	a. Lymphatic Filariasis	05

Source: Based on an interview of 2000 users of the LV by the Author during 1993 (Marothia,1995a, 1997c, 2004d)

and other agencies are equally detrimental issues. To resolve these conflicts, it is necessary to restructure the property rights and institutional arrangements which control the individual users' actions as well as the actions of the RMC and other agencies engaged in monitoring LV's ecological parameters. Understanding the interactions between various policies may assist policy makers to design appropriate institutional arrangements to sustain LV's ecosystem and the ecosystems of other wetlands of the state.

The critical issues in case of LV are the nature and extent of the rights of individual or groups of individual resource users, as compared to the rights of society, over a valuable multifunctional and multiuse wetland, and the collective actions that are needed for the conservation and management of the total lake ecosystem, including its buffer zone.[4]

During the course of the study, LV changed to an open-access regime from a state property regime (RMC). As a result, LV suffered from overuse, abuse, congestion, unchecked degradation, widespread growth of water hyacinths, and the loss of ecosystem services and use and non-use values.

Consider the case of sewage disposal from up-lake inhabitants. As a renewable resource, LV has a limited purification capacity. Sewage disposal has a constant and widespread health, chemical, biophysical, and ecological impact, and third-party effects, with high transaction costs in terms of polluted water, water-borne diseases, reduction in aquatic life, and deterioration in water-centric recreation activities, make improper sewage disposal a matter of great concern. The nature of its impact can be irreversible if the waste load affects the self-purifying capacity of the lake or damages the critical renewable resource limits[5] (see Table 5.2 for exceeding values compared to permissible limits of many physio-chemical parameters).

Take another example. A group of washer folk use the LV water to clean clothes. Hutment households, located up-lake, dump human, domestic, and animal waste into the lake, thereby increasing the cost of cleaning clothes. These costs are external to the hutment household's budget estimates. Households, as a group, think LV is a waste-disposal resource. Both the washer folk and the hutment households feel that their location on the lake gives them the right to use its waters in their own way.

Waste disposal by these hutments also adversely affects other third parties, like the bathers, swimmers, fishermen, recreationists, and themselves, too. It is important to note that hutment households located on the west-east boundary of the lake, which had initially encroached on the lake area some years ago, claimed ownership of occupied plots. Their activities have led to inter-temporal or inter-generational interferences or externalities.

Rather than repairing the aforementioned ecological damages, a move was made between January and April 2000 to create urban amenities on an atoll, and a park, by converting the water area into land. A massive movement against such an ecological alteration of LV was organized by various civil societies, the press, the media, environmentally concerned citizens, and scientists (see Table 5.11). I also joined the movement and my extensive interviews appeared on the front page of the leading newspaper in Central India on April 20th, 2000, giving strength to the agitation – and, as a result, LV was saved.[6]

In the course of this movement to save LV, I had realized that it is extremely important to estimate monetary and non-monetary values and WTP by different users and non-users for the conservation of LV. Further, I was quite aware that policymakers may possibly be convinced by the estimates of the economic benefits being provided by LV. Such information may improve the thinking of public policymakers, helping them in making informed decisions on the wise use of LV and many other multiuse water bodies in the state. After this movement, I decided to make an attempt to estimate the use and non-use values of LV to create awareness among a large section of stakeholders about wetland wealth in general and LV in particular.

5.4 Valuation of LV and WTP for Conservation

For the economic valuation of LV, both direct and indirect techniques were used. Indirect techniques are based on the expressed preferences of visitors, or the market approach in which the visitation rate to LV to use the park or to engage in boating and other activities is taken as an indirect measure of the value that people attribute to LV's ecosystem. Such a valuation approach is known as the Travel Cost Method (TCM). The Contingent Valuation Method (CVM), or stated preference method, uses a direct approach to elicit WTP. Respondents express their WTP assuming a market exists for an improved environment, better water quality, and a well-managed park within the physical boundaries of LV. To estimate the demand function and to capture the recreation value, the TCM can be used. Two approaches, namely, Zonal Travel Cost Method (ZTCM) and Individual Travel Cost Method (ITCM), have been used by researchers to estimate the value of a particular recreation site. However, their calculations depend on the unit of observation, the nature of dependent and independent variables, and the sample size (see Cesario, 1969; Clawson and Knetsch, 1978; Marothia, 2001a; Chopra, 2002, for a comparison of the two methods).

5.4.1 Database and Valuation Methods

Keeping in view the individual-specific characteristics of LV, the ITCM approach has been used herein to capture the recreation value and to estimate demand functions. CVM was also used to estimate the WTP of visitors to LV. The WTP by the local population, scientists, and environmentally conscious citizens was also estimated using a CVM to assess the non-use or existence value of LV. For an economic valuation of LV, two interview schedules were prepared. The first interview schedule was designed for visitors and had two parts. Part one dealt with ITCM questions and part two dealt with contingent valuation questions. The second interview schedule was designed for non-users (environmentally conscious citizens, scientists). In addition to these two interview schedules, a third interview schedule was also designed to assess the perceptions of value among the households living around the periphery of LV – those who have the maximum stake in its ecosystem. A close-ended questionnaire format was used to carry out a contingent valuation survey. The respondents were asked whether or not they would be willing to pay a single specified sum. In other words, in a close-ended format, the respondent's decision involves a dichotomous choice between 'yes' and 'no'.

The ITCM approach was applied to determine the influence of various socio-economic characteristics and the level of demand for recreation activities on LV. Since LV is visited frequently by local residents for the park and water-recreation facilities, the ITCM approach seems to be more appropriate. Also, as mentioned, it can be meaningfully used for a relatively small sample size. A visitor survey was conducted to collect data about the number of visits made by each visitor per time period and the individual, site-specific characteristics of 60 visitors. The respondents were also asked how much they would be willing to pay for the better management and preservation of LV.

In order to assess the scientists' perception of the relationship between existence and use value to humans of the LV ecosystem, and to estimate the WTP for upkeep of the LV, 29 scientists were interviewed using a pre-structured interview schedule. Fifteen of the scientists were either involved in teaching or research relating to water bodies, and their area of specialization was fisheries, limnology, ecology, and biology. Six scientists were working in the area of social science, and the remaining eight were working in agriculture, animal science, and forestry. Similarly, to assess the non-users' perception of the value of LV, 32 environmentally conscious citizens were interviewed. To assess the perceived value of LV by the households living around the lake's periphery, 90 households were interviewed using questions related to existence, value, and WTP. All the respondents were interviewed face to face.

The interviews were conducted by nine trained interviewers. These nine interviewers completed about 230 interview schedules. Incomplete interviews and those judged insincere by the interviewers were rejected. There were 19 such cases among the visitors', who were interviewed for estimating the travel cost and contingent valuation method for LV.

Finally, the sample contained 211 respondents. The visitors to LV and households living around the periphery of LV were selected randomly. However, the scientists and environmentally conscious citizens were chosen purposely, keeping in mind their involvement in the preservation of the environment in general and LV in particular. The survey was conducted during March-April 2001.

Data on visitors' profiles, attitudes towards environmental preservation, expenditure on their trip, reasons for their visit, the frequency of their visits, and their opinions on the management aspects of LV were collected to estimate the ITCM model. Information gathered for contingent valuation related to the perception of visitors regarding the value of LV for different uses (recreation, aesthetic and scenic beauty, biological diversity, education, and exclusiveness). Their opinions were obtained on uniqueness, attitudes, and concerns about the degradation of environmental quality, perceptions of conservation, alternative management systems to maintain LV, proposed entrance fees, and WTP and its mode of payment.

Specification of Travel Cost Model and Variables

Individual travel cost method (ITCM) was applied to determine the value of recreational activities and the nature of demand for LV. The following demand function was estimated using ITCM:

$$V = f\,(DIST, LC, TC, REA, INC, OCT, EDU, VH, CON, MAG) \quad (1)$$

Where:

V = visits made by each visitor per period (number)

DIST = distance from residence to LV (Km)

LC = local cost incurred on parking, boating, and eating (Rs.)

TC = total travel cost from place of residence and expenses of parking, boating, and eating within the surroundings of LV. It is defined on a per capita basis (Rs.)

REA = reason for visiting LV. Information for this variable was elicited under the headings of "combination of water and park recreation", "educational interest", and "presence of LV". Most of the visitors mentioned recreation value as their main reason. This variable was normalized

INC = household income per family (Rs.)

OCT = per capita household income per day corrected (normalized) for number of hours spent in LV for park and water recreation (opportunity cost of time in Rs.)

EDU = education of the respondent (five-point scale) is defined in five levels (1: Illiterate, 2: Primary, 3: High school, 4: College, 5: University) (Score)

VH = time spent in LV per visit (hours)

CON = concern for upkeep of LV (dummy variable: Yes = 1, No = 0)

MAG = present management (dummy variable: satisfied = 0, unsatisfied = 1)

Specification of WTP Function and Variables

To estimate the WTP of the visitors to LV, the contingent valuation method was used. The WTP function is expressed in the following form.

$$W = f\,(V, REA, INC, OCT, EDU, VH, CON, MAG, DIST, ENT) \quad (2)$$

Where W is WTP for upkeep of LV (Rs.), and other variables are as defined previously, and ENT is proposed entrance fee (Rs.).

For the estimation of demand function for LV visitation rate (equation 1) and individual WTP for a visit to LV (equation 2), an OLS specification with linear and non-linear forms (semi-log with log of dependent variable) of function were selected. Both linear and non-linear regression equations were also estimated for the two sets – one with TC and the other with LC. Both forms of the functions provide almost similar results for equation 1.

5.4.2 Empirical Evidence

The results of the interviews are discussed in three sections. The first section discusses the results of use value of LV to visitors. In the second section, the results of the WTP by visitors to LV are discussed. The third section analyses the perception of the non-use value accruing from LV to the local population living around the periphery of LV, environmentally conscious citizens, and scientists.

A. Demand Functions for LV Visitation Rate

The estimated demand functions for the visitation rate to LV indicate its use value to visitors (Tables 5.4, 5.5, and 5.6). It can clearly be seen from Table 5.4 that the visitation rate is affected by distance, total travel cost, and purpose of visit, in linear and semi-log demand functions. Education variable has a positive and significant coefficient in the semi-log functions. Level of education thus has an impact on the visitation rate. Since most of the visitors are local, the opportunity, cost of time, and hours spent per visit, probably indicate their preference for LV as a recreation site. This is also reflected in the coefficient of concern for the upkeep of the LV and the existing management

Table 5.4 Demand Function for LV Visitation Rate

Independent Variables	*Linear*	*Semi-Log*
DIST	-21.405 (2.212)**	-0.086 (2.345)**
LC	1.332 (0.896)	0.009 (1.180)
TC	-2.341 (2.815)***	-0.007 (2.305)**
REA	-30.335 (3.356)***	-0.113 (3.267)***
INC	-0.003 (0.422)	0.000 (0.788)
OCT	0.328 (0.262)	0.003 (0.691)
EDU	21.883 (1.393)	0.097 (1.631)*
VH	23.933 (0.470)	0.154 (0.792)
CON	-21.252 (0.612)	-0.040 (0.301)
MAG	82.217 (1.196)	0.251 (0.959)
DIST	130.68	1.65
R^2	0.4374	0.4448
F-Value	3.81 (10, 49) df	3.93 (10, 49) df

Note: The figures in brackets are t – values of the regression coefficients, df – degree of freedom
*** Significant at 1 percent level, ** Significant at 5 percent level, *Significant at 10 percent level

Table 5.5 Demand Function for LV Visitation Rate (Total Travel Cost)

Independent Variables	*Linear*	*Semi-Log*
DIST	−21.469 (2.223)**	−0.087 (2.366)**
LC	−1.799 (3.156)***	−0.004 (2.604)**
TC	−30.696 (3.406)***	−0.118 (3.399)***
REA	−0.003 (0.549)	−0.000 (0.882)
INC	0.655 (0.549)	0.005 (0.978)
OCT	18.677 (1.223)	0.081 (1.390)
EDU	24.785 (0.488)***	0.147 (0.752)
VH	−24.095 (0.698)	0.046 (0.344)
CON	71.705 (1.060)	0.200 (0.772)
MAG	154.87	1.77
R^2	0.4282	0.4291
F-Value	4.16 (9,50) df	4.18 (9,50) df

Note: The figures in brackets are t – values of the regression coefficients, df – degree of freedom
*** Significant at 1 percent level, ** Significant at 5 percent level

Table 5.6 Demand Function for LV Visitation Rate (Local Cost)

Independent Variables	*Linear*	*Semi-Log*
LC	−1.589 (1.526)	−0.006 (1.593)
REA	−31.808 (3.464)***	−0.118 (3.421)***
VH	27.345 (0.728)	0.209 (1.477)
OC	0.154 (0.266)	0.001 (0.546)
DIST	−26.601 (2.732)***	−0.102 (2.774)***
MAG	60.289 (0.876)	0.206 (0.797)
CONSTANT	198.24	1.90
R^2	0.3365	0.3645
F-*Value*	4.48 (6,53) df	5.07 (6,53) df

Note: The figures in brackets are t – values of the regression coefficients, df – degree of freedom
***Significant at 1 percent level

system. It is interesting to note from Table 5.5 that the linear as well as semi-log functions based on total travel cost (excluding LC variable) indicate an inverse relationship between distance, total travel cost, and reasons to visit and visitation rate. However, education in the case of the linear function has a positive and a highly significant relationship with the visitation rate. The demand function based on local cost (excluding TC variable) explains the inverse relationship between reasons to visit and distance with visitation rate (Table 5.6). The level of variation explained by each of the models is not very high (less than 50 percent), although it is better than many other studies that adopted the Travel Cost Method to assess the value of recreation sites in India and elsewhere. Similarly, the level of significance of a few estimated parameters are somewhat weak but consistent with earlier studies (Cesario, Jr., 1969; Chopra, 2002; Isangkura, 1998 reviewed a few such studies with similar variables). The estimated area under the demand curve and the consumer surplus relating the visitation rate to total travel cost, local cost, and distance, indicate WTP for recreation services provided from LV. It can also be considered as WTP for the preservation of the LV ecosystem (Table 5.7).

B. WTP

CVM was used to estimate the WTP of 60 respondents (the same respondents were also interviewed for ITCM) for the upkeep of LV. The linear form of the WTP equation provides better results in terms of the level of variation seen. However, the semi-log function indicates that education, distance, and proposed entrance fee variables have a significant positive relationship with WTP. The proposed entrance fee also has a significant positive relationship with WTP in the linear function. These results clearly indicate that the respondents are willing to pay if entrance fees could be introduced to avoid congestion and nuisances in the park and the boating area. Congestion and nuisance involve a psychological cost to the respondents. The other variables were not found to have a significant effect on WTP (Table 5.8).

Table 5.7 Area Under Demand Curve and Consumer Surplus

Variables	*Area*		*Consumer Surplus*
	Linear	*Semi-Log*	
TC	122.811	53.223	25.87
LC	122.808	54.882	42.35
DIST	122.804	55.454	46.43

Table 5.8 Multiple Regression Estimate Results for WTP

Independent Variables	*Linear*	*Semi-Log*
V	-0.012 (0.973)	0.000 (0.881)
REA	-0.053 (0.058)	-0.050 (1.282)
INC	0.000 (0.898)	0.000 (0.410)
OC	-0.067 (0.628)	0.004 (0.935)
EDU	-0.571 (0.399)	0.222 (3.648)***
VH	6.044 (1.291)	-0.096 (0.484)
DIST	1.377 (1.444)	0.069 (1.698)*
ENT	0.995 (83.206)***	0.002 (4.495)***
CON	-1.030 (0.323)	0.020 (0.145)
MAG	-6.662 (1.063)	-0.263 (0.988)
CONSTANT	1.97	0.71
R^2	0.9939	0.5434
F-Value	792.55 (10,49) df	5.83 (10,49) df

Note: The figures in brackets are t – values of the respective coefficients, df – degree of freedom
*** Significant at 1 percent level, * Significant at 10 percent level

C. Use and Non-Use Values

In this section, an analysis is presented of the perception of use value and non-use values accruing from LV as assigned by visitors for recreation, the local population living around the periphery (PLIPLV), ECCs, and scientists. As stated earlier, the analysis of WTP clearly indicates that visitors are willing to pay if entrance fees could be introduced to avoid congestion and nuisances in the park and boating area. During the study period, there was no entrance fee for the park (Marothia, 2001a). The non-use values relate to the existence or intrinsic values expressed by respondents such that those values are unrelated to present and future uses of the LV ecosystem, by them or their generation. The results of WTP by visitors, scientists, ECCs, and the local population living around the periphery (PLIPLV), and their WTP for an alternative management system are also presented.

The reasons chosen for the upkeep of LV by these four groups of users and non-users are presented in order of importance. Ranks assigned to different reasons by respondents are given in Figure 5.1 to Figure 5.4. Aesthetic value, biological diversity, education, and future value were the main reasons expressed by scientists for the conservation of LV. The ECCs were concerned about aesthetic value, future value, and biological diversity, followed by other important reasons listed in Figure 5.2. It is interesting to note that more than 55 percent of the respondents of the PLIPLV category were concerned about the rarity/uniqueness of LV and would like to conserve it for future generations due to its aesthetic value (Figure 5.3).

The responses of visitors, scientists, ECCs, and PLIPLV for the upkeep of the park are presented in Table 5.9. A majority of the respondents in all the four groups stated that they would be willing to pay for the upkeep of the LV. Interestingly, a large majority of ECCs and PLIPLV seemed to prefer

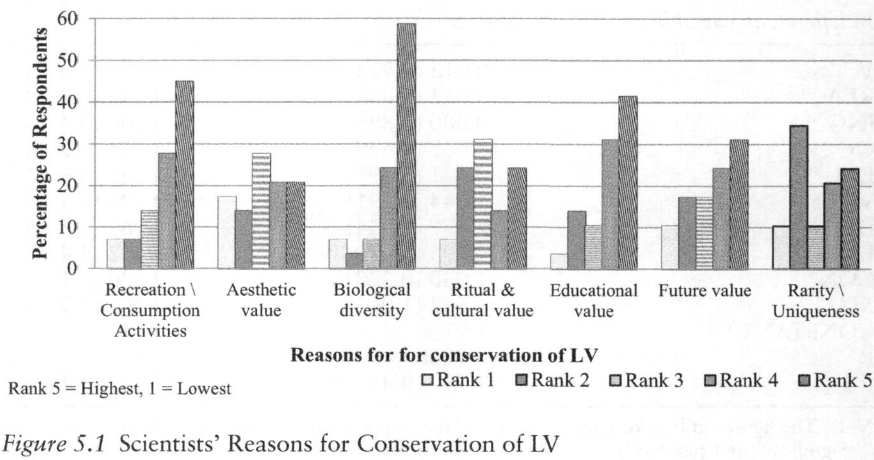

Figure 5.1 Scientists' Reasons for Conservation of LV

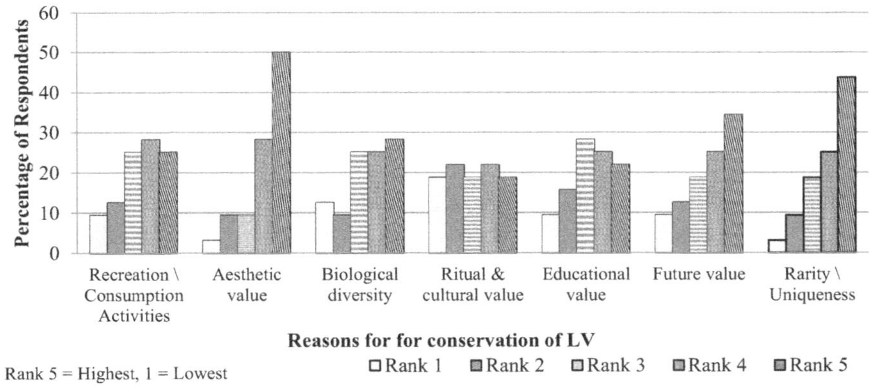

Figure 5.2 ECCs' Reasons for Conservation of LV

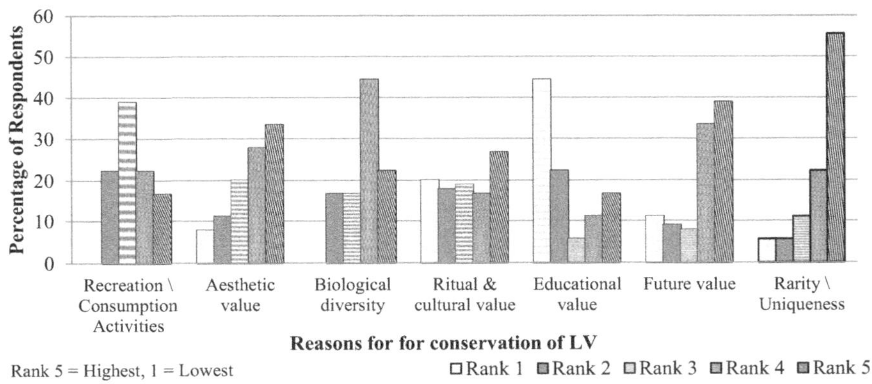

Figure 5.3 PLIPLVs' Reasons for Conservation

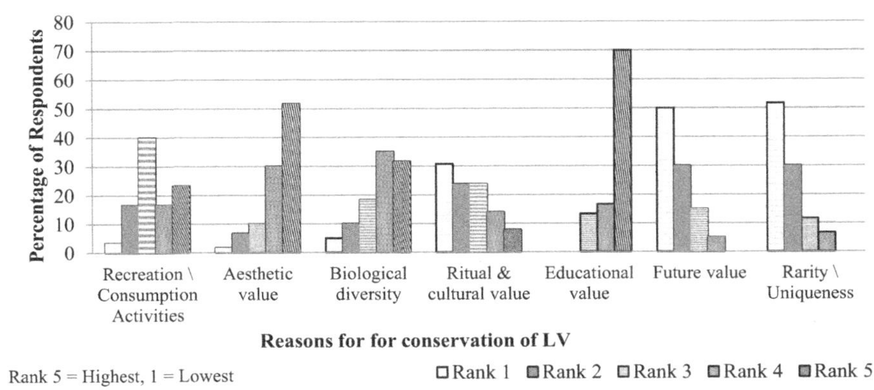

Figure 5.4 Visitors' Reasons for Conservation of LV

management by an autonomous government body, despite the fact that, currently, the RMC manages LV. A large number of the scientists were not satisfied with the present management. However, their WTP was highest among the four categories of respondents (Table 5.9). Scientists were willing to pay for LV if the management was in the hands of the local people, a local NGO, or a national NGO (Table 5.10).

During the course of the survey, we spoke with the members of the FCSs, who had expressed serious concerns about the quality and availability of

Table 5.9 WTP of Visitors, Scientists, Environmentally Conscious Citizens, and Population Living on the Periphery of LV

S. No.	*Resource*	*No. of Responses*			
		Visitors	*Scientists*	*ECCs*	*PLIPLV*
1.	No. of responses to WTP				
	(i) Yes	55	16	29	65
	(ii) No.	5	13	3	25
2.	No. of WTP respondents due to				
	(i) Income constraints	2	2	2	20
	(ii) Dissatisfaction with present management	3	11	1	5
3.	WTP				
	(i) By Lump sum payment	28	6	1	55
	(ii) By annual payment	32	10	28	10
5.	Average lump sum payment	Rs. 45.00 per capita	Rs. 510.00 per capita	Rs. 50.00 per capita	Rs. 62.00 per capita
6.	Average annual payment	Rs. 48.00 per capita	Rs. 235.00 per capita	Rs. 100.00 per capita	Rs. 75.00 per capita

ECCs = Environmentally Conscious Citizens, PLIPLV= Population Living on the Periphery of LV

Table 5.10 Perception of Visitors, Scientists, Environmental Conscious Citizens, and Population Living on the Periphery of LV for WTP to Alternative Management

S.No.	*Alternative management*	*Visitors %*	*Scientists %*	*ECCs %*	*PLIPLV %*
1.	Autonomous govt. body	48.33	10.34	56.25	40.62
2.	Local interest groups	11.67	34.48	21.00	6.25
3.	Local NGOs	23.33	20.69	12.00	9.37
4.	A nationally managed NGO	6.67	20.69	6.25	3.12
5.	Local NGOs with monitoring by a national body	10.00	13.29	3.12	3.12

ECCs = Environmentally Conscious Citizens
PLIPLV = Population Living on the Periphery of LV

water, in the summer month particularly. As a group, the fishermen were willing to pay almost 20 to 25 percent more lease rent to the LV for providing pollution-free water. During the period of the investigation, the lease rent was Rs. 300,000 per year, and every year the RMC increased the rent by 15 percent. The fishermen would prefer the LV to be managed by the district collector or fisheries administrative officer, rather than RMC, to minimize conflict between them and other users responsible for polluting LV.

From the point of view of valuation (values attached by users and non-users) and the management of a MW site like LV, this study provides some meaningful feedback to the RMC and other agencies engaged in urban wetland conservation and governance. The findings in this section may also enhance the understanding of environmentally conscious citizens, scientists, and civil society about multifunctional nature and the various uses of LV. With such data, awareness across different sections of society can be increased to conserve and sustainably use the wetland wealth of the state. In fact, the use and non-use values may contribute significantly to the state's GDP, if GDP is measured in a green income accounting framework.

5.5 Restoration: Ecological Gain or Loss

Over the years, particularly after the formation of Chhattisgarh, few efforts have been made by the RMC, the Department of Urban Development, and the Chhattisgarh Tourism Board (CGTB) to reverse the ecological decline of LV. In the last four years, however, Raipur Smart City Limited (RSCL) has made efforts to restore LV and create a few water-centric recreation activities. I have been tracking restoration efforts at LV since 1993 and analysed the socio-ecological impact at different points of time. Based on a synthesis of my LV research (Marothia, 1995a, 1997c, 2001a, 2004d), coupled with recent developments, this section describes the purpose, process, outcome, and social-ecological impacts of restoration efforts aimed at reversing the degradation of LV's ecosystem. The recovery of key ecosystem services reflects the direct and indirect value of wetland restoration and can provide useful insights for sustainable management. Major initiatives taken between 1977 to 2020 are described in Table 5.11.

5.5.1 First Phase of Restoration

The RMC had made sporadic efforts between 1977 to 2000 to eradicate water hyacinths, but major restoration initiatives were only taken during 2004–06. During this period, the following actions have been taken by the RMC to improve the LV ecosystem. (i) A large number of washer folk have been using a substantial amount of detergents. Most of these detergents contain phosphate, which adversely affects the water quality. The RMC therefore shifted the washer folk to a nearby pond which was already in use for

Table 5.11 Restoration Activities and Their Socio-Ecological Impact

S. No.	*Restoration Activities and Process*	*Year of Activities*	*Implementing Agencies*	*Property Right Regimes*	*Social-Ecological Impact*
1.	Construction of an approach road of about 300 metres from the main entrance to atoll. Plantation of trees along approach road and around atoll	1972	Govt. of MP.	State property (Urban local body)	Public space created for visitors and nature lovers, but for a road-construction area of 300(l) x 6(W) metres area of LV reclaimed resulted in irreversible loss of water-spread area. Waterways option would have saved the LV area and added value to water recreation.
2.	Manual eradication of water hyacinth	1977	RMC	State property (Urban local body)	Improved overall ecosystem services of LV.
3.	Budhatalab named as Vivekananda *Sarovar*	1984	RMC	State property (Urban local body)	Some efforts initiated in subsequent years to improve the ecosystem of LV.
4.	LV leased out by RMC to FCS for aquaculture development	1985	RMC	Usufructuary Rights to FCS	Improved food and nutritional security of fishers' community. Fish was available to local communities living around LV, Aquaculture continued till 2016–17.

5.	Environment planning and coordination organization of the Environmental Department Govt. of M.P. initiated a project to improve LV ecosystem and beautification. Improved bathing spots, lighting facilities and some plantations. Diversion of sewage was part of the initiative but never took off.	1992–93	RMC	State property (Urban local body)	Improved basic water-centric activities for local population, but inflow of sewage water in LV continued. Biological, chemical, ecological, physical and health damages persisted
6.	First single lane road construction was done, in the Name of Beautification of Budhatalab, around its periphery, reducing the area of Budhatalab and adjoining area of Madhav Rao Sapre School (Established 1913) and Dani Girls School (Established 1944) grounds	1999–2000	RMC	State property (Urban local body)	Reduction in the water-spread area and buffer zone lead to irreversible loss.
7.	Attempt to alter ecological landscape of pond and garden and atoll.	2000	RMC	State property (Urban local body)	With the massive public protest and environmental conscious citizens campaign alteration was stopped*.

(Continued)

Table 5.11 (Continued)

S. No.	*Restoration Activities and Process*	*Year of Activities*	*Implementing Agencies*	*Property Right Regimes*	*Social-Ecological Impact*
8.	Cleaning water hyacinth, fencing of LV, removal of encroachment (huts), washing spots and restriction on cattle tending, In April 2005, a statue of Swami Vivekananda was installed in the centre of the garden. But immersion of Ganesh and Durga idols continued	2004–05	RMC	State property (Urban local body)	Shifting washer folks, unauthorized occupied huts, restriction of cattle tending, fencing periphery of LV improved LV ecosystem, use and non-use values, chemical contamination continued due to immersion of Ganesh and Durga idols. Immersion stopped sometime between 2011–12.
9.	Second time road widening was proposed in the name of Beautification of Vivekananda Sarovar. This time the ground of Sapre school, and construction of roads on two sides, was reduced from both the sides. Two Buildings of Primary school, Scouts and guide office, and *Karmchari Bhavan* (class 3 employees organization) were demolished.	2013–14	RMC	State property (Urban local body)	Irreversible damage created by reducing the area of a historical school.

10.	CGTB signed an agreement for 30 years (PPP Mode) with Modern Maintenance Products India Pvt. Ltd (MMPI) where in CGTB had made a three phased project of Integrated Leisure Destination at LV and its periphery. In 2017–18, CGTB planned to develop facilities in the LV. The site was handed over to the company in August 2017. Under the contract, the company was empowered to build 12 shops near the Budheshwar Temple outside the pond with beautification of the pond, water games, boating, floating, food zone, light and laser show. All these facilities were to be developed in 18 to 36 months. Not much work of restoration was carried out except boat facilities introduced.	25 October 2016–02 May 2020	MMPI (PPP Mode)	RMC transferred ownership to CGTB. In turn Tourism Board transferred to MMPI	The company had started work but soon got embroiled in controversies. Actually the company started bridging the water body at Nilabh Garden- might have ended irreversible loss. But due to strong mass opposition in the city proposed activities stopped. Due to disconnect between RMC, CGTB and MMPI some activities proposed by MMPI might have created persistent serious social-ecological interference.

(*Continued*)

Table 5.11 (Continued)

S. No.	*Restoration Activities and Process*	*Year of Activities*	*Implementing Agencies*	*Property Right Regimes*	*Social-Ecological Impact*
11.	RSCL started Phase 1 of Renovation and Beautification of Vivekananda Sarovar and garden, cleaning of water hyacinths and silt clearance was done. 2 Poclain machines, 85 sanitary workers, 35 fishermen were engaged in cleaning operations.	11 May 2020	RMC	RSCL	Cleaning of water hyacinths was almost completed. However, a mechanical solution to the ecological problem damaged the buffer zone and might have destroyed the food web for wetland based habitats.
12.	Destruction of play ground, made for senior citizens in 2018 by spending Almost Rs. 2cr. Demolished one Class Room (Chemistry Lab) of Dani Girls school, almost 50 Trees demolished	25 May 2020	RMC	RSCL, RMC	RMC and RSCL have revived the MMPI and CGTB project with a more tourism centric portfolio. Even at this time, there has been a lot of opposition to the shortening of the grounds in the name of road widening and beautification and alteration of LV ecology. Irrational trade-off between beautification and reducing area of educational institutions and uprooting trees (nesting refuge for local and migratory birds).

13.	RSCL put hoardings displaying the scheme/ project at different prominent places of the city without information related to measurement of road widening, how much land is being acquired?	3 June 2020	RMC	RSCL	Avoided public hearings and voices of civil societies, environmental conscious citizens.
14.	Uprooting of trees in the buffer zone of LV	10–20 August 2020	RSCL	RSCL	Ecological base for wetland habitation was lost.
15.	Sewerage treatment plant, and methane gas chamber not completed	2018–2020	NEERI, BHEL	RMC/CG State Urban Development Department	Without a sewage treatment plant, the LV ecosystem may continue to degrade and create Biological, chemical, ecological, physical and health damage. Frequency of impact of these damages may be constant and widespread with significant third-party effects (externalities), habitat loss, high transaction cost, and irreversibility.

Source: Information listed in the table is based on author's research and supplemented with information sought through RTI from a civil activist (with gratitude)

* A massive movement was launched by civil society, press, media, and environmentalists during January-May 2000 against the ecological alteration of LV, atoll, and park for creating urban amenities (see Marothia, April 24 2000, interview on LV crisis published in *Deshbandhu* Hindi newspaper)

washing purposes. (ii) People living in the hutments situated at the mouth of the lake were evicted and moved to an alternative location. (iii) Cattle tending was stopped in the lake area to reduce a substantial amount of animal waste. (iv) The lakeshore was suffering because of water hyacinths, which had a negative effect on the fish biomass, choked waterways, increased the mosquito population, and hampered recreational boating. Water hyacinths were removed only periodically. (v) A nominal entry fee was introduced. Crowding in the park was perceived by many users as a leading factor in the deterioration in the quality of environmental services rendered by the park (all these institutional arrangements were suggested by Marothia, 1997c, 2001a, 2004d, and shared with major stakeholders). (vii) To commemorate Swami Vivekananda, a statue of him was installed in the centre of the garden in April 2005.

However, the problem of waste and sewage inflow from up-lake colonies still exists today. Between 2006–2019, many cleaning drives had been conducted to improve the ecosystem of LV. The most important action was to enforce a total ban on the immersion of clay statues (idols) of Lord Ganesh and Goddess Durga, sometime between 2011–12, but the lake degeneration continued.

5.5.2 Second Phase of Restoration

A massive cleaning and beautification drive was initiated in May 2020. RSCL started Phase 1 of the renovation and beautification of LV, in which the water hyacinths and silt were cleared. Over 85 sanitation workers, 30 additional contractual labourers, and 50 local fishermen, with the help of heavy construction equipment, removed over 1,050 truck loads of silt and filth from LV. Political leaders and senior administrative functionaries, from cabinet ministers to the chief secretary, participated in the cleaning operation. The authorities also had a plan to beautify and develop the surroundings of the lake after the cleaning. The cleaning operation was achieved through mechanical cleaning, and it was almost completed by the end of May 2000. After the cleaning operation, the following water-centric recreation facilities were developed[7] between May and October 2020:

(i) The musical fountain, 120 metres long, claimed to be the world's longest musical fountain.
(ii) A tunnel fountain, which gives a warm welcome to the tourists-yet to be developed.
(iii) A high-jet fountain.
(iv) Decorative LED lights.
(v) A chase board in the garden.
(vi) A children's sports space.
(vii) 2-D and 3-D art by accomplished artists on the boundary wall of LV.
(viii) Gym equipment for fitness lovers.
(ix) Modern boating facilities.

5.5.3 Urban Amenities at the Cost of Ecological Chaos[8]

All these facilities and decorations added value to LV as an outdoor recreation site. The renovation of LV is a welcome step towards making it more attractive for tourists. The RMC and the Government of Chhattisgarh deserve to be congratulated. However, the beautification of the lake was more in the nature of a cosmetic improvement. In the process, almost 50 old trees and vegetation in the buffer zone were uprooted with heavy construction equipment to develop a concrete walking plaza along the shoreline. The following are the socio-ecological consequences of destroying old trees, shrubs, herbs, and vegetation, and the concretization of LV's shoreline:

(a). The impact of human activity on the local environment, including plant and animal life, cannot be ignored. This medieval lake has got immense value in terms of hydrology, water-quality maintenance, the food chain, the habitat for wildlife, food production, and heritage. These values are tied to the different ecosystem services. Ecological alterations have already reduced provisioning services (fisheries activities were banned as LV is now designated as a tourism site), and may minimize the availability of regulatory, supporting, and social ecosystem services.

(b). During beautification, the hydrodynamic aspects of the lake were overlooked. The water level in the lake has maintained equilibrium with the groundwater of the area. Any change (decrease) in the lake's water level will result in lowering of the groundwater level. Such a decrease in the lake's water level may occur if the catchment is further encroached.

(c). In regard to aquatic life, installing fountains is a welcome move. The fountains may result in better circulation of water and increase the dissolved oxygen (DO) level of the water. The increased DO will, however, help aquatic life, only if wastewater and solid waste are not allowed to mix with the lake water.

(d). Boating activities should not be encouraged inside a relatively small water body like LV. This activity endangers the existence of aquatic fauna. The water sports facilities will have an adverse impact on aquatic flora and fauna habitat due to careless fuelling and vibrations, and might become a cause of degradation in future.

(e). The presence of vegetation not only beautifies the water body but also acts as a carbon sink. It creates a habitat, along with a food chain, which sustains insects, aquatic animals, amphibians, avifauna, and mammals. A water body bare of vegetation is lifeless. Ever since the vegetation has been removed, local breeding birds like cormorants, white egrets, snipes, bitterns, and jacanas have lost their habitat. This is also true of resident bleeder ducks like the lesser whistling ducks and the Indian spot-billed ducks. During every winter season, more than 100 birds of myriad shapes, sizes and colours, like the red headed duck, the northern pintail duck, the pheasant-tailed Jacana, and the little ringed plover, make LV their home. They have attracted many birdwatchers and nature photographers. Due to the loss of bird habitats around LV, migratory birds did not visit the lake during the winter of 2020. Further, even local birds are not seen due to the loss of trees and vegetation in and around LV.[9] The biggest threat to aquatic and avian fauna in this area is sound and light pollution – there are floodlights all around the lake. This is an unnatural environment for them to properly breed and rear young as their biological clocks get disturbed. Further, the removal of mud from the lake-bottom during the process of cleaning means removing the aquatic creatures that regenerate the soil to make it more conducive for aquatic plants. The ecology changes accordingly. The restoration process ended up making LV an amusement park for city dwellers and tourists, and unsuitable for birds and biodiversity conservation.

(f). Uprooting big trees and bushes by the RMC will adversely affect the habitat and niches of local fauna and will compel them to migrate away from the area. The old, big trees will slowly be replaced by some unwanted species of plants and this will deter migratory birds because the flowers, fruits, insects, and shelter of big trees will no longer be available. This will certainly reduce the carrying capacity of the lake's complete ecosystem. Hundreds of thousands of organisms inhabit a single tree and interact with each other. They are interdependent. Therefore, uprooting even a single tree is disastrous, and might unknowingly drive a species to extinction. Ruthlessly uprooting trees will also alter the micro-climate of the town, leading to higher temperatures during summer and flooding during the monsoon.

(g). Riparian higher plants are also important for the health of the pond's ecosystem in terms of bank stability, minimizing soil erosion, oxygen balance, natural shade, and nesting, and providing perching places for aquatic birds. Preventing soil erosion in the catchment directly influences water turbidity or transparency. Water transparency is vital for light penetration and pond respiration, especially in nutrient-rich water bodies. The natural shade provided by a green canopy along the shores of the pond controls water temperature and reduces the growth of phytoplankton that prevents algal blooms. Frequent and sustained algal bloom in such water bodies often causes anoxic conditions during night hours, which may turn the pond into a biological desert. The marginal higher plants are important for the wetland birds that rely on the pond's ecosystem for food, for perching, and for nesting. In return, these birds enrich the ecosystem with their droppings, thus contributing to maintaining the nutrient cycle and the ecological balance.

(h). Another serious threat to LV is the concretization of its shoreline and the direct loss of habitat this causes. The land-water transition zone or shoreline provides a unique habitat for several aquatic and terrestrial animals, such as turtles, frogs, water snakes, crabs, and snails. The loss of shoreline results in the habitat destruction for shoreline-wading

> birds like the stilts and sandpipers. Construction of guard walls along the shoreline has many ecological implications. The loss of gradient due to such construction halts the periodic drying and wetting needed for mineralization of nutrients. This also prevents the growth of marginal hydrophytes and their associated fauna. Many planktonic animalcules, snails, insects, and crustaceans grow around aquatic plants of the littoral zone, attracting a variety of fishes and other vertebrates. This micro-environment is also considered as a suitable habitat for breeding and hiding, and as a resting place for diverse fish species. Such areas exhibit higher biodiversity and an abundance of fish species compared to deeper open-water zones. Aquatic birds also prefer such areas because of the availability of a variety of food items, making the area a rich grazing field for them. Shoreline concretization will therefore prevent the occurrence and growth of aquatic plants and associated fauna, including fish and birds, causing the extinction of fish species that prefer marginal, shallow, muddy-bottom areas.

The restoration process adopted by the RMC and RSCL for LV was not aligned with "Ramsar Resolution X.27: Wetlands and Urbanization". Resolution emphasized that "wetlands in urban and peri-urban environments can deliver many important ecosystem services to people's physical and mental health and well-being . . . can play important role in communication, education, participation and awareness for urban communities about wetlands, as well as the value of establishing education and visitor centres in such places" (X.27.4,5). Further concern is expressed in the Resolution about "recognition of urban and peri-urban wetlands values and importance by both decision-makers and urban communities" (X.27.7). The relevance of various provisions of the Resolution is discussed in Chapter 9 to conserve and achieve the wise use of urban and peri-urban wetlands.

5.6 Issues and Lessons Learned

My three and a half decades of research on the LV ecosystem, at different points of time, has revealed that the extent of pollution, externalities, impact of interdependent activities, use and non-use values, and WTP for conservation of this unique wetland are all important. But the critical issues are a tendency to convert usufruct rights to legal property rights, to grant political patronage to lake encroachers, the ineffective exercise of property rights, entitlement systems, and the role of the RMC authority.

In addition to these findings, some key issues and lessons learned from the LV restoration journey are worth sharing to shape the thinking of various agencies responsible for the conservation and management of the LV ecosystem. After the formation of Chhattisgarh State, LV became an epicentre for restoration experiments. The RMC and RSCL have planned about 13 projects, but a sizable number are still incomplete despite adequate financial allocation.[10] The main reason for non-completion of environmental and restoration works was the frequent shifts in the property rights regime and management system of LV, from the RMC to other agencies. Another equally important factor was a lack of coordination among various agencies, namely, CGPCB, CGTB, RMC, RSCL, the State Ministry of Urban Development, the State Ministry of Forest, Environment and Climate Change, and the State Wetland Authority (SWA).

A lack of resilience in thinking about sustaining the ecosystem services in a socio-ecological system is also a crucial cause for the loss of the ecological landscape and biodiversity of LV. Thanks to massive movements and protests organized by civil society and environmentally conscious citizens, some of the projects and activities that might have created irreversible loss to LV have been stopped.

The state government should build up collective expertise, including wetland scientists, NGOs and practitioners, environmentally conscious citizens, members of nature clubs and nature journalists, to identify barriers to wetland restoration and to plan how to achieve ecologically friendly goals and to avoid ecological losses, based on the experience at LV. This should form part of restoring the degraded and depleted wetlands of Raipur city and the Chhattisgarh State.[11] To avoid the ecological chaos suffered during LV's restoration, responsible federal and state authorities can sustainably manage many Ramsar wetland sites and other peri-urban and urban wetlands of the country, which are facing multiple challenges to their existence. Restoration and conservation efforts have also been made to urban water bodies in Asian countries aligned with the Ramsar Convention Resolution. Lessons from these case studies may be useful to restoring India's urban wetlands.[12]

Notes

1 These illustrations are from Maratha Bhosales Basta's records of Chhattisgarh. They show the history of different cities of Chhattisgarh, with its forts and ponds. The illustration is not only of Budhatalab but other ponds were also mentioned in the 1838 AD record. (*Chhattisgarh Ka Itihas* by Ramendra Nath Mishra, 2015, pp. 93–99). Occasionally the *Gond* tribal community claimed that it was constructed by *Gondwana* kings for the collective use of the *Gond* community. They also claimed that the lake was named after the *Gond* dynasty's god *Budha Dev ji*. Their claim is based on eco-cosmology (an indigenous knowledge system, particularly connected with Shamanic dreams: human-non-human-animal transformation).

2 See Thakur (2002); Tripathi and Tripathi (2012); Shukla and Shukla (2018) for the historical background and multiple uses of these seven ponds and their connection to LV.

3 Personal communication from Debashish Sircar.

4 Using Bromley (1991) and Calabresi and Melamed (1972) taxonomy of property rules, liability rules, and inalienability rules, I argued on behalf of the society and users of the lake, that the RMC can use the inalienability rule to resolve intrusion into LV.

5 See Bromley, 1991, Chapter 3, for critical observations on such complex situations of environmental resources. See also Kneese (1971); Bromley (1976, 1978).
6 See Marothia, 24th April, 2000, interview on LV crisis published in *Deshbandhu*, a Hindi newspaper.
7 S.K. Sundrani, Joint Director, Urban Administrative Department, GoCG, provided details of the development plan of Lake Vivekananda. Profuse thanks to him. Support from D.S. Misra in this regard is also gratefully acknowledged.
8 I am indebted to A.K. Pati, Ninad Bodhankar, M.A. Hassan, B.P. Nonhare, Mohit Sahu, and Ajit Degwekar for providing their critical input for this section. A.K. Gauraha feedback on restoration operations of phase two of LV was also helpful. Thanks to him.
9 Interaction with LV Protection Youth Groups and people living on the periphery of LV.
10 See *Dainik Bhaskar* Star Investigation Report of LV Restoration Projects Status (*Dainik Bhaskar*, Sunday, 24th May, 2020, Chhattisgarh edition).
11 See endnote 7 of Chapter 4.
12 See Ghosh and Sen (1987); Euisoon et al. (1997); Huryn et al. (2003); Galbraith and Burns (2007); Khai and Yabe (2014); Bachmann (2016); Gevers et al. (2018); WISA (2018) for restoration efforts of urban wetlands.

6 Tanks and Tank Irrigation

Challenges for Institutions and Collective Action Restitution

Recognizing the utility of tanks as a traditional technology for sustaining today's farming needs, an attempt is made in this chapter[1] to examine the history and context of tanks and tank irrigation in the rainfed ecosystems of Indian states, including Chhattisgarh. Details about study area, the methodology, and the data base are discussed in section two. The current hydraulic, hydrological, and agricultural status of selected tanks is described in section three. The socio-economic profile and land and water resource availability of sample farmers is presented in section four. Decision-making arrangements and patterns of interaction under the state management system is discussed in section five. Comparative economics of crop establishment methods under perennial and seasonal tanks, incremental benefits of tank irrigation at farm level, and financial gains to main beneficiaries are detailed in sections six, seven, and eight respectively. Farmers' reflections on the existing tank-irrigation systems and the WTP higher charges by different user groups for improved water quality and services are assessed and presented in section nine. In the last section, institutional and policy interventions are suggested to sustain tanks and improve the efficiency of tank irrigation.

6.1 Introduction: History and Context

India derives its irrigation water from canals, reservoirs, tube wells, open wells, ponds, and tanks. Ponds and tanks, a crucial category of multiuse wetlands, are small reservoirs constructed across a slope to store rainwater runoff for irrigation. Tank irrigation is a centuries-old practice for rainwater harvesting in South Asian countries, and Indian states, including Chhattisgarh. Recognizing the socio-economic, cultural, and ecological relevance of traditional water-harvesting systems to meet the varied requirements of rural communities, several ponds and tanks were constructed during the British colonial regime in different parts of Chhattisgarh, particularly in the Central Plains. British rulers acquired land for tanks under the Land Acquisition Act of 1898. Prior to the British regime, the construction of ponds and tanks for irrigation and different community uses was the utmost priority of kings and

DOI: 10.4324/9781003456742-6

village landlords (*Malgujars*) in different parts of Chhattisgarh. Many ponds and irrigation tanks are centuries old and continue to provide multiple ecosystem services (see Chapters 3 and 4). This trend of constructing ponds and tanks has been continuing in the post-independence era under different rural water-creation schemes (see Chapter 3 and 7).

Irrigation policymakers since independence have recognized that the supply of water, more than any other single factor, controls the production of field crops, and a dependable and controlled water supply is therefore a prerequisite to achieving the full potentiality of farm technology (Marothia, 1986, 1992b, 1993b, 2002b, 2003b, 2005a, 2018). The existing canal irrigation system, though it supports farm and off-farm production activities, has distributional implications for water use. In addition, there are long-term problems, such as siltation, salinity, and water-logging due to canal seepage (Joshi and Agnihotri, 1985; Marothia et al., 1987, 2010a; Marothia, 2005a). Canal irrigation also tends to be expensive per unit service due to high construction costs and delays (Marothia et al., 1987; Marothia, 1992b, 2002b, 2005a). The potential for expanding well irrigation is constrained due to the geological characteristics of aquifer formation and problems associated with exploration, development, testing, pump design, and maintenance, in addition to limited opportunities for new capital investment by farmers (Marothia, 1992b). Similarly, reservoirs have limited scope for further expansion. Also, reservoirs adversely affect forest ecosystems, change water regimes, and increase the problems of siltation, salinity, and water-logging (Marothia, 1997a; Marothia et al., 2010). The agro-climatic and topographical characteristics of the Chhattisgarh Central Plain region also justify the construction of irrigation tanks to take full advantage of wider geographical distribution and to provide positive interdependencies within the agro-socio-economic situation of village life (Marothia, 1992b). In-situ utilization of precipitation through irrigation tanks may improve micro-ecological balance; mitigate climate change effects; improve the timeliness and reliability of the water supply for stable food-grain production; generate employment opportunities and reduce the distress village people (Marothia, 1992b). Research on irrigation tanks in Chhattisgarh (Marothia, 1986, 1992b, 2002d, 2003b, 2004b; 2010a, Marothia et al., 2010) and elsewhere (Von Oppen and Subba Rao, 1980a, 1980b; Von Oppen, 1985; Palanisami and Easter, 1984; Rao and Chandrakant, 1984; Palanisami et al., 1997; Mosse, 1997; Vaidyanathan, 1997; Palanisami and Balasubramanian, 2002; Shah et al., 2002; Sakthivadivel et al., 2004; Shah, 2003; Raju et al., 2003; Shah and Raju, 1999) has shown that tank-irrigation efficiency can be improved with effective institutional reform.

To ensure last-mile delivery in canal and tank (connected with the distributionary network) command areas, the Government of Chhattisgarh (GoCG) under the Farmers Participation in the Management of the Irrigation System in Chhattisgarh Act (CG FIMS Act 2000), transferred management of the irrigation system to FOs in 2002 (see Chapter 4 for a detailed description

FIMS provisions). After the WUAs had become defunct, this irrigation system fell under the control of the DWR. Most of the tanks and tank-irrigation water-carrying structures are in a degraded condition due to ineffective local institutions (WUAs and village Panchayats) and the DWR management (Marothia, 2002b, 2005a).[2]

As mentioned in Chapter 3, rainfall in Chhattisgarh is heaviest during June to October, and irrigation tanks serve as reservoirs for storing rainwater. Many tanks in the Central Plains have traditionally been canal fed and are being used for supplementary irrigation. Canal-fed tanks are generally perennial sources of irrigation, whereas rainfed tanks are seasonal in nature and can provide irrigation only at critical crop-growth stages. The number and area of irrigation ponds/tanks across districts of the state are shown in Figure 6.1. Tank water is basically used for irrigating paddy crops. Intermittently, one irrigation from perennial tanks is also available to irrigate rabi crops. Paddy accounts for 85 to 90 percent of the total cropped area. It is important to note here that nearly 75 percent of the cropped area in the state is under a rainfed farming system and subject to the vagaries of the monsoon.

6.2 Profile of the Study Area, Database, and Methodology

Out of 1,350 minor irrigation tanks located in the Arang and Dharsiwa blocks of Raipur district of Chhattisgarh (see Map 6.1), this study was

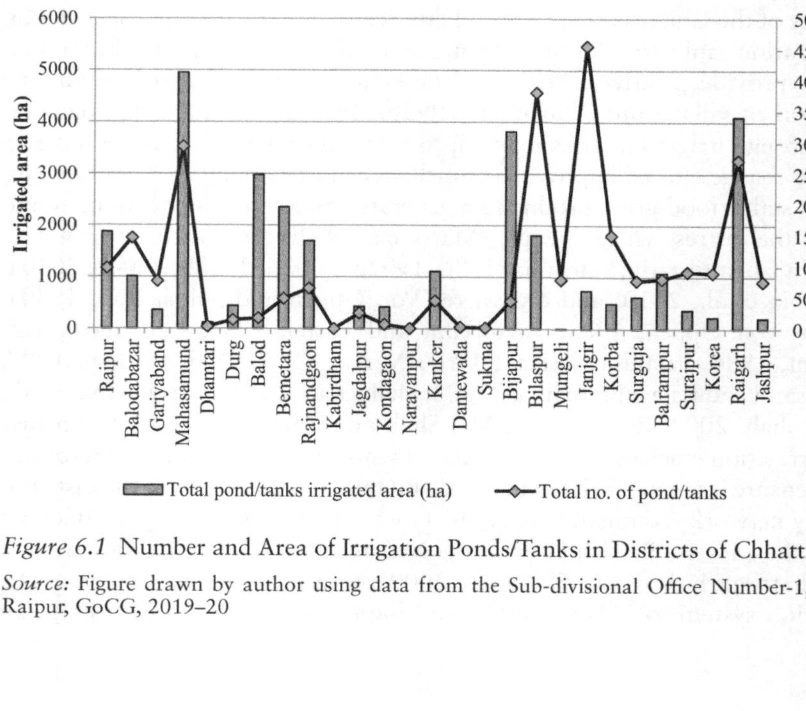

Figure 6.1 Number and Area of Irrigation Ponds/Tanks in Districts of Chhattisgarh

Source: Figure drawn by author using data from the Sub-divisional Office Number-1, DWR, Raipur, GoCG, 2019–20

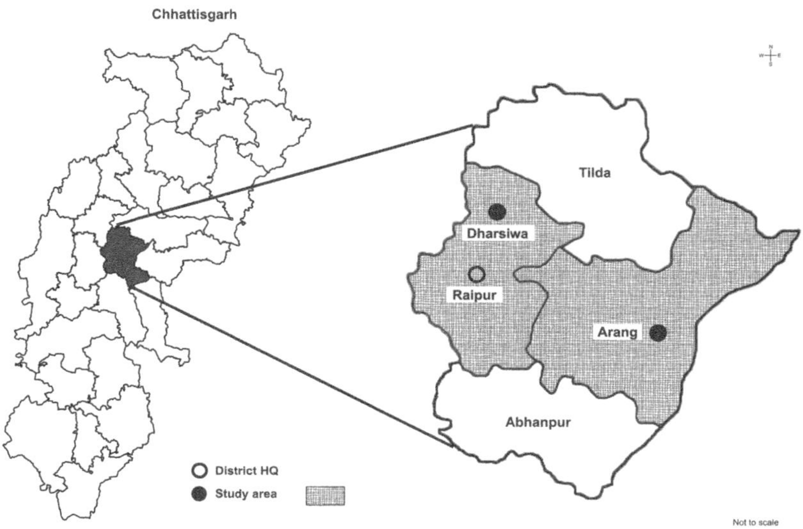

Map 6.1 Location Map of Tank Study Area (Devised by the Author)

confined to eight irrigation tanks for detailed investigation. These eight tanks, including four perennial tanks and four seasonal tanks, were selected from the Arang and Dharsiwa blocks. Perennial tanks have an adequate water supply throughout the year, while seasonal tanks just have adequate water, collected through runoff, to irrigate one kharif crop. Out of 320 farmers selected for economic analysis, 210 were from the command area under tanks and the remaining 110 were from the non-tank areas, representing rainfed agricultural conditions. Farmers were selected from the villages under the command area of tanks. The selection of tanks, the number of villages, and the farmers under irrigated and rainfed situations are shown in the sampling design (Figure 6.2). Information regarding operational holdings, land-use patterns, cropping patterns, crop production, input-utilization patterns, off-farm and non-farm employment, and farmers' perceptions of the existing tank-irrigation system was collected and analysed for both irrigated and rainfed farming systems. From the selected farmers, the required information was collected through well-structured, brief surveys, personal interviews, and focused group-discussion methods. Information pertaining to fish culture, farmers' WTP for equitable and efficient tank management was also collected. The data were collected during 2017–18 and refer to the same crop year.

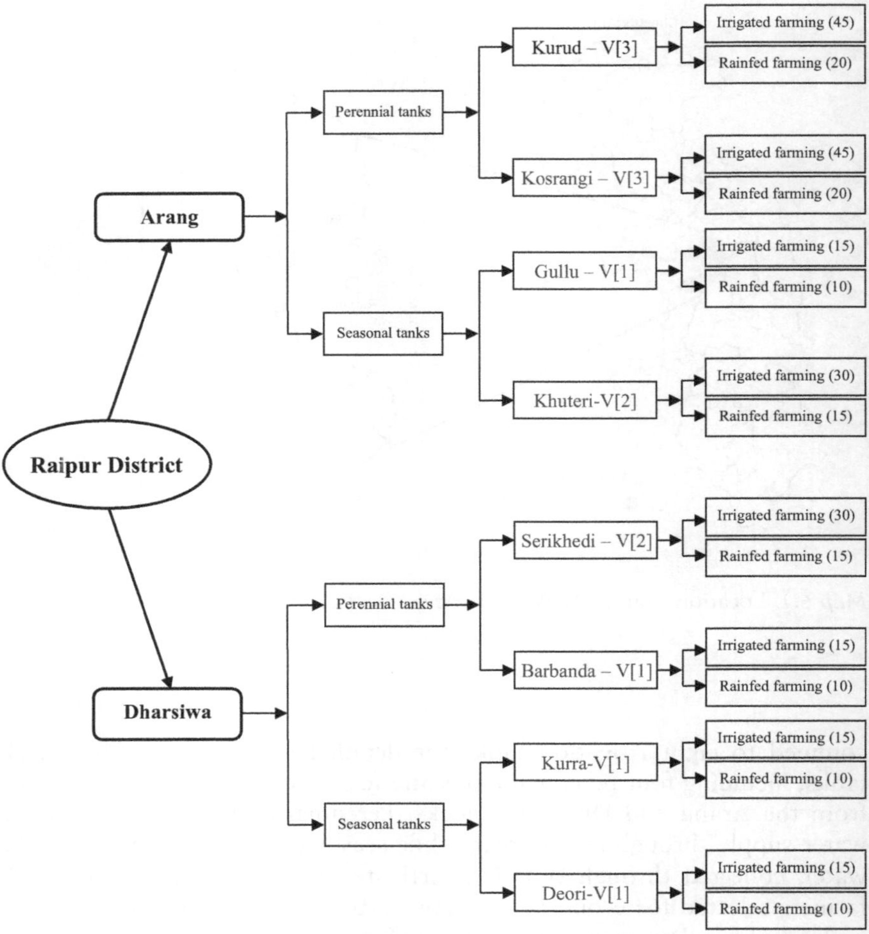

Figure 6.2 Sampling Structure for Data Collection in Tank Study Area

Source: Devised by the author

Note: Figures in parenthesis indicate number of selected farmers in irrigated and rainfed farming situation in perennial and seasonal tanks; V within tank box indicates number of villages selected [in brackets] from respective tanks.

6.3 Hydraulic, Hydrological and Agriculture Attributes of Selected Tanks

Before discussing the physical and technical attributes of selected tanks, we will briefly mention agro-climatic conditions and the availability of irrigation sources. Arang and Dharsiwa blocks constitute nearly 53.69 percent of the total geographical area of the district. These blocks contribute 31 percent to the district's cropped area. Canal water is the main source of irrigation in Dharsiwa and

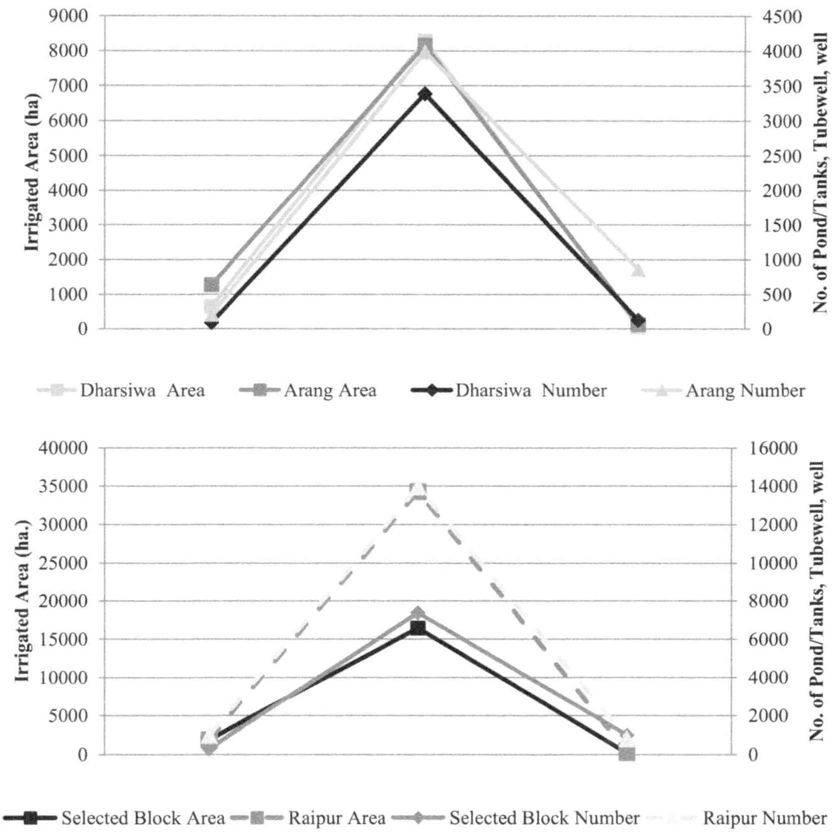

Figure 6.3 Irrigation area under ponds/tanks, tubewell and wells in Selected Blocks and Raipur District

Source: Figure drawn by author using data from the Sub-divisional Office Number-1, DWR, Raipur, GoCG, 2019–20

Arang blocks. Tanks contribute nearly 19 percent to irrigation (see Figure 6.3 for irrigated area under ponds, tanks, tubewell and wells with their numbers).

The net sown area of the blocks in which the selected tanks are located, is 18 percent of the net sown area of the district. The percent of double-cropped area or area sown more than once is marginally higher than the district. Four general classes of soil are found: gravelly sand, sandy loam, loam, and loamy clays, corresponding to the local names *Bhata, Matasi, Dorsa* and *Kanhar*. The topography is undulating. *Bhata* soils have great potential for agro-forestry (Marothia, 1988, 1989a; Marothia and Nandi, 1994).

The climate, including rainfall and weather conditions, the soil and the topography of the selected blocks and villages are similar to the Central Plain region of the state. Most of the cultivated fields or farms in the selected blocks are bunded to store rainwater for distributing (*biasi*) and transplanting for paddy cultivation.

The hydraulic, hydrological, and agricultural features of the tanks[3], and the years of construction, construction costs, gross-storage capacity with dead-storage capacity and live-storage capacity, catchment areas, command areas, cultivable and irrigated areas, water-spread areas, length of dam earth, length of main canal, head discharge, and number of villages under command area are presented in Tables 6.1 and 6.2. The perennial tanks are generally canal fed and can irrigate kharif as well as rabi crops. However, the proportion of irrigation in rabi season is quite low as compared to kharif. Seasonal tanks have proportionately less gross-storage capacity than perennial tanks, which indicates that seasonal tanks provide life-saving irrigation to the crops in kharif season only. The live-storage capacity of any tank has great importance because irrigation potentiality is based on it. The live-storage capacity for perennial tanks varies from 5.09 mcft (Barbanda) to 200.73 mcft (Kurud) among the sample tanks.

However, the live-storage capacity of seasonal tanks varies from 2.64 to 45.00 mcft. Live-storage as a proportion of design-storage capacity varies from 0.83:1 to 0.99:1 in perennial tanks, whereas it varies from 0.75:1 to 0.94:1 in seasonal tanks (Tables 6.1 and 6.2). All irrigation tanks (perennial and seasonal) have an estimated life span of 100 years. Kurud, Kosrangi, and Serikhedi are meeting irrigation requirements even 100 years after their construction. The tanks are largely constructed in sandy loam and loamy clay soils. The command area of all the tanks also has these two types of soils. Most of the tanks are situated on upland topography, whereas some tanks are also situated on the midland topographic situation. However, tanks that are situated on midland topography are considered to be perfect for irrigation. The water-spread area in perennial tanks when they are full varies from 8 ha to 200 ha, whereas in seasonal tanks it varies from 14 ha to 65 ha. Field-to-field is the most common system of irrigation (Table 6.1). The operation of sluice gates depends on the requirement of the irrigation. The charges of irrigation are fixed by the DWR at Rs. 230 per hectare for kharif and Rs. 600 per hectare for rabi crops. The water is supplied to the farmers subject to its availability in the tanks and its supply from the canal.

6.4 Socio-Economic and Resource Profile of Sample Farmers

The socio-economic profile of the sample farmers is presented in Table 6.3. The distribution of family size and the available labour force in irrigated and rainfed farming situations indicates that economically active family members are largely engaged on their farms, and during the crop off-season, they seek employment in off-farm and non-farm work. Information about social caste structure, education level, and distribution of farm size in irrigated and non-irrigated farming situations is given in Table 6.3.

Table 6.1 Hydraulic and Hydrological Attributes of Tanks

Tanks	*Block*	*Year of Construction*	*Construction Cost (Rs. in lakhs)*	*Length of dam Earth (ft)*	*Top Bank level (TBL) (mtr)*	*Maximum Tank Level (MTL) (mtr.)*	*Full Tank Level (FTL) (mtr.)*	*Lowest Still Level (LSL) (mtr.)*	*Catchment Area (sq miles)*	*Mean Monsoon Yield (mcft)*	*Gross-storage capacity (mcft)*	*Live-Storage Capacity (mcft)*	*Dead-Storage Capacity (mcft)*	*WSA Maximum Tank Level (MTL) (ha.)*	*WSA Full Tank Level (FTL) (ha.)*
Kurud	Arang	1909	1.53	6500	100	95	93	75	6.7	202.8	201.8	200.73	1.07	250.00	200.00
Kosrangi	Arang	1909	0.96	5850	104	100	97	80	5.88	174.46	160.4	157.3	3.1	150.00	125.00
Serikhedi	Dharsiwa	1919	0.43	3500	982	978	977	964	0.53	16.03	46.23	45.93	0.3	240.00	200.00
Barbanda	Dharsiwa	1981	1.42	2200	105	100	99	79	0.3	9.07	6.1	5.09	1.01	10.20	8.50
Kurra	Dharsiwa	1964	7.03	800	106	103	102	90	0.95	56.77	33.55	31.46	2.09	17.40	14.50
Deori	Dharsiwa	1965	7.78	320	104	101	99	87	2.2	113.5	6.23	5.19	1.04	28.45	20.00
Gullu	Arang	1971	1.00	1830	104	100	98	87	1.15	34.78	13	11.82	1.18	67.80	65.00
Khuteri	Arang	1984	17.35	2204	981	978	974	961	1.12	33.88	49.3	45	4.29	62.45	46.85

Source: Sub-divisional Office Number-1, Department of Water Resource, GoCG Raipur, 2019–20 WSA – Water-spread area (compiled by author)

Table 6.2 Agricultural Attributes of Tanks of Chhattisgarh

Tanks	*Block*	*L of M* **Canal**	*Head Q* **Cusecs**	*Nos. of* **Village**	*Nos. of Farmers in Command Area*	*Nos. of* **Sluice Gate**	*Duty*	*Total Command Area (ha)*	*Total Cultivable Area (ha)*	*Kharif % of Total Cultivable Area (ha)*	*Rabi % of Total Cultivable Area (ha)*	*Net Cropped Area as % of Total Cultivable*	*Irrigated Area as % of Total Command*	*No of Waterings (kharif)*	*No of Waterings (rabi)*	*% Recovery of Irrigation Fee*
Kurud	Arang	414	12	6	1474	1	32	1232.10	1183.10	50.37	3.67	54.04	17.25	3	2	100
Kosrangi	Arang	239	30.1	5	1740	1	28	1413.60	1648	77.33	2.43	79.76	28.30	3	2	75
Serikhedi	Dharsiwa	234	16	3	576	1	32	480	700	68.57	6.4	74.97	66.67	2	1	40
Barbanda	Dharsiwa	30	2	1	30	1	18.4	120	160	87.50	12.5	100	50	2	1	60
Kurra	Dharsiwa	141	5.08	1	148	1	32	314.4	280	57.14	0	57.14	50.89	2	1	90
Deori	Dharsiwa	19	6	1	211	1	32	171.2	120	66.67	0	66.67	46.73	2	1	70
Gullu	Arang	96	4.8	1	144	1	32	288	100	95.2	0	95.20	21.53	2	1	70
Khuteri	Arang	59	6	2	204	1	28	124	124	82.47	17.53	100	61.86	3	1	25
								4043.30	**4315.10**							

Source: Sub-divisional Office Number-1, Department of Water Resource, GoCG Raipur, 2019–20 WSA – Water-spread area (compiled by author)

Table 6.3 Socio-Economic Characteristics of Sample Farmers

S. No.	*Particulars*	*Perennial Irrigation Tanks*		*Seasonal Irrigation Tanks*	
		Irrigated	*Rainfed*	*Irrigated*	*Rainfed*
1.	No. of farmers	135	65	75	45
2.	Family size (no.)	10.72	9.56	8.35	9.42
3.	Active adult family member (no.)				
	a. Male	3.29	2.89	2.73	2.63
	b. Female	3.19	2.48	2.40	2.35
4.	Working full time on farm (no.)	3.20	2.60	2.50	2.40
5.	Off farm (no.)[1]	1.80	3.10	1.60	3.40
6.	Non-farm (no.)[2]	0.90	4.54	0.84	1.86
7.	Social structure of farm households (% of total sample farmers):				
	a. General	4	2	6	4
	b. OBC	70	67	72	68
	c. SC	22	27	20	25
	d. ST	2	4	2	3
8.	Education (%):				
	a. Illiterate	16.67	14.58	6.25	16.67
	b. Primary to Secondary School	72.92	79.17	85.42	70.83
	c. Graduate	10.42	6.25	8.33	12.50
9.	Farm size (%):				
	a. Marginal (less than 1 ha.)				
	(i) No.	12.50	56.25	20.83	66.67
	(ii) Area	5.51	13.71	5.46	6.41
	b. Small (1–2 ha)				
	(i) No.	29.17	29.17	41.67	16.67
	(ii) Area	11.52	32.06	11.40	15.60
	c. Medium (2–4 ha)				
	(i) No.	39.58	14.58	31.25	14.58
	(ii) Area	25.76	54.23	21.38	29.71
	d. Large (above 4 ha)				
	(i) No.	18.75	0	6.25	2.08
	(ii) Area	57.20	0.00	61.77	48.28

Source: Based on field data of the author

1. Agriculture labour
2. Small-scale village industry/processing/repairing activity, construction work, shop keeping

Table 6.4 Land and Water Resource Availability on Sample Farms

Particulars	*Perennial Irrigation Tanks*		*Seasonal Irrigation Tanks*	
	Irrigated Farming Situation	*Rainfed Farming Situation*	*Irrigated Farming Situation*	*Rainfed Farming Situation*
Land resource per farm				
Operated area (ha)	3.27	1.28	2.70	1.29
Cropped area (ha)	3.70	1.40	2.84	1.35
Cropping intensity %	113.15	109.38	105.16	104.65
Water resource per farm				
Irrigated area (ha)	2.80	0.90	2.78	0.50
Irrigated area to cropped area (%)	75.68	64.29	97.88	37.04
Source-wise irrigated area (%)				
Canal	6.12	0.00	13.89	62.50
Tanks	86.88	0.00	80.18	36.50
Tube well	7.00	1.00	5.93	1.00
No. of Parcel	2.33	1.75	1.83	1.40
Location of farmer in the command area (%)				
Head	33	-	41	-
Middle	35	-	35	-
Tail	32	-	24	-

Details of each farm's average size of operated area, cropped area, irrigated area, and irrigated area as a percentage of cropped area, are presented in Table 6.4. Per farm operated area varies significantly between rainfed and irrigated conditions. Irrigated area as a percentage of total cropped area is higher in seasonal tanks because the area irrigated under seasonal tanks is also supplemented with tube wells. Cropping intensity is higher in perennial tank-irrigated areas. Tanks are the main source of irrigation and cover the maximum (83.53%) area. Nearly 75.68 percent of the area was irrigated by the sample farmers, irrespective of the farm size groups, through perennial tanks. The area irrigated through seasonal tanks by the sample farmers was around 97.88 percent. It is evident from these facts that both types of tanks are providing irrigation to the crops during the dry season. Therefore, tanks make a major contribution to mitigating water scarcity. Tube wells also contribute about 12.95 percent of irrigation in the command area.

6.5 Decision-Making Arrangements and Patterns of Interaction Under State Management System

To analyse the performance of tanks and tank irrigation under state property regimes, we have used four attributes: namely, the physical and technical attributes of a resource; the characteristics of the resource users' community;

the external and internal institutional arrangements in place; and patterns of interaction and outcomes of the institutional framework presented in Chapter 2. The institutional framework used here has assessed the physical and technical attributes of the tanks, the characteristics of the farmers' community in relation to other stakeholders (fisheries and the village community for domestic uses), the external and internal institutional arrangements, the impact and outcome in terms of productivity levels, the income and cropping intensity under different sowing methods, the patterns of interaction among farmers due to rule conformance, the adequacy of water flow, and the maintenance of tanks and tank-irrigation structures (see Table 6.5 for the inter relationship between four major attributes of the institutional analysis of tanks and tank-irrigation systems).

The physical and technical characteristics of perennial tanks and seasonal tanks are presented in Table 6.1 and Table 6.2. The socio-economic profile of sample farmers is presented in Table 6.3. The physical and technical attributes of tanks and the farming community's socio-economic characteristics jointly affect the overall performance of the tank-irrigation system. For example, as shown in Table 6.5, most of the perennial tanks and seasonal tanks, as well as the irrigation supply system, suffer from continuous degradation due to poor maintenance above and below the water outlets, and the encroachment of catchment areas. The poor physical and technical health of tanks results in the reduction of the command area and impedes the water flow. Poor irrigation flow may create conflict among irrigators when water is scarce, particularly during critical crop-growth stages when all the farmers located at the head, middle, and tail-end of a water course require adequate water at the same time.

Both perennial tanks and seasonal tanks were leased out by Panchayats, according to water-spread area, to local fisheries with an understanding that adequate water for crops should be available and that other users cannot be excluded from using tank water, as these water bodies are multipurpose and multifunctional in nature. Leasing arrangements are being followed according to the state's fisheries cooperatives policy. Based on the water-spread area, Kurud, Kosrangi, and Serikhedi tanks (above 200 ha) were leased out by the District Panchayat at Rs. 500/- lease per ha of water-spread area. Kurra, Deori, Gullu, and Khuteri (between 10 and 200 ha) tanks were leased out at Rs. 1000/- per ha water-spread area by the Janpad Panchayat. Only Barbanda tank (below 10 ha) was leased out by the village Panchayat at Rs. 500/- per ha water-spread area (see Chapter 8 for leasing policy under the three-tier FCS).

Although all irrigators in the command area of selected tanks were part of defunct Water User Associations (WUAs) and had full knowledge of the decision-making arrangements, operational rules, and rules protecting water structures, in the absence of WUAs, most of them engage in exploiting the maximum water from total flow as free riders, and in the process create multi-party

Table 6.5 Tanks and Tank Users' Attributes, Decision-Making Arrangements, and Patterns of Interaction Under the State Management System

	Attributes	*Description*
A.	**Physical and technical Resource attributes**[*]	
	1. Original property rights regime	Private land acquired during British rule (three tanks constructed in pre-independence period came under state property rights), remaining tanks constructed on revenue lands in post-independence era, property regime (DWR)
	2. Current management and property rights regime	WUAs under PIM System-A Decentralized governance regime (although currently default)
	3. Size and boundary	Water-spread area and physical boundary are technically demarcated in perennial tanks and seasonal tanks
	4. Number of irrigators in command area of perennial tanks and seasonal tanks	Registered under agreement with DWR
	5. Irrigation supply system	Perennial tanks and seasonal tanks are linked with canal irrigation system
	6. Alternative water sources available to farmers	During the last ten years, the density of tube wells increased in perennial tanks' and seasonal tanks' command areas for supplementary and conjunctive use of tank and tube well irrigation. Also, construction of *anicuts* and stop dams in a few villages covered under perennial tanks and seasonal tanks
	7. Current status of perennial tanks and seasonal tanks, and irrigation supply system	• 90% of water-carrying channels are damaged due to weeds and siltation • 50–60% of water outlets are damaged • 50% of distributaries, 90% of minor, sub-minor, and feeder canals damaged • 100% of tanks are silted, no regular desilting of tank bed • 80% of outlets are distorted by farmers to get maximum possible water in their fields • Reduction in command area due to poor maintenance above and below water outlets • Encroachment of catchment area impeding water flow

(*Continued*)

Table 6.5 (Continued)

	Attributes	*Description*
B.	**Attributes of community**[**]	
	1. Social structure of tank users	Heterogeneous social class: farmers and fisheries are the main water users, households of villages use water to meet domestic requirement and tending cattle
	2. Irrigators main income and livelihood sources	Agriculture, on-farm, and off-farm activities
	3. Legal status of irrigators	Agreements with DWR for availing irrigation from perennial tanks and seasonal tanks
	4. Variability in farm size and numbers	Not significant in both types of tanks
	5. Dependency of irrigators	Majority of farmers depend on tank irrigation
	6. Relationship among farmers located at head, middle, and tail-end of water-carrying channel	Due to flood-irrigation system, conflicts occasionally observed, particularly during water scarcity and concurrent requirement of water at critical crop-growth stage
	7. Conflict between farmers and fishers over water use	Occasional conflict over water sharing during water scarcity
C.	**Decision-making arrangements**	
	1. Legal and administrative relationship with the state	Under CG FIMS Act 2000, management of irrigation systems in Chhattisgarh was transferred to FOs under three tiers of management. FOs include WUAs at the primary level, distributary committees (DCs) at distributory level, and project committees (PCs) at project level. Institutional structure of the FO was as follows: (i) minor irrigation scheme: single tier i.e. only (WUAs up to 2,000 ha), (ii) medium scheme: two tier, i.e. WUAs and PCSs (2,001 to 10,000 ha), (iii) major schemes, three tier i.e. is WUAs, DCs and PCs (10,001 ha and over)
	2. Functional jurisdiction of WUAs and TCs	In all the above categories, the water users' area for an individual WUA is restricted to a maximum of 2,000 ha. The area of a WUA is subdivided into territorial constituencies with a minimum of four and a maximum of ten, depending upon the size of the water users' area. The delineation of water users' area and its sub-division (territorial constituencies) is based on hydraulic parameters and must be hydraulically viable. Every WUA is called by its local, district name.

	According to the above norm, the water users' area for individual WUA is restricted to a maximum of 2,000 ha. Command area of all perennial and seasonal tanks included in the study fall under this rule of the Act. WUAs of Kurud and Kosrangi perennial tanks should have ten Territorial Constituencies (TCs) of command area farmers representatives, Serikhedi Perennial tanks – 8, Barbanda, Deori, and Khuteri – 4 TCs Gullu and Kurra WUAs -6 TCs
3. Organization setup	One president, four to ten members from TCs nominated. 3 Ex-officio members–1 sub-engineer+1 admin. of WRD+1 from Agr. or Ayacut. All the beneficiaries are member of the hydraulic unit of WUAs
4. Current status of WUA	The 2006 election of WUA was not held and WUA has not been functioning since then. All tanks and tank-irrigation systems are managed by the sub-engineer for irrigation and Amin (field staff). Financial allocation is being made through the DWR. Depletion of social relationship and informal institutional arrangements between traditional village head and elected head (*Sarpanch*).
5. Method of irrigation	Field-to-field should be rotational.
6. Percentage of recovery of irrigation fee	Very poor to 100% recovery
7. Maintenance of tank	• Routine maintenance of outlet, feeder, and minor branch canal by the DWR (not regularly). • Due to lack of participation of farmers and Panchayats, most of the tanks and irrigation structures are in bed shape
8. Ability to influence other Panchayat/ government activities	Due to non-functional WUAs, farmers are not in a position to ensure adequate water supply or maintain tanks. Since Panchayats have no control over these tanks, farmers are unable to insist upon village Panchayat for the proper maintenance of the tanks Continuous reduction in number of field workers at the DWR at village and command area levels is also a major constraint

(*Continued*)

Table 6.5 (Continued)

	Attributes	*Description*
	9. External organizational and technical assistance	Non-existence of WUA resulted in absence of coordination with DCs and PCs. This leads to poor management and continuous degradation of tanks and irrigation structures Resistance to change in existing management of tank irrigation water in the command area from the DWR Mixed opinions on restructuring tank-management systems with respect to WUAs, possibility of formation of tank users association (decentralized governance and tank user Panchayats).
D.	**Patterns of interactio**	
	1. Membership, legitimacy, and outsiders	According to CGFPIM Act, all farmers in the command area of the tanks are members. Elections every five years. Election norms, rights, and duties of the members and executives of irrigation Panchayats are well-defined. Outsiders of the tank command areas are excluded from irrigation use and management. However, due to non-functional WUAs, the system of collective irrigation management has collapsed
	2. Reciprocal interaction	Not contributing to induce collective behaviour
	3. Protection	Due to the absence of institutional arrangements related to maintenance and distribution systems, tanks are not protected by irrigators. Alterations of water outlets by farmers to release water into their fields is a common problem
	4. Use regulations	Water distribution rules are non-operative, and as a result, conflict over water use among farmers located at the head, middle and tail-ends has become a common phenomena. Poor operation of sluice gates leads to wastage of water
	5. Consequence	Emergence of individual interests of farmers, each irrigator made every effort to maximum water for his field from the total water supply/available, poor excludability due to unregulated irrigation supply, water insecurity and uncertainty due to ineffective rules and authority systems

Source: Author's analysis based on field work

* Details given in Table 6.1 and 6.2.

** Details given in Table 6.3

externalities. Further, poor maintenance, lack of field channels, poor water control, inadequate labour, and poor recovery of irrigation fees are the most serious problems the tanks under investigation are confronted with. The impact of not implementing operational rules, and internal and external institutional arrangements have resulted in a distorted pattern of interaction in terms of negative reciprocity, protection, and use regulations, as can be seen in Table 6.5.

Despite the degraded condition of many tanks, non-functional WUAs, failures of institutional arrangements, and poor monitoring of water resources beyond outlets, being a complex irrigation system, these tanks play a significant role in mitigating the effects of droughts and sustaining rainfed agriculture. To this end, agronomical and economic gains are discussed next.

6.6 Comparative Economics of Crop Establishment Methods Under Perennial and Seasonal Tanks

Almost 91 percent area was allocated by the sample farmers to rice cultivation during kharif season, whereas lathyrus, wheat and gram were allocated 5.7 percent 1.95 percent and 1.3 percent area respectively during rabi season under perennial tank-irrigation system. System of Rice Intensification (SRI) cultivation methods have yet to be introduced under seasonal tanks. Lathyrus was grown under the *utera* (in utera cultivation a crop is grown in paddy fields before harvesting of paddy crop) system as a follow-up crop (Table 6.6). Information regarding yield levels, and cost and returns incurred under different crop establishment methods has been estimated and presented in Table 6.7.

Table 6.6 Distribution of Cropped Area Under Sowing Methods (in percentages)

Particulars	*Perennial Tanks*		*Seasonal Tanks*	
	Irrigated Farming Situation	*Rainfed Farming Situation*	*Irrigated Farming Situation*	*Rainfed Farming Situation*
Kharif season				
Paddy:				
a. Improved broadcast biasi	49.21	100.00	72.48	100.00
b. Transplanting	32.71	0.00	14.73	0.00
c. Line sowing	12.31	0.00	12.79	0.00
d. SRI	5.77	0.00	0.00	0.00
total (paddy crop)	100.00	100.00	100.00	100.00
Rabi season				
Lathyrus	5.70	0.00	2.25	0.00
Gram	1.30	0.00	0.00	0.00
Wheat	1.95	0.00	0.00	0.00
Total cropped area	100.00 (499.50)	100.00 (84.00)	100.00 (213.00)	0.00 (67.50)

Source: Author's analysis based on field work

Note: Figures in parenthesis are total cropped area in ha

Table 6.7 Paddy Yield and Income Under Different Crop Establishment Methods in Perennial and Seasonal Tanks (per ha.)

Particulars	*Irrigated Farming Situation*				*Rainfed Farming Situation*
	IBB	*TPR*	*Line Sowing*	*SRI*	*IBB*
Perennial tanks					
Yield (q)	27.68	33.77	37.50	45.70	20.00
Total variable cost (Rs.)	29800.00	33852.00	34500.00	32990.00	23500
Gross return (Rs.)	48440.00	59097.50	65625.00	79975.00	35000.00
Net return (Rs.)	18640.00	25245.50	31125.00	46985.00	11500.00
Input-output ratio	1:1:62	1:1:74	1:1:90	1:2:42	1:1:49
Seasonal tanks					
Yield (q)	23.90	25.50	34.40	40.87	15.00
Total variable cost (Rs.)	24995.60	25475.56	32648.00	30652.50	22720.00
Gross return (Rs.)	41825.00	44625.00	60200.00	71522.50	26250.00
Net return (Rs.)	16829.40	19149.44	27552.00	40870.00	3530.00
Input-output ratio	1:1:67	1:1:75	1:1:84	1:2:33	1:1:16

Source: Based on field data of the author

Note: TPR – Transplanting; IBB – Improved Broadcast Biasi; SRI – Rice Intensification System; Cost of production include CostA_2+FL

Returns over variable cost are shown in Table 6.7. Rice cultivation using SRI methods provides higher net returns over total variable costs in perennial and seasonal tanks. In an earlier study, based on a large sample of paddy growers (102 farmers with SRI and non-SRI fields) in Raipur district of Chhattisgarh and Balaghat district of MP respectively, it was revealed that paddy yield and gross return using a SRI over a non-SRI system were 24.5 percent and two percent higher respectively, with 29 percent lower cost of cultivation (Marothia, 2012b). The availability of irrigation water through tanks was a major determinant of returns.

6.7 Incremental Benefits of Tank Irrigation

The direct costs and benefits of tank irrigation at farm level are presented in Table 6.8, which reveals that variable cost (cash cost), gross, and net returns are associated with the degree of reliable and adequate water availability. Returns per rupee invested in input cost also indicate that farmers under the command area of perennial tanks have an advantage over those using seasonal tanks or rainfed situations. It is obvious that the incremental benefits/costs are related to the degree of water availability.

6.8 Financial Gains to Main Beneficiaries of Tanks

Tanks are multipurpose and multifunctional in nature, and managed under different property rights regimes. They are used for irrigating crops,

Table 6.8 Incremental Benefits Over Cash Cost in Perennial and Seasonal Irrigation Tanks (Rs. per ha)

Particulars	*Direct Benefits*			*Incremental Gains Due to Tank Irrigation*		
	PTs	*STs*	*Rainfed*	*PTs-STs*	*PTs-RF*	*STs-RF*
Variable cost	32785.50	28442.92	23110.00	4342.58	9675.50	5332.92
Gross return	63284.38	54543.13	30625.00	8741.25	32659.38	23918.13
Net return	30498.88	26100.21	7515.00	4398.67	22983.88	18585.21
Return per rupee invested in input cost	1:1.93	1:1.92	1:1.33	1:2.01	1:3.38	1:4.48

Source: Field survey by the author

Note: PTs – Perennial Tanks, STs – Seasonal Tanks, RF – Rainfed

Table 6.9 Financial Gains to Main Beneficiaries of Tanks

Particulars	*PTs*	*STs*	*Total*
Average size of tanks (HWSA)*	145.88	36.59	182.46
Net crop income (Rs):			
a. Per hectare	30498.88	26100.21	56599.09
b. Per tank	4449176.61	955006.68	5404183.30
Net income from total command area (A)	95937558.60	23427548.50	119365107.10
Fisheries net income per HWSA (Rs.)	34500.00	23000.00	57500.00
Net income per tank	5032860.00	841570.00	5874430.00
Net income for total water-spread area (B)	20131440.00	3366280.00	23497720.00
Income to Panchayats from lease of tanks for fisheries			
Gram Panchayat	4250.00	0.00	4250.00
Janpad Panchayat	0.00	136350.00	136350.00
District Panchayat	260000.00	0.00	260000.00
Total income to Panchals (C)	264250.00	136350.00	400600.00
Income to State Department of Fisheries (D)	125000.00	0.00	125000.00
Income to DWR (irrigation fee (Rs) (E)	707782.00	201960.00	909742.00
Total income (A+B+C+D+E)	117166030.60 (81.20)	27132138.50 (18.80)	144298169.10 (100.00)

Source: Field survey by the author. * Hectare water-spread area

Note: Figures in parenthesis indicate percentage of total gains under perennial tanks and seasonal tanks

freshwater aquaculture, and to meet the domestic requirements of villagers (see Marothia, 2004c, 2005c, 2006b). We have also calculated the financial benefits to various stakeholders, which are presented in Table 6.9, showing that these tanks contribute significantly to the income of beneficiaries. Similarly, if all tanks of the state are used for irrigation and fisheries, the gains may be quite substantial.

6.9 Farmers' Responses to Existing Tank-Irrigation Systems and WTP

The responses of farmers on existing tank irrigation systems are presented in Table 6.10. Out of 210 farmers, 75 percent were under perennial tanks and 80 percent of farmers were under seasonal tanks, and reported that they had been using tank irrigation for the past 20 years. However, 20 percent and 35 percent of farmers located in the command area of perennial

Table 6.10 Responses of Farmers' Reflecting on Existing Tank-Irrigation System (Percentage Farmers Agreed)

Particulars	*Perennial Tanks*	*Seasonal Tanks*
1. How long have you been using tank water for irrigation?		
(i) 20 years	65	70
(ii) Above 20 years	75	80
2. Have you used tank water every year during the last five years?	80	65
3. Is there an inadequate and uneven supply of water?	30	20
4. Do you have water supply security during crop seasons?	40	25
5. Do you receive water at various/critical stages of crop growth? Is water available in the tank:	70	45
(i) In kharif season?	70	40
(ii) In rabi season?	5	0
(iii) In summer season?	0	0
6. Do you have any alternative source of water? (a) tube wells (b) wells (c) canal	5	8
7. Is the field-to-field irrigation system an effective system in your opinion?	50	45
8. Do you think that more socially and economically powerful farmers affect water distribution?	25	10
9. Are you satisfied with water charges?	70	55
10. Are you satisfied with the present system of irrigation management?	55	40
11. How effective are the following organizations for managing tanks?		
(i) Water Users Associations (WUAs)*	90	60
(ii) Department of Water Resource	5	20
(iii) Panchayat**	85	70

Source: Field survey by the author

* *The subject to impartial and apolitical elected body takes over the tank management.*

** *Panchayats were given authority to run the irrigation system with adequate funds and use the collected irrigation fee for maintaining the irrigation system beyond the main outlet.*

tanks and seasonal tanks respectively had not been able to use irrigation water from tanks for the last five years as they are tail-enders. All the farmers reported that they get water in turn during most of the year. Nearly 32 percent of farms under perennial tanks and 24 percent of farms under seasonal tanks are located at the tail-end of the command area of tanks (refer Table 6.10). Only five percent (perennial) and eight percent (seasonal) of farmers have alternative sources of irrigation. Irrigation water at different stages of crop growth was available in comparatively adequate quantities through perennial tanks than through seasonal tanks. Most of the farmers felt that the supply of water depends totally on the amount of rains. About 50 and 45 percent of farmers respectively, were of the opinion that field-to-field irrigation is an effective system. However, the remaining farmers did not think so, because they suffered from an unreliable and inadequate supply of water to their plots.

In farmers' opinions, the existing irrigation system affects soil fertility, erodes the upper layer of soil, results in seepage and evaporation, and increases water loss and incidence of diseases. Nearly 65 percent of perennial-tank users and 50 percent of seasonal-tank users were satisfied with the present irrigation system. However, nearly 70 percent and 55 percent respectively were willing to pay higher irrigation fees if water distribution services were improved. The majority of farmers across both types of tanks were of the opinion that WUAs should be revitalized to manage the tanks so that conflicts among water users regarding water distribution can be resolved through well-defined and workable institutional arrangements. Sluice gates are operated by field staff of the DWR, and the department is responsible for the collection of irrigation charges. Recovery of irrigation charges varies from 25 to 100 percent (see Table 6.2). In the state of Chhattisgarh, WUAs were created under the PIM system implemented in 1999 in the undivided Madhya Pradesh (including present-day Chhattisgarh) (Marothia, 2002b, 2005a). The WUAs were formed by an election process. All farmers were members of the WUA. If anybody violates the rules of water distribution, and/or does not contribute labour for maintaining and repairing tanks, there is a provision under the PIM that the WUA may impose monetary fines, or sometimes violators must do labour for repairing tanks. Since 2006, WUA elections have not been conducted and WUAs no longer function.

Since WUAs are defunct, the DWR has ownership of the tanks. Irrigation tanks, besides supplying water to crops, are used for fish culture, bathing, tending cattle, and washing clothes (Marothia, 2004b, 2004c, 2006b). We have carried out a series of quick but well-structured focus-group discussions to assess the WTP of various stakeholders for better and improved irrigation services. The information is summarized in Table 6.11. It can be seen that a sizable percentage of total respondents are willing to pay higher prices for better services than charges fixed by the DWR for irrigating kharif and rabi crops, and the SDF for leasing rights.

Table 6.11 WTP Higher Charges by Different Users Groups for Improved Water Quality and Services

Stakeholders	*Total Respondents (No.)*	*Percentage of Total Respondents WTP*	*Range of WTP*
Farmers			
Perennial	210	85	50–60
Seasonal	110	75	40–50
Fishermen			
Perennial	80	90	100–200
Seasonal	80	70	50–80

Source: Field survey by the author

6.10 How to Restore Institutions and Collective Action?

Tank irrigation in Chhattisgarh is an economically, ecologically, socially beneficial, and time-tested technology. However, irrigation tanks have been deteriorating rapidly. Despite the deteriorating conditions of many tanks, they have played a significant role in mitigating the effects of droughts and sustaining rainfed agriculture. However, the performance of seasonal tanks in mitigating drought depends on rainfall patterns. With 60 percent of normal rainfall, these tanks can be filled to a water level that can provide life-saving irrigation (Marothia, 2002b, d; Marothia et al., 2010a). Even canal-fed tanks do not generally experience water-scarcity problems unless there is extremely low rainfall. In addition to storing water for crop irrigation, tanks provide ecosystem services such as the recharge of groundwater used by adjacent communities, and fertilizing silt soil that allows sand and soil to be used by small-scale village industries. In a large number of tanks (including the sampled tanks) social, cultural, human, and livestock activities are part of the tank-based ecosystem.

Taking into account all these benefits, tanks can certainly be considered as socio-ecological, community, and economic assets in the context of Chhattisgarh. If tanks are to be used as traditional technology to minimize the negative effects of drought and to contribute to the rural economy of Chhattisgarh, they need to perform three basic functions: better water allocation and distribution; water fee collection; and infrastructure maintenance and repair. Currently, in Chhattisgarh, none of these functions are being performed satisfactorily due to absence of effective internal and external institutional arrangements.

Soon after the birth of Chhattisgarh, the state government introduced several reforms for rehabilitating irrigation tanks to improve the allocation of water, maintenance, and collection of irrigation fees (water tax, irrigation councils). However, these strategies did not work for several reasons such as conflict between farmers located at tail and head ends, lack of participation, poor collection of water fees, and lack of coordination among different levels of the water-management bureaucracy. Under the new reforms of PIM,

WUAs were formed in Chhattisgarh around a cluster of five or six large and small tanks with territorial constituencies based on the size of their command areas. WUAs were supposed to do a better job of performing the three essential functions mentioned previously. However, the PIM is experiencing power-sharing conflicts between traditional owners of water resources (DWR) and members of WUAs, arising from the allocation of financial resources for repair and maintenance work (see Marothia, 2002b, 2005a, 2016 for a detailed discussion on PIM). Unless WUAs are empowered in a real way and supported with irrigation management experts, the three tasks listed previously cannot be fulfilled (see chapter 4). Post 2006, elections of WUAs have not been conducted and so these bodies are not functional. It is extremely important to revive WUAs with an apolitical election process (Marothia, 2002b, 2005a, 2016). Tanks involve multiple stakeholders with conflicting interests (fishermen, irrigators, lift irrigators, and domestic users). Tanks have traditionally been generating significant revenue from fisheries and irrigation. A large number of tanks have been operating under an open-access management regime due to ineffective institutional arrangements and authority systems (local governance/WUAs). This open access has to be replaced with a common property management regime with well-empowered WUAs and the engagement of Panchayati Raj Institutions.

After the implementation of the 73rd Amendment Act in 1992, Panchayat Raj Institutions (PRIs) have been given constitutional status for managing local resources. The WUAs under PIM and FCSs have to work under PRIs for the management of the irrigation tanks to minimize conflicting interests (see Marothia, 1992b, 1993b, 2004b, 2004c, 2005c, 2006b for successful management of tanks with multiple uses and interests). Also, before the rehabilitation of old tanks, DWR, WUAs and Panchayats must take a closer look at all the current activities of an ecosystem and their values, which extend beyond irrigation.

Evaluating tanks solely from an irrigation perspective will deprive Chhattisgarh's rural communities of the valuable ecosystem services and functions that these tanks provide today. The best strategy for policy makers, irrigation authorities, and donor agencies involved in rehabilitation and transfer of tank irrigation to the farmers is to view tanks as a complex socio-ecological MWs system with multiple stakeholder groups. Without re-crafting institutional arrangements it will be difficult task for management authorities and farmers to sustain tank's ecosystem services in years to come.

Notes

1 This Chapter is a revised version of Marothia (2002d) and Marothia et.al. (2010) in terms of recent field survey and datasets, collected by me, and analysis of performance of tank irrigation systems using a synthesis of institutional framework developed in Chapter 2. Thanks to V.K. Choudhary and Sushil Pandey for their support during 2000-1 for project work (Marothia, 2002d). A series of conversations with V.K. Vachhani, Superintendent Engineer Bureau of Design and Hydrological Investigation, helped me to keep track of field level realities. Sincere thanks to him.

2 Based on a study of the participatory irrigation system in Chhattisgarh, Marothia (2002b, 2005a, 2010a, 2016) observed that successful adoption of the PIM requires a complete change in mindset, and in several cases the officers of the DWR, members of managing committees of WUAs, and farmers are still part of a learning process. Periodic training for officials and nominated and elected members of the committee and farmers is required to educate them about the importance of institutional arrangements to achieve self-governance in canal-irrigation systems and canal-fed tanks. Sustainability of PIM largely depends on political and bureaucratic will to share power with farmers and create an apolitical environment for the smooth functioning of WUAs. (See also note 1 of Chapter 4.)

3 Sunil Kumar Ambast's, ICAR, input on hydraulic and hydrological parameters is duly acknowledged.

7 Analysing Self-Organized and Self-Governed *Dabri* Irrigation System

7.1 Introduction and Context

In the undivided state of Madhya Pradesh (MP), including present-day Chhattisgarh state, the State Department of Agriculture (SDA) introduced *Dabri Yojana* (a micro-minor irrigation tanks scheme) through its Soil Conservation Wing in 1977–78, to provide irrigation facilities within the shortest possible time, increase food-grain production, generate employment, and minimize distress migration. The scheme was largely confined to rice-growing districts of the state. All the districts of present-day Chhattisgarh were part of the scheme (Marothia, 1986, 1992b). *Dabries* were constructed on village common land or revenue land across the natural water courses to store rainfed runoff. The command area of *dabries* was generally below 40 ha. The Agriculture Department constructed *dabries*. After construction, the Panchayats were given ownership rights of *dabries* were responsible for their management. It was expected that these *dabries* were to be managed under a common property regime by the groups of farmers within the command area of a particular *dabri* (Marothia, 1986, 1992b).[1,2]

In MP, nearly 10,000 *dabries* were constructed in rice-growing areas between 1977 and 1985. In an earlier in-depth study Marothia (1992b) examined factors which were responsible for the degradation of *dabries* in MP (including areas of Chhattisgarh). His study revealed that *dabries* constructed under *Dabri Yojana* were supply driven and largely target-oriented. Soil Conservation Officers (SCOs) of the Agriculture Department would make decisions about the locations for constructing *dabries*. The *dabries* scheme was introduced in MP with the traditional assumption that farmers would contribute with different types of resources – labour, materials, and maintenance – once they realized that the *dabries* would be of significant benefit to them. This assumption has largely been proven false in the case of *dabries* at village level due to their poor management by those who use them and a lack of financial resources and administrative powers of Panchayats to manage them under common property regime. Further, Panchayat Raj Institutions (PRIs) were not effectively empowered to manage local commons and did not have the financial resources

DOI: 10.4324/9781003456742-7

to maintain and manage *dabries* in the undivided MD. One reason for poor management is that *dabries* have extremely poor excludability i.e. it is difficult to prevent non-contributors of labour or capital resources from taking advantage of the water in the absence of a well-defined structure of rights, laws, and duties. The Panchayat system broke down due to a lack of well-defined rules, regulations, and laws for regulating water use, and the common property (*res communis*) regime of *dabries* ultimately degenerated into an open-access (*res nullius*) resource. Marothia (2004b), in another case study of Bastar villages, clearly highlighted that small irrigation structures (*dabries*) created using target-oriented approaches have largely failed to create any significant demand for irrigation in tribal-dominated societies. Most *dabries* were constructed simultaneously by many departments, namely, Agriculture, Water Resources, Rural Development, Panachayat and Tribal Development using relief funds. Most of the *dabries* did not hold water even during rainy days. These *dabries* served as quick percolation ponds. Departmental officers have not amply discussed with Sarpanch, members of the Panchayat, farmers, or villagers, issues of financial allocation, maintenance, the transfer of the irrigation system to Panchayats, the process for forming users groups, the mechanism for collecting irrigation fees, the distribution of water along watercourses, and other attributes of internal and external institutional arrangements. The design and location of *dabries* were often not appropriate. These water structures were handed over to the Panchayats on paper, but the Panchayats were not willing to take any responsibility for maintaining or managing them, as this irrigation system involves a large number of water users and the Panchayats did not, at that point in time, have the financial resources or administrative powers needed. It is interesting to note that users of *dabries* had no knowledge or understanding that these resources belonged to them as community resources. Just creating water structures in tribal areas can fulfill the targets of water resource development programmes, but may not ensure the active participation of tribal farmers in using and managing the structures (Marothia, 2004b). This was also true in other parts of the state (Marothia, 1992b). It was necessary to develop economic and social incentives, sanctions, and capacity-building programmes that promote long-term participation and involvement of water users and Panchayats in the maintenance and management of *dabries*. Such incentives and sanctions should have constituted the major focus for the revival of *dabries* (Marothia, 1992b, 2004b).

After splitting from MP, the Government of Chhattisgarh (GoCG) continued rehabilitating old *dabries*, and constructing new ones under *Dabri Yojana*, with a focus on community participation.[3] Policy makers and managers understood that there are large tracts of land in Chhattisgarh with an annual rainfall of around 1,200 mm, yet these areas suffered with frequent droughts, large-scale distress migration from villages to the cities, labourers forced to become bonded to large and better-off farmers in order to earn their livelihoods, and low productivity under a rainfed and mono-cropped agriculture system. This situation has been and is still prevalent in many villages of the State of Chhattisgarh. In-situ conservation of rainwater in small ponds, village tanks, and other small water structures (*dabries*) provides capacity to cope with such a situation. An urgent need was felt to create irrigation facilities at the command of farmers to increase the production of crops and cropping intensity, introduce diversification, and minimize distress migration. In this context, small water structures were considered as the most appropriate technology for the tribal, scheduled caste, and resource-poor farmers of Chhattisgarh who could collectively operate the such an irrigation system with the support from local bodies or village Panchayats. It has been established in some parts of the state that runoff water stored in small-scale water-harvesting structures helped both diversification and intensification of agriculture. Besides, these structures are an appropriate and centuries-old, established technology for soil and water conservation, helping in groundwater recharge, reduction of soil erosion, and mitigating climate change effects (Marothia, 1986, 1992b, 1993b, 1997a, 2005a).

In view of the above facts in Chhattisgarh, focus shifted from a supply-driven to a demand-driven approach from 2001–2002 onwards. If farmers of any village demand the construction of a *dabri*, the Panchayat of that particular village has to approach the Department of Agriculture and request the construction of a *dabri*. The Agriculture Department constructs *dabries* with a command area below 40 ha after seeking technical approval for the design from the DWR. The DWR is responsible for designing the physical and technical parameters of *dabries*. Management responsibility of *dabries* is assigned to the Panchayats after their construction. After the formation of Chhattisgarh State, Panchayats were assigned reasonable legislative, administrative, and financial resources. The flow of financial resources directly to Panchayats has also been initiated after the recommendations of the 14th Finance Commission.

In this chapter, an attempt has been made to take stock of the construction of *dabries* at block, district, and ACZs levels after the formation of Chhattisgarh State. Using physical and technical attributes of *dabries*, the characteristics of the irrigators' communities, and institutional arrangements, it examines how self-organized and self-governed institutional mechanisms can sustain *dabries* and improve the income and livelihoods of farmers, and how institutional chaos can be avoided with self-organized and self-governed institutional arrangements to sustain *dabries*.

7.2 Steps and Norms for Construction of *Dabries*

Information regarding the steps and norms used for the construction of *dabries* was obtained from SCOs and is presented in Table 7.1.

Table 7.1 **Steps and Norms for Construction of *Dabries***

1.	Village Panchayats demand the construction of *dabries* – they submit formal application to the Assistant Soil Conservation Officer, District Office of Deputy Director Agriculture (DDA), SDA.
2.	Discussion with sarpanch, members of elected village body, and farmers of the village.
3.	Discussion with farmers who will be part of the command area of the proposed construction.
4.	Engineers from the DWR and SCOs assess the technical parameters (catchment area, slope of up and down streams, topography, hydrology, counter mapping, reservoir capacity, tank bund.
5.	*Dabri* should be constructed only on government land.
6.	Site with less requirements for earthworks may be preferred.
7.	A safe spillway site, i.e. a natural vegetative spillway, may be preferred.
8.	The soil of the site must have water-holding capacity.
9.	The site should not cause excessive seepage losses.
10.	The catchment area of a *dabri* normally ranges between 0.16 to 0.25 sq. km.
11.	Slope – not specified but toposheet used for calculating the water flow of the area and water storing ability of the *dabri*.
12.	Gully area should be preferred to reduce the cost of construction and maximize water storage in the *dabri*.
13.	Command area varies from *dabri* to *dabri*, but it is generally within 50–150 ha.
14.	The selection of *dabri* should be determined on the basis of rainwater that is estimated to flow out to farmers' fields.
15.	Minimum size of *dabri*: length – 20 m. width – 20 m. depth- 3 m. maximum size 40 ha.
16.	Construction cost is estimated by the DWR; it is around 100,000 rupees for one ha.
17.	The DWR approves the technical design of the *dabri*, which is checked and approved by the Executive Engineer (DWR).
18.	Planning to connect the *dabri* to the canal irrigation system before construction begins is based on technical and physical feasibility.
19.	The number of *dabries* to be constructed depends on the catchment area and other physical and technical parameters, as well as the concentration of farmers' fields in the command area.
20.	After construction, the *dabri* should be transferred to village Panchayats for repair, maintenance, and management – a prerequisite in the *debri* scheme.
21.	After taking control of a *dabri*, Panchayats handle all disputes over water use through *Sinchai Samitee,* water user groups or directly.
22.	*Dabri* should have an element of a community irrigation governance system.

Note: Complied by the author based on information collected from SCOs of the respective blocks and the Deputy Director Agriculture of respective districts.

7.3 Coverage of *Dabries* Construction

Details about construction of *dabries* and irrigated areas created under *Dabri Yojana* in districts and ACZs of the state from the year 2000–01 to 2019–20 are presented in Table 7.2, Figure 7.1 and 7.2. Current construction cost is estimated at Rs. 100,000 per ha, which is fully borne by the SDA. If any *dabri* is connected with an irrigation canal, irrigation charges as fixed

Table 7.2 Construction of *Dabries* Under *Dabri Yojana* (Year 2000–01 to 2019–20)

S.No.	*Zone*	*District*	*Total No. of* Dabries	*Total Irrigated Area (ha)*	*Average Irrigated Area (ha)*
1.	Northern Hills (NH)	Korea	146	3090	21.16
		Surajpur	21	580	27.62
		Surguja	277	5955	21.5
		Balrampur	46	1335	29.02
		Jashpur	158	3723	23.56
2.	Central Plains (CP)	Raigarh	173	3883	22.45
		Korba	73	1693	23.19
		Janjgir	129	2698	20.91
		Bilaspur	151	3446	22.82
		Mungeli	8	235	29.38
		Kabirdham	81	1685	20.8
		Rajnandgaon	58	1299	22.4
		Durg	137	2969	21.67
		Balod	6	160	26.67
		Bemetara	4	120	30
		Raipur	210	4494	21.4
		Balodabazar	18	510	28.33
		Gariyaband	40	1140	28.5
		Mahasamund	56	1210	21.61
		Dhamtari	99	2034	20.55
		Kanker	58	1343	23.16
3.	Bastar Plateau (BP)	Jagdalpur	115	2570	22.35
		Narayanpur	0	0	0
		Kondagaon	13	360	27.69
		Dantewada	54	1220	22.59
		Bijapur	13	355	27.31
		Sukma	12	330	27.5

S.No.	*Zone-wise*	*Total no. of dabries*	*Total Irrigated Area (ha)*	*Average Irrigated Area (ha)*
1	NH	648 (30.06)	14683 (30.31)	22.66
2	CP	1301 (60.34)	28919 (59.70)	22.23
3	BP	207 (9.60)	4835 (9.98)	23.36
State		**2156** (**100**)	**48437** (**100**)	**22.47**

Source: Compiled from data provided by the State Department of Agriculture, Government of Chhattisgarh

Note: Figure in brackets indicate percentage of state total

up by the DWR are applicable and have to be paid by farmers. Otherwise, irrigation water is provided by Panchayats without any irrigation fee. Panchayats can lease out *dabri* for fisheries provided it does not have a negative effect on farmers' water requirement. If leased out for fisheries, Panchayats may follow leasing rules and lease rents as per the State Department of

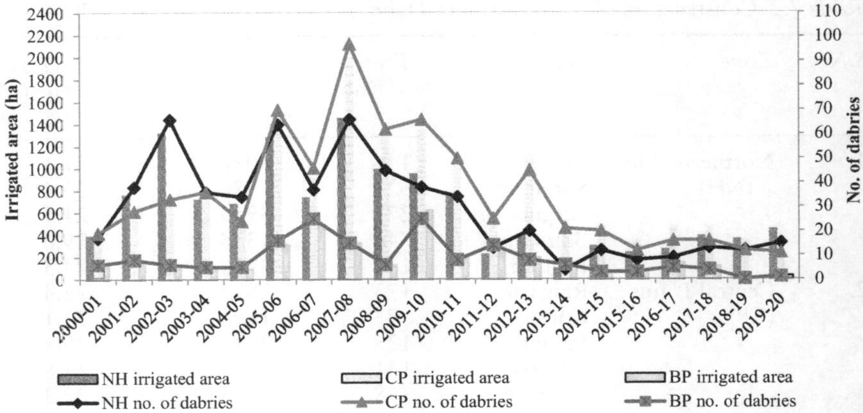

Figure 7.1 ACZs-Wise Construction of *Dabries* Under *Dabri Yojana* (Year 2000–01 to 2019–20)

Source: Drawn using data from Joint Director, State Dept. of Agriculture, Government of Chhattisgarh

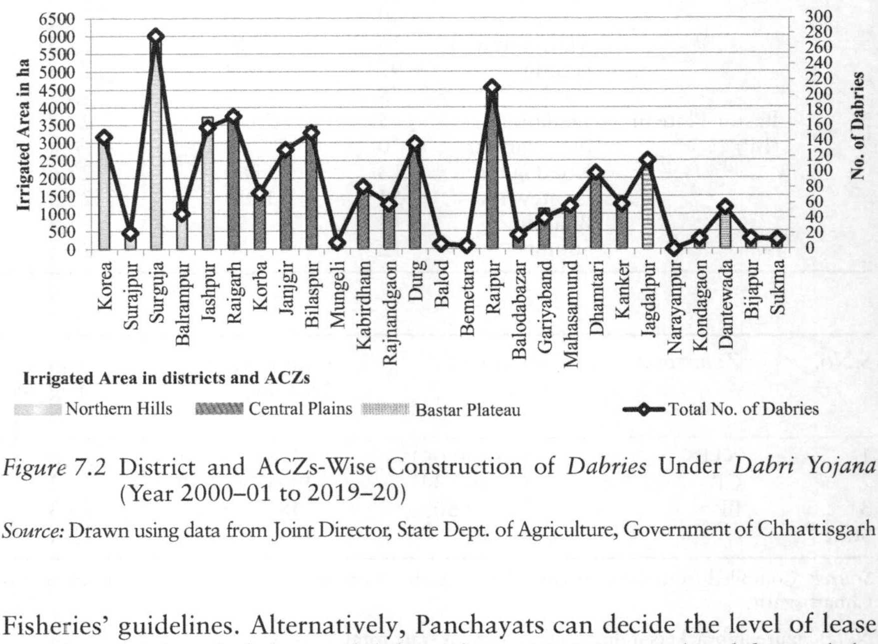

Figure 7.2 District and ACZs-Wise Construction of *Dabries* Under *Dabri Yojana* (Year 2000–01 to 2019–20)

Source: Drawn using data from Joint Director, State Dept. of Agriculture, Government of Chhattisgarh

Fisheries' guidelines. Alternatively, Panchayats can decide the level of lease rent, keeping in mind the availability of water in a particular *dabri* and the collective decision of *Sinchai Samitee* and farmers' groups. In most of the cases, *dabries* have been leased out to locals or fishermen or fish cooperatives, if functional.

7.4 Profile of the Study Area, Database and Application of the Synthesized Institutional Framework

After the formation of Chhattisgarh State, 2,156 *dabries* were constructed between 2000–2001 and 2019–2020. The Assistant Soil Conservation Officer (ASCO) has records of the constructed *dabries* at village and block levels. District and state-level information is available from the Deputy Director and the Joint Director Agriculture. This information includes the physical and technical parameters of each *dabri* constructed in different years in different villages in respective blocks under the jurisdiction of a district. As mentioned, after construction of *dabries*, these water bodies are transferred to Panchayats for management. Many Panchayats directly manage *dabries* although in several villages these structures are managed by self-organized irrigators to make rational use of available water. Assistant Soil Conservation Officers, Rural Agriculture Extension Officers, and scientists of Krishi Vigyan Kendra (KVKs) generally remain in touch with sarpanch, members of Panchayats, and farmers, to transfer technology in agriculture and allied sectors.

In view of these facts, this study was carried out to understand the performance of *dabries* that are administered and controlled by Panchayats and irrigators' groups. After a series of discussions with the Joint Director of the Department of Agriculture, the Deputy Director of Agriculture, SCOs, rural development officers, and Panchayats stationed at district level, and after collecting some baseline information and speaking with some Sarpanches and irrigators' groups, a set of four questionnaires was designed and pre-tested to collect information about: (i) criteria for construction of *dabries* and the physical and technical attributes of *dabries*, (ii) the socio-economic profile of irrigators, (iii) decision-making arrangements, (iv) the impact of institutional arrangements on productivity levels, incomes, cropping intensity, diversification, availability of adequate water, and the maintenance of *dabries* under different management systems. For this study, out of 188 *dabries* constructed between 2000–01 and 2019–20 in the villages of selected blocks and districts of Chhattisgarh, 95 *dabries* from 85 villages of 11 blocks of 11 districts spread over three ACZs were selected in consultation with SCOs and the Deputy Director of Agriculture. Aggregate data pertaining to year-wise construction, technical and physical parameters, and socio-economic characteristics, management systems, yields, and income levels for selected *dabries* were provided by SCOs of respective blocks. Details of the coverage of *dabries* and the beneficiary farmers included in this study at ACZ, district, block and village levels are given in Figure 7.3. Further, details about the area and number of selected *dabries* in the sample of villages from selected blocks are given in Figure 7.4. Information provided by SCOs was re-checked by KVK scientists and data was cleaned accordingly. A series of quick Focus Group Discussions (FGDs) were conducted with the farmers of sampled *dabries* to understand the institutional arrangements and collective choices governing them. The

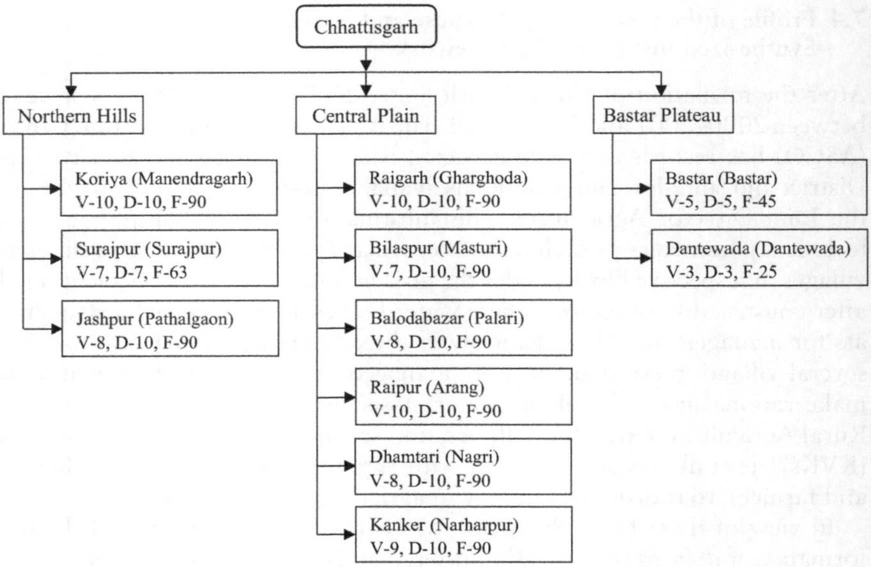

Figure 7.3 Coverage of *Dabries* and Beneficiary Farmers in ACZs, Districts, Blocks, and Villages

Source: Devised by the author

Note: Figures in brackets indicate name of selected blocks, V, D, and F indicate number of villages, *dabries*, and total beneficiary farmers respectively. Total number of farmers is given in Table 7.3.

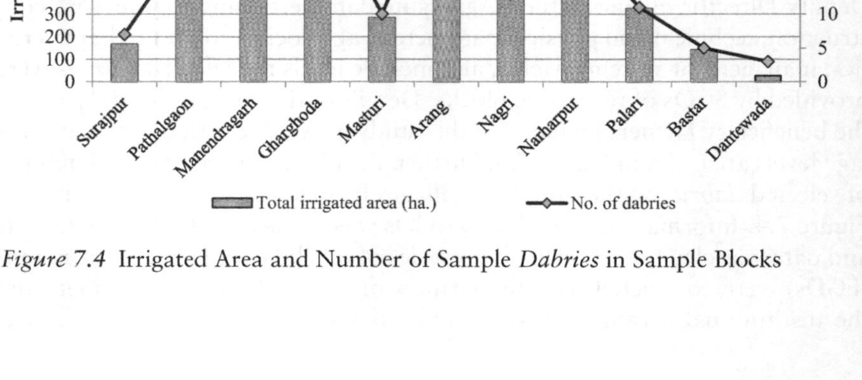

Figure 7.4 Irrigated Area and Number of Sample *Dabries* in Sample Blocks

primary criteria for selecting districts, blocks, and villages were continuity in the construction of *dabries* and agro-climatic parameters. Further, for collection of data pertaining to yield, income from important crops, and outcome parameters, nine farmers from each *dabri* (except in case of Dantewada, where all the 25 farmers were selected) were randomly selected to ensure a fair representation of farmers along the water course (head, middle, and tail). This was done to further cross-check aggregate data obtained from the office of the assistant SCOs of respective blocks. The required information was gathered from mid-March 2020 to February 2021. Data collected refer to the year 2019–20. The synthesis of institutional analysis frameworks developed in Chapter 2 is used here to analyse performance of *dabries*. The details of resource and community attributes, institutional arrangements, patterns of interaction, and outcomes are presented in the next four sections.

7.5 Physical and Technical Attributes of *Dabries*

The physical and technical characteristics of constructed *dabries* in a sample of villages of selected blocks in different districts across three ACZs are presented in Table 7.3. Year-wise data related to the construction of the sample *dabries* and irrigation areas created in selected villages and blocks are given in Appendix 7.1. Catchment areas, water-spread areas, depth when tank is full, size and boundary, source of water, command area, physical condition of water structures, main uses of the water, duration of availability of water, number of sluice gates, and watering adequately are the physical and technical parameters of a typical *dabri*. Normally, one *dabri* is constructed in a village. However, in five of the sample villages, more than one *dabri* was constructed (Table 7.3). The physical condition of most of the *dabries*, except Masturi and Palari, are in very good condition (Table 7.3). All these factors jointly affect the overall performance of the *dabri* irrigation system. For example, if the catchment area of a *dabri* is encroached, it may have static and inter-temporal externalities in terms of water flow and scarcity, which in turn increases water conflicts.

Water from *dabries* is provided for irrigating kharif paddy and rabi crops. Paddy is the main crop and occupies almost 90 percent of the kharif crop area. Due to higher MSP over and above CACP's recommended price (state-fixed Rs. 2,500 per quintal from 2018 onwards and likely to continue till 2024), farmers have also increased the area under summer paddy. During summer, subsidized groundwater and power are other reasons for the expansion of the area under summer paddy. This is affecting efforts to increase diversity that have been initiated in recent years. Also, a few years ago, the state government took the decision to discourage summer paddy. Currently, the state has been contributing a substantial quantity of paddy to the central pool, compared with the situation that prevailed before formation of the state.

Table 7.3 Physical and Technical Attributes of Sample *Dabries*

S. No.	*Particulars*	*Northern Hills*			*Central Plains*						*Bastar Plateau*	
	Districts	*Surajpur*	*Jashpur*	*Koriya*	*Raigarh*	*Bilaspur*	*Raipur*	*Balodabazar*	*Dhamtari*	*Kanker*	*Jagdalpur*	*Dantewada*
	Selected Blocks	*Surajpur*	*Pathalgaon*	*Manendragarh*	*Gharghoda*	*Masturi*	*Arang*	*Palari*	*Nagri*	*Narharpur*	*Bastar*	*Dantewada*
1.	Sample villages in which *dabries* constructed	Karva, Ketka, Podi, Lanchi, Pahadgao, Beltekri, Haripur	Kadro, Pandripani, Farsatoli-1, Farsatoli-2, Goliyagarh, Jhimki-1, Jhimki-2, Karrabevra, Lakchhar, Minjpur	Paradol, Saraula, Bundeli, Lalpur, Bhauta, Putadand, Shankargarh, Chaughada, Nagma, Bauridand	Gharghodi, Baronakunda, Chharratangar, Kathrapali, Faguram, Tenda, Kaya, Bulekera, Karrikachhar Bhalumar	Navagaon, Kuli, Karichhapar, Madai, Jondhara-1, Jondhara-2, Pachpedi-1, Pachpedi-2, Loharabor-1, Loharabor-2	Gukhera, Nardaha, Ghumarabhata, Khapari, Banarsi, Borid, Malidih, Khorasi, Baihar, Jarod	Charauda, Galakusmi, Gaitara, Datan, Sasaha-1, Sasaha-2, Zara, Sankari-1, Sankari-2Kosmandi	Charra, Padampur, Ghutkel, Garhdongari, Dongardula, Parsapani, Rajpur, Kaspur-1, Kaspur-2, Bhalupara Kaspur-3	Devdongar, Devgaon, Basanwahi, Jamgaon (Imlipara), Bhainsmudi, Bansagar, Gaharinbandha, Thanabodi, Navadbari-1 (Shitlapara), Navadbari-2	Junapani Bakel, Godiyapal Kadamunda, Cherakur Banvajhori, Piringmara Cherakur, Kavadgaon	Balud, Muraki, Teknar
2.	Nos of villages covered under sample *dabries*	7	8	10	10	7	10	8	8	9	5	3
3.	Nos of sample *dabries*	7	10	10	10	10	10	10	10	10	5	3
4.	Average catchment area (sq km)	1.56	0.006	1.46	0.89	0.73	1.27	1.00	0.51	0.49	0.39	0.26
5.	Average water-spread area at full tank level (ha)	1.58	0.80	0.63	1	4.39	5.41	3.068	5.833	1.77	3.2	2.00

6. Average Depth at full tank level (mtr.)	4.75	4.70	4.70	4.00	4.01	4.76	4.60	3.50	4.00	4.75	4.72
7. Total Irrigated Area (ha.)	166.41	229.363	140	177.83	282.62	341.512	334.44	276.86	273.84	140.63	26
8. Average irrigated area per dabri (ha)	23.77	22.94	14.00	17.78	28.26	34.15	33.44	27.69	27.38	28.13	8.67
9. Size and boundary parameters	Size and numbers of irrigators in the command area are well demarcated and identified	Size and numbers of irrigators in the command area are well demarcated and identified	Size and numbers of irrigators in the command area are well demarcated and identified	Size and numbers of irrigators in the command area are well demarcated and identified	Size and numbers of irrigators in the command area are well demarcated and identified	Size and numbers of irrigators in the command area are well demarcated and identified	Size and numbers of irrigators in the command area are well demarcated and identified	Size and numbers of irrigators in the command area are well demarcated and identified	Size and numbers of irrigators in the command area are well demarcated and identified	Size and numbers of irrigators in the command area are well demarcated and identified	Size and numbers of irrigators in the command area are well demarcated and identified
10. Source of water	Rainfed	Rainfed	Rainfed	Rainfed	Rainfed	Rainfed	Rainfed	Rainfed	Rainfed	Canal and rainfed	Rainfed
11. Physical condition of *dabries* and water delivery structure (% of good *dabries*)	97	90	90	97	70	80	60	90	95	97	95
11. Main uses of *dabries'* water	Irrigation for paddy crop	Irrigation for paddy crop and domestic uses	Paddy, fisheries & domestic uses	Paddy & fisheries	Paddy	Paddy & fisheries	Paddy	Paddy, fisheries & domestic uses	Paddy, fisheries & domestic uses	Paddy, fisheries & domestic uses	Paddy, fisheries & domestic uses

(Continued)

Table 7.3 (Continued)

S. No.	*Particulars*	*Northern Hills*			*Central Plains*						*Bastar Plateau*	
	Districts	*Surajpur*	*Jashpur*	*Koriya*	*Raigarh*	*Bilaspur*	*Raipur*	*Balodabazar*	*Dhamtari*	*Kanker*	*Jagdalpur*	*Dantewada*
	Selected Blocks	*Surajpur*	*Pathalgaon*	*Manendragarh*	*Gharghoda*	*Masturi*	*Arang*	*Palari*	*Nagri*	*Narharpur*	*Bastar*	*Dantewada*
12.	Availability of water in *dabries*	July to March	July to February	July to June	July to April	July to November	July to March	July to November	July to January	July to March	July to February	July to April
13.	No. of sluice gate per *dabri*	1	1	1	1	1	1	1	1	1	1	1
14.	No of waterings (kharif)	2	1	2	2	1	2	2	1	1	2	2
15.	No of waterings (rabi)	2	2	2	0	0	2	0	2	1	1	1
16.	Sample villages with more than one *dabri*	1	1–2	1	1	1–3	1	1–2	1–3	1–2	1	1

7.6 Profile of Irrigators' Community

The socio-economic characteristics of irrigators covered by the sample of *dabries* are given in Table 7.4. Scheduled Tribe (ST) irrigators dominate the social structure in the majority of the *dabries* studied. In the Northern Hills and Bastar Plateau, besides crops, NTFPs and fisheries also contribute to the total income of irrigators. Income levels and standards of living depend on the extent of earning potential from activities other than paddy and crops. In the majority of *dabries*, a positive relationship was observed between irrigated area and cropping intensity. Higher cropping intensity in some *dabries* has also been observed, which may be due to supplementary irrigation being available to irrigators. Since the size of the command area is below 40 ha, the location of irrigators and the operated area owned by them along the water course do not show any visible difference. In all the *dabries*, irrigators had plots at multiple locations along the water course. Although irrigators depend heavily on *dabries* for irrigation water, the majority of them have supplementary sources of water, either owned or hired.

7.7 Decision-Making Arrangements and Pattern of Interactions

To analyse the self-organized and self-governed *dabri* irrigation system, the institutional framework developed in Chapter 2 is used here. This institutional framework assesses the physical and technical attributes of *dabries*, the characteristics of irrigators' communities in relation to other stakeholders (fishermen, and the village), the external and internal institutional arrangements, and the impact and outcomes in terms of productivity levels, income, cropping intensity, diversification, patterns of interaction among irrigators due to rule conformance, adequacy of water flow, and maintenance of *dabries*. As mentioned in Chapter 2, these attributes are usually combined in similar configurations. Applying this configuration to evaluate the effects of the physical and technical conditions of a *dabri*, it is important to understand what effects the physical condition (for example, a well-maintained or degraded *dabri*) may have on the behaviour of irrigators and, in turn, on decision-making arrangements, patterns of interaction, and outcomes in terms of productivity levels, cropping patterns, and the process of diversification. Further, the interrelationship between the internal and external arrangements involved in managing *dabries* and irrigators may change positively or negatively. Further, the physical and technical attributes of *dabries*, the irrigators' socio-economic characteristics, the institutional arrangements, the alternative property regimes or management regimes, and the role of Panchayats) collectively affect external and internal institutional arrangements, interaction patterns, and outcomes and impacts (see Tables 7.4 and 7.5). The characteristics of a *dabri* and irrigators/water users can affect outcomes directly or through patterns of interaction between external and internal institutional arrangements.

Table 7.4 Attributes of *Dabries* Irrigators' Community

Particulars / *Blocks*	*Surajpur*	*Pathalgaon*	*Manendragarh*	*Ghargboda*	*Masturi*	*Arang*	*Palari*	*Nagri*	*Narharpur*	*Bastar*	*Dantewada*
1. Total Farmers in *dabri* command area	141	190	141	165	561	402	542	265	226	96	25
2. Average number of farmers per *dabri*	20	19	14	17	56	40	54	27	23	19	8
3. Social structure of beneficiaries (% of total beneficiaries)	General (60) ST (37) and SC (3)	ST (63) General (27) and SC (10)	ST (50) General (42) and SC (8)	ST (67) General (25) and SC (8)	General (60) SC (28) and ST (12)	General (70) SC (27) and ST (3)	General (79) SC (15) and ST (6)	ST (61) General (34) and SC (5)	ST (61) General (32) and SC (7)	ST (66) General (33) and SC (1)	ST (73) General (26) and SC (1)
4. Irrigators Literacy %	55.52	54.65	66.50	50	55	50	50	74.92	71.95	50	42.23
5. Main income and livelihood sources of farmers	Crops, NTFPs collection, and non-farm activities	Crops, small ruminants, NTFPs	Crops, livestock, largely small ruminants	Paddy Fisheries, NTFPs, and Silkworm collection	Paddy and non – farm employment	Crops, Fisheries	Paddy, small ruminants	Crops, Fisheries & NTFPs	Crops, Fisheries & NTFPs	Crops, Fisheries & NTFPs	Crops, Fisheries & NTFPs
6. Variability in income and living standards of irrigators	Not significant, only in case of those farmers who are engaged in NTFPs collection	Income difference among irrigators due to supamitetary income from small ruminants	Not significant	Some variation due to employment in non form sector	Not significant	Not significant	Income difference among irrigators due to supamitetary income from small ruminants	Significant if irrigators also engaged in fish culture and NTFPs collection	Significant if irrigators also engaged in fish culture and NTFPs collection	Significant if irrigators also engaged in fish culture and NTFPs collection	Significant if irrigators also engaged in fish culture and NTFPs collection
7. Operated area (ha)	0.966	1.710	1.110	1.600	0.758	1.018	0.871	1.113	1.180	1.689	4.676
8. Cropped area (ha)	1.265	1.892	1.261	1.924	0.807	1.389	1.139	1.261	1.359	1.955	6.557
9. Irrigated area as % of cropped area	85.00	80.00	95.00	90.00	70.00	80.00	95.00	82.00	90.00	90.00	98.00
10. Cropping intensity (%)	130.957	110.666	113.585	120.279	106.525	136.456	130.665	113.309	115.147	115.728	140.214
11. Location of irrigators along watercourse (%)											
a. Head (H)	34.05	34.38	37.60	33.00	33.57	33.66	33.91	33.90	34.49	32.99	35.71
b. Middle (M)	50.73	35.79	23.40	36.08	35.91	28.43	38.37	25.71	27.30	25.77	25.72
c. Tail (T)	15.22	29.83	39.00	30.92	30.52	37.91	27.71	40.40	38.22	41.24	38.58

12. Command area under H, M, T (%)											
a. Head (H)	34.56	34.31	38.46	31.25	34.35	33.29	34.10	32.14	33.48	32.14	36.63
b. Middle (M)	39.26	23.46	38.46	43.75	31.00	29.00	41.62	39.29	29.10	25.00	38.02
c. Tail (T)	26.17	42.23	23.08	25.00	34.65	37.70	24.28	28.57	37.42	42.86	25.35
13. % of irrigators owned plots at multiple locations (H,M,&T)	10	3	8	5	15	10	7	10	5	12	2
14. Dependency of irrigators on *dabries*	Heavy	Heavy	Heavy	Heavy	Heavy	Heavy	Heavy	Heavy	Heavy	Heavy	Heavy
15. Alternative/ supplementary irrigation	Wells, tube wells	Wells, tube wells	Wells, tube wells	Wells, tube wells	Pond, Check dam	Wells, tube wells	Wells, tube wells	Tube wells, check dam	Wells, tube wells	Tube wells, electric pump	Canal, pond/ tanks

The analysis of decision-making mechanisms adopted by Panchayats and irrigators' groups in managing *dabries* structure and irrigation flow and their effects on patterns of collective interactions is presented in Table 7.5. Basically, three management regimes are used for distributing irrigation water from *dabries*, namely, Panchayats manage directly, Panchayats manage through *Sinchai Samitee*, or they are controlled and managed by irrigators' groups. Irrigators or local fisheries have been given usufruct rights for fish culture with an understanding that adequate water for crops should be available and other users cannot be excluded from using the *dabries'* water, as they are needed for many purposes in many villages. The irrigators of six *dabries* had full information about decision-making arrangements. The irrigators of the remaining *dabries* also had fair understanding of institutional arrangements. Access to information does not only help in the decision-making process but also minimizes transaction costs. The impact of effective implementation of operational rules, internal and external institutional arrangements, and continuing patterns of interaction in terms of reciprocity, protection, and use regulations is presented in Table 7.5.

It is clear from Table 7.5 that decision-making arrangements with respect to the type of irrigation method, operational rules, and conditions of collective-choice arrangements differ significantly. The combined impact of these can be seen in the patterns of interaction. Further, the impact of patterns of interaction was measured in terms of high, medium, and low impact, as indicated in Table 7.5.

To this end, the impact of the decision-making mechanism on outcomes is examined in terms of productivity and income levels of important crops grown by the farmers in *dabries* across the selected villages.

7.8 Outcomes of Workable Institutional Arrangements

The effective interrelationship among *dabries*' physical and technical attributes, irrigators' community characteristics, decision-making arrangements, and patterns of interaction can result in economic and equitable outcomes. The outcomes of workable institutional arrangements in terms of changing from direct/broadcasting of seeds to improved methods of sowing paddy, increases in cropping intensity, increases in yield of main crops, average fish production, and diversification are presented in Table 7.6. The rational use of available water promotes diversification. In *dabries* located in tribal-dominated villages, the value of the Simpson Diversity Index moving towards 1 indicated diversification. It is worth noting that tribal farmers have traditionally been practicing diversification to ensure maximum income and minimize the risk of crop failure due to uncertain rains. Farmers in other *dabries* are in the transition phase, moving away from mono-crop systems dominated by paddy production.[4] According to farmers' assessment, there has been a many-fold increase in crop yields. Except in Gharghoda *dabries,* fish yields are equal or higher than the yields of the state's inland fisheries.

Table 7.5 Decision-Making Arrangements, Patterns of Interaction, and Pattern of Collective Outcome Under Common Property Regime

Particulars / Blocks	*Surajpur*	*Pathalgaon*	*Manendragarh*	*Gharghoda*	*Masturi*	*Arang*	*Palari*	*Nagri*	*Narharpur*	*Bastar*	*Dantewada*
A. Decision-making arrangements											
1. Ownership over *dabries*	Village Panchayats	Village Panchayats	Village Panchayats	Village Panchayats	Village Panchayats	Village Panchayats	Village Panchayats	Village Panchayats	Village Panchayats	Village Panchayats	Village Panchayats
2. Legal and administrative relation with state-local body legally structured*	Village Panchayat is legal unit of three-tier Panchayats institutions	Village Panchayat is legal unit of three-tier Panchayats institutions	Village Panchayat is legal unit of three-tier Panchayats institutions	Village Panchayat is legal unit of three-tier Panchayats institutions	Village Panchayat is legal unit of three-tier Panchayats institutions	Village Panchayat is legal unit of three-tier Panchayats institutions	Village Panchayat is legal unit of three-tier Panchayats institutions	Village Panchayat is legal unit of three-tier Panchayats institutions	Village Panchayat is legal unit of three-tier Panchayats institutions	Village Panchayat is legal unit of three-tier Panchayats institutions	Village Panchayat is legal unit of three-tier Panchayats institutions
3. Management system	Farmer's Group under CPR Regime*	Sinchai Samitee (Irrigation committee)	Farmers group	Water users committee of irrigator farmers representing head, middle, tail, ender	Panchayats directly managing tanks & tank irrigation	Sinchai Samitee manage tanks & tank-irrigation system	Panchayats directly managing tanks & tank irrigation	Farmers group	Panchayats directly managing tanks & tank irrigation	Sinchai Samitee manage tanks & tank-irrigation system	Sinchai Samitee manage tanks & tank-irrigation system
4. Irrigators awareness of collective-choice rules (%)	85	85	100	100	70	85	80	100	90	100	100
5. Operational rules related to: a. Size and boundary rules	a. Well demarcated size and a closed set of landowners in command area of *dabries*, b. exclusion of outsiders collectively by irrigators	a. Well demarcated size and a closed set of landowners in command area of *dabries*, b. exclusion of outsiders collectively by irrigators	a. Well demarcated size and a closed set of landowners in command area of *dabries*, b. exclusion of outsiders collectively by irrigators	a. Well demarcated size and a closed set of landowners in command area of *dabries*, b. exclusion of outsiders collectively by irrigators	a. Well demarcated size and a closed set of landowners in command area of *dabries*, b. exclusion of outsiders collectively by irrigators	a. Well demarcated size and a closed set of landowners in command area of *dabries*, b. exclusion of outsiders collectively by irrigators	a. Well demarcated size and a closed set of landowners in command area of *dabries*, b. exclusion of outsiders collectively by irrigators	a. Well demarcated size and a closed set of landowners in command area of *dabries*, b. exclusion of outsiders collectively by irrigators	a. Well demarcated size and a closed set of landowners in command area of *dabries*, b. exclusion of outsiders collectively by irrigators	a. Well demarcated size and a closed set of landowners in command area of *dabries*, b. exclusion of outsiders collectively by irrigators	a. Well demarcated size and a closed set of landowners in command area of *dabries*, b. exclusion of outsiders collectively by irrigators

(Continued)

Table 7.5 (Continued)

Particulars / *Blocks*	*Surajpur*	*Pathalgaon*	*Manendragarh*	*Ghargboda*	*Masturi*	*Arang*	*Palari*	*Nagri*	*Narharpur*	*Bastar*	*Dantewada*
b. Allocation rules	Water withdrawal method from *dabri* is field-to-field, To avoid uncertainty and conflict fixed timeslot and water flow is regulated, in two *dabri* irrigators (H,M,T) withdraw water simultaneously through diesel pump of 5HP under fixed timeslot rule	Water withdrawal method from *dabri* is field-to-field, To avoid uncertainty and conflict fixed timeslot and water flow is regulated	Water withdrawal Rotational arrangement. To avoid uncertainty and conflict fixed timeslot and water flow is regulated	Rotational arrangement. To avoid uncertainty and conflict fixed timeslot and water flow is regulated	Water withdrawal method from *dabri* is field-to-field, To avoid uncertainty and conflict fixed timeslot and water flow is regulated	Water withdrawal method from *dabri* is field-to-field, To avoid uncertainty and conflict fixed timeslot and water flow is regulated	Rotational arrangement. To avoid uncertainty and conflict fixed timeslot and water flow is regulated	Rotational arrangement. To avoid uncertainty and conflict fixed timeslot and water flow is regulated	Water withdrawal method from *dabri* is field-to-field, To avoid uncertainty and conflict fixed timeslot and water flow is regulated	Water withdrawal method from *dabri* is Field-to-field, To avoid uncertainty and conflict fixed timeslot and water flow is regulated	Water withdrawal method from *dabri* is field-to-field, To avoid uncertainty and conflict fixed timeslot and water flow is regulated, Panchayat and Sinchai Samitee coordinates with canal local managers
c. Input rules (maintenance of tank)	From village Panchayat fund and labour contribution of irrigators	Village Panchayat fund	Through MGNREGA activities and by irrigators	Village Panchayat fund and labour contribution of irrigators committee	Village Panchayat	By irrigators groups	Village Panchayat	Village Panchayat fund and labour contribution of irrigators committee	Village Panchayat	Voluntary labour contribution for maintaining irrigation system	Voluntary labour contribution for maintaining irrigation system
d. Penalty rules	Nil	Nil	Nil	Nil	Nil	Nil	Nil	Nil	Nil	Nil	Nil
6. Irrigation fees (Rs/ha)	Nil	Nil	Nil	250/-	Nil	Nil	Nil	Nil	Nil	Nil	Nil
7. Percentage of recovery of irrigation fees	-	-	Nil	95	Nil	Nil	Nil	Nil	Nil	Nil	Nil
8. Ability to raise fund for *dabri*	Panchayat's fund	Panchayat's fund	*Dabries* leased out for fish cultivation, collection of lease rents	Irrigation fee & lease rent from fish cultivation	Panchayat's fund	MGNREGA & lease rent from fish cultivation	Panchayat's fund	Through MGNREGA	Panchayat's fund	Panchayat's fund	Panchayat's fund

9. Ability of irrigators to influence Panchayat	Farmer's Group collectively influence Village Panchayats if major maintenance issue arise	Irrigators are members of Sinchai Samitee, through Sinchai Samitee irrigators influence to Panchayat to ensure adequate and timely supply water and overall maintenance of irrigation system has	Farmers group is quite active can effectively engaged Panchayat if there is any conflict over water issue between fishers and irrigators	Effective control of the Water Users committee on overall management. In case of any issue related to Panchayat, users committee collective influence Panchayat	Irrigators influence for adequate water supply for critical growth stage of paddy	Irrigators are members of Sinchai Samitee, through Sinchai Samitee irrigators influence to Panchayat to ensure adequate and timely supply water and resolve conflict, if any, between irrigators and fishers	If any issue in rotational irrigation, Panchayat intervene on request of irrigators	Farmers group is quite active can effectively engaged Panchayat for maintain the system	Irrigators influence for adequate water supply for critical growth stage of paddy	Since irrigators regularly contribute labour for maintaining *dabries*, irrigators engage Panchayat for any issue related to adequate water supply	Since irrigators regularly contribute labour for maintaining *dabries*, irrigators engage Panchayat for any issue related to adequate water supply
10. External organizational and technical assistance	Soil Conservation officer, Rural extension officer and KVK provide technical guidance to farmers	Soil Conservation officer, Rural extension officer and KVK provide technical guidance to farmers	Soil Conservation officer, Rural extension officer and KVK provide technical guidance to farmers	Soil Conservation officer, Rural extension officer and KVK provide technical guidance to farmers	Soil Conservation officer, Rural extension officer and KVK provide technical guidance to farmers	Soil Conservation officer, Rural extension officer and KVK provide technical guidance to farmers	Soil Conservation officer, Rural extension officer and KVK provide technical guidance to farmers	Soil Conservation officer, Rural extension officer and KVK provide technical guidance to farmers	Soil Conservation officer, Rural extension officer and KVK provide technical guidance to farmers	Soil Conservation officer, Rural extension officer and KVK provide technical guidance to farmers	Soil Conservation officer, Rural extension officer and KVK provide technical guidance to farmers
B. Patterns of interaction											
1. Reciprocal interaction:	Individual co-irrigators make reasonable efforts to avoid over use of water and negative externality	Since Sinchai Samitee has representation from head, middle and tail, irrigators make efforts collectively to maintain equitable water flow	Farmers group collectively regulate use of available water in *dabries* to ensure productive and equitable use	Farmers group collectively regulate use of available water in *dabries* to ensure productive and equitable use	Individual co-irrigators make reasonable efforts to avoid overuse of water. Due to water availability, in *dabries*, is up to November, few cases of water subtraction was observed.	Since Sinchai Samitee has representation from head, middle and tail, irrigators make efforts collectively to maintain equitable water flow	Farmers group collectively regulate use of available water in *dabries* to ensure productive and equitable use	Farmers group collectively regulate use of available water in *dabries* to ensure productive and equitable use	Individual co-irrigators make collectively efforts to avoid over use of water under overall supervision of Panchayat	Individual co-irrigators make reasonable efforts to avoid over use of water and negative externality	Individual co-irrigators make reasonable efforts to avoid over use of water and negative externality

(*Continued*)

Table 7.5 (Continued)

Particulars \ Blocks	*Surajpur*	*Pathalgaon*	*Manendragarh*	*Ghargboda*	*Masturi*	*Arang*	*Palari*	*Nagri*	*Narharpur*	*Bastar*	*Dantewada*
2. Protection	Panchayats protect *dabries* and restrict free ridding behaviour with irrigators support	Sinchamiteee protects and maintain *dabries*	Farmers group collectively protects *dabries*	Farmers group collectively protects *dabries*	Panchayats protect *dabries* and restrict free ridding behaviour with irrigators support	Sinchamiteee protects and maintain *dabries*	Farmers group coamitely protects *dabries*	Farmers group collectively protects *dabries*	Panchayats protect *dabries* and restrict free ridding behaviour with irrigators support	Panchayats protect *dabries* and restrict free ridding behaviour with irrigators support	Panchayats protect *dabries* and restrict free ridding behaviour with irrigators support
3. Use regulations	To enforce use regulations Panchayats have played critical role to minimize water substrative behaviour of irrigators through engaging irrigators as limited amount of flow is available for limited months in *dabries*	To enforce use regulations Panchayats through Sinchai Samitee have to played critical role to minimize water substrative behaviour of irrigators through engaging irrigators as limited amount of flow is available for limited months in *dabries*	To enforce use regulations farmers groups have played critical role to minimize water substrative behaviour of irrigators through engaging irrigators as limited amount of flow is available for limited months in *dabries*	To enforce use regulations farmers groups have played critical role to minimize water substrative behaviour of irrigators through engaging irrigators as limited amount of flow is available for limited months in *dabries*	To enforce use regulations Panchayats have played critical role to minimize water substrative behaviour of irrigators through engaging irrigators as limited amount of flow is available for limited months in *dabries*	To enforce use regulations Panchayats through Sinchai Samitee have to play critical role to minimize water substrative behaviour of irrigators through engaging irrigators as limited amount of flow is available for limited months in *dabries*	To enforce use regulations farmers groups have played critical role to minimize water substrative behaviour of irrigators through engaging irrigators as limited amount of flow is available for limited months in *dabries*	To enforce use regulations farmers groups have played critical role to minimize water substrative behaviour of irrigators through engaging irrigators as limited amount of flow is available for limited months in *dabries*	To enforce use regulations Panchayats have played critical role to minimize water substrative behaviour of irrigators through engaging irrigators as limited amount of flow is available for limited months in *dabries*	To enforce use regulations Panchayats have played critical role to minimize water substrative behaviour of irrigators through engaging irrigators as limited amount of flow is available for limited months in *dabries*	To enforce use regulations Panchayats have played critical role to minimize water substrative behaviour of irrigators through engaging irrigators as limited amount of flow is available for limited months in *dabries*
4. Impact of pattern of interaction on performance of management system **	Medium	Medium	High	High	Low	Medium	Medium	High	High	Medium	Medium

Source: Author's analysis based on field work

* Uniform organizational structure of village Panchayat: Sarpanch (1), Pach (4), and Sinchai Samitee (elected irrigator – head, middle, and tail-ender)

**level of performance = 3 (high), 2 (medium), 1 (low).

Table 7.6 Outcomes of Workable Institutional Arrangements

S.N.	Particulars / Blocks	*Surajpur*	*Pathalgaon*	*Manendragarh*	*Gharghoda*	*Masturi*	*Arang*	*Palari*	*Nagri*	*Narharpur*	*Bastar*	*Dantewada*
1.	Shift from direct/broadcasting to improved method of sowing paddy	60+	70	65+	80	50	80++	60	40	40++	30	20
	a. % of farmers adopted transplanting											
	b. % area covered transplanting	75+	80	60+	90	45	85++	55	50	30++	20	15
2.	% increase in cropping intensity	15–25	5–10	5–10	10–20	2–5	10–20	5–15	5–10	10–15	10–15	10–20
3.	Diversification (value of Simpson Diversification index)*	0.630	0.446	0.896	0.558	0.191	0.567	0.273	0.336	0.531	0.571	0.761
4.	% increase in yield of main crops											
	Paddy	25	11	20	20	15	10	18	18	12	18	20
	Maize	18	15	15	20			20	20	6	15	15
	Pigeon pea	29	10	10				10	20	10		
	Niger		8	8								
	Wheat	40		11	20		20	11			11	11
	Chickpea pea			10			20	10		10	10	10
	Mustard	35										
	Vegetable		25	22	20		20	20	40		20	22
5.	Average fish production (kg/ha/yr)	Nil	Nil	3000	1000	-	3300	-	2000	3000	2000	2500

Source: Author's analysis based on field work

Note: + and ++ indicate the adoption of line sowing and SRI methods of sowing respectively

* SID = 1-Σ(a/A) where SID is the Simpson Index for Crop Diversification, a is the area planted to the jth crop in i^{th} year. Index value ranges from 0 to 1. If the value moves towards 0 or 1 it refers to specializations or diversification respectively.

Performance of selected *dabries* in relation to irrigators' implicit goal of enhancing their livelihood and generating income through self-organized and self-governed irrigation systems is evaluated in terms of crop yield per ha of water-spread area and net income. It can be seen from Table 7.7 that higher yields can even be achieved under flood/field-to-field irrigation, if operational rules are effectively and collectively implemented with support from Panchayats. Also, the increase in productivity levels of the majority of irrigators depends on the simultaneous and adequate availability of water at critical crop-growth stages in the head, middle, and tail-end of water-carrying channels. In a few *dabries* with better management, where irrigation is also available in rabi season, wheat and other rabi crops have been introduced into the cropping pattern.

7.9 Approaches to Sustain Institutional Arrangements for Self-Governance of *Dabries*

To recap, with the formation of the Chhattisgarh State, *Dabri Yojana* was re-launched with a package of institutional arrangements. Focus shifted from target-orientation to a demand-driven approach with the commitment of Panchayats to manage *dabries* sustainably under community-based management systems. It has adequately been documented that, like other village commons, *dabries* in the sample villages are small-scale, CPRs being managed under common property regimes. The advantage of having small-scale CPRs is that the process and performance of self-organized and self-governed structures are easy to evaluate by the users themselves, in order to redefine operational rules, if needed. As shown in this study, in an interdependent situation, the irrigators under 70% of *dabries* have self-organized and developed functional rules for self-governance to reap the joint benefits, when, without enforceable institutional arrangements, they might all be tempted by the lure of becoming free riders.

Due to well-defined institutional provisions in terms of operational rules and their effective implementation by Panchayat and irrigator groups, free-riding behaviour could not spread among irrigators. The effective interrelationships of *dabri* and irrigators' community attributes, decision-making arrangements, and patterns of interaction provided economic and distributive outcomes for irrigators in terms of efficiency (crop or fish output and income) and equity (distributive gains in terms of equitable distribution of water among irrigators). Besides equity and efficiency criteria suggested by Oakerson, two additional criteria, namely, sustainability and expression of satisfaction (Berkes, 1992), are also used in this chapter to assess the outcomes of *dabries* as CPR-management systems. The system is sustainable if it "meets the needs and aspirations of the present without compromising the ability of future generations to meet their own needs" (IUCN/UNEP/WWF, 1991). Using this definition of sustainability, it can fairly be said that decision-making arrangements adopted by the Panchayats and water users'

Table 7.7 Yield, Gross, and Net Returns of Crops *(per ha)*

Crops	**Particulars** / *Blocks*	*Surajpur*	*Pathalgaon*	*Manendragarh*	*Gharghoda*	*Masturi*	*Arang*	*Palari*	*Nagri*	*Narharpur*	*Bastar*	*Dantewada*
Paddy	MSP (Rs./q)	1815	1815	1815	1815	1815	1815	1815	1815	1815	1815	1815
	Yield (q)	32.5	22	35	30	18	20	35	34	22.77	35	22
	Total cost (Rs.)	26700	38250	25000	27500	25000	28000	28000	32500	30000	26950	16940
	Gross Return (Rs.)	58987.50	39930.00	63525.00	54450.00	32670.00	36300.00	63525.00	61710.00	41327.55	63525.00	39930.00
	Net Return (Rs.)	32287.50	1680.00	38525.00	26950.00	7670.00	8300.00	35525.00	29210.00	11327.55	36575.00	22990.00
	BC Ratio	2.21	1.04	2.54	1.98	1.31	1.30	2.27	1.90	1.38	2.36	2.36
Maize	MSP (Rs./q)	1760	1760	1760	1760			1760	1760	1760	1760	1760
	Yield (q)	31	18.2	20	26			30	18	24.65	32	25
	Total Cost (Rs.)	18900	30000	20000	22181			25200	28000	27000	25000	16000
	Gross Return (Rs.)	54560.00	32032.00	35200.00	45760.00			52800.00	31680.00	43384.00	56320.00	44000.00
	Net Return (Rs.)	35660.00	2032.00	15200.00	23579.00			27600.00	3680.00	16384.00	31320.00	28000.00
	BC Ratio	2.89	1.07	1.76	2.06			2.10	1.13	1.61	2.25	2.75
Pigeon pea	MSP (Rs./q)		5800	5800				5800	5800			
	Yield (q)		5.03	10				10	12			
	Total Cost (Rs.)		19200	25000				27000	28000			
	Gross Return (Rs.)		29174.00	58000.00				58000.00	69600.00			
	Net Return (Rs.)		9974.00	33000.00				31000.00	41600.00			
	BC Ratio		1.52	2.32				2.15	2.49			
Niger	MSP (Rs./q)		5940	5940								
	Yield (q)		3.2	5								
	Total Cost (Rs.)		9600	12000								
	Gross Return (Rs.)		19008.00	29700.00								
	Net Return (Rs.)		9408.00	17700.00								
	BC Ratio		1.98	2.48								
Wheat	MSP (Rs./q)	1925		1925	1925		1925	1925			1925	1925
	Yield (q)	16		10	22		10	12			9	12
	Total Cost (Rs.)	21500.00		13000	19859		15000.00	16000.00			15000.00	20000.00
	Gross Return (Rs.)	30800.00		19250.00	42350.00		19250.00	23100.00			17325.00	23100.00
	Net Return (Rs.)	9300.00		6250.00	22491.00		4250.00	7100.00			2325.00	3100.00
	BC Ratio	1.43		1.48	2.13		1.28	1.44			1.16	1.16

(*Continued*)

Table 7.7 (Continued)

Crops	Particulars / *Blocks*	*Surajpur*	*Pathalgaon*	*Manendragarh*	*Gharghoda*	*Masturi*	*Arang*	*Palari*	*Nagri*	*Narharpur*	*Bastar*	*Dantewada*
Chickpea	MSP (Rs./q)			4875	4875		4875	4875		4875	4875	4875
	Yield (q)			8	11		5	7		7	6	8
	Total Cost (Rs.)			21000	22024		20000.00	23000		24000	22000	22000
	Gross Return (Rs.)			39000	53625		24375.00	34125.00		34125.00	29250.00	39000.00
	Net Return (Rs.)			18000	31601		4375.00	11125.00		10125.00	7250.00	17000.00
	BC Ratio			1.86	2.43		1.22	1.48		1.42	1.33	1.77
Pea	MSP (Rs./q)				4875			4875				
	Yield (q)				15			7				
	Total Cost (Rs.)				31812			20000				
	Gross Return (Rs.)				73125			34125				
	Net Return (Rs.)				41313			14125				
	BC Ratio				2.30			1.71				

Source: Author's analysis based on field work

Note: Cost of production includes $CostA_2$+FL, MSP fixed by CACP for the year 2019–20 considered for calculation. MSP – Minimum Support Price, BC Ratio – Benefit Cost Ratio

groups under the common property regime reflect long-term sustainability. The criterion of 'expression of satisfaction' may provide a composite measure of outcomes as perceived by farmers. Under both situations, the resource users were satisfied with the existing management systems and the flexibility in redefining institutional arrangements, if needed.

Success in managing *dabries* for crop production under common property regimes does not depend solely on the technical-physical nature of the *dabries* or the decision-making arrangements or behaviour of the irrigators, but on a combination of four key attributes. It may safely be concluded from this analysis that village commons in general and *dabries* in particular, can effectively be managed under common property regimes. This study has also documented that, while degradation of many village commons, including dabries, incorrectly attributed to community management systems, actually originates in the dissolution of village-level working rules whose very purpose is to manage and safeguard the use of natural, rural resources efficiently, equitably, sustainably, and satisfactorily for the rural community as a whole. Similar findings are also reported in CPRs case studies of Nepal, Philippines, Sri Lanka, and African countries, highlighting the robustness of internal governance.[5]

Appendix 7.1

Details of *Dabries* Construction in Villages of Selected Blocks

District	Dabries *Block*	*Dabries Village*	*Year of construction*
Surajpur	Surajpur	Karva	2007–08
		Ketka	2008–09
		Podi	
		Lanchi	
		Pahadgao	2009–10
		Beltekri	
		Haripur	2015–16
Jashpur	Pathalgaon	Kadro	2007–08
		Pandripani	2008–09
		Farsatoli-1	2010–11
		Goliyagarh	2011–12
		Farsatola-4	2012–13
		Jhimki-1	2013–14
		Jhimki-2	2014–15
		Karrabevra	2015–16
		Lakchhar	2016–17
		Minjpur	2017–18
Koriya	Manendragarh	Paradol	2000–01
		Saraula	2001–02
		Bundeli	2002–03
		Lalpur	2003–04
		Bhauta	2004–05
		Putadand	2005–06
		Shankargarh	2006–07
		Chaughada	2008–09
		Nagma	2010–11
		Bauridand	2013–14
Raigarh	Gharghoda	Gharghodi	2006–07
		Baronakunda	2007–08
		Chharratangar	2008–09
		Kathrapali	2009–10
		Faguram	2010–11
		Tenda	2011–12
		Kaya	2013–14
		Bulekera	2015–16
		Karrikachhar	2018–19
		Bhalumar	2019–20

District	Dabries *Block*	*Dabries Village*	*Year of construction*
Bilaspur	Masturi	Navagaon	2006–07
		Kuli	
		Karichhapar	2011–12
		Madai	
		Jondhara-2	2012–13
		Pachpedi-1	2013–14
		Loharabor-2	
		Loharabor-1	
		Jondhara-1	
		Pachpedi-2	2015–16
Raipur	Arang	Gukhera	2006–07
		Nardaha	2007–08
		Ghumarabhata	2008–09
		Khapari	2009–10
		Banarsi	2010–11
		Borid	2011–12
		Malidih	2012–13
		Khorasi	2014–15
		Baihar	2016–17
		Jarod	2018–19
Balodabazar	Palari	Charauda	2007–08
		Gadakusmi	
		Gaitara	2008–09
		Datan	2010–11
		Sasaha	2012–13
		Jara	
		Sankari-1	2016–17
		Sankari-2	2017–18
		Sasaha	2018–19
		Kosmandi	2019–20
Dhamtari	Nagri	Charra	2006–07
		Kaspur-1	
		Padampur	
		Ghutkel	2007–08
		Garhdongri	2008–09
		Dongardula	
		Parsapani	2010–11
		Rajpur	
		Kaspur-2	2011–12
		Bhalupara Kaspur-3	2012–13
Kanker	Narharpur	Devdongar	2005–06
		Devgaon	2006–07
		Basanwahi	2007–08
		Jamgaon (Imlipara)	2008–09
		Bhainsmudi	2009–10
		Bansagar	2010–11
		Gaharinbandha	2011–12
		Thanabodi	2012–13
		Navadbari-1 (Shitlapara)	2013–14
		Navadbari-2	2015–16

(*Continued*)

(Continued)

District	Dabries *Block*	*Dabries Village*	*Year of construction*
Dantewada	Dantewada	Balud	2009–10
		Muraki	2010–11
		Teknar	2011–12
Bastar	Bastar	Junapani Bakel	2007–08
		Godiyapal Kadamunda	2013–14
		Cherakur Banvajhori	2015–16
		Piringmara Cherakur	2016–17
		Kavadgaon	2018–19

Notes

1 In the initial phase (between the 1980s and the mid-1990s) of *Dabri* research, I had the unique opportunity to interact and travel extensively with the late Deodutta Shukla, the then Additional Director Agriculture, to understand the socio-ecological aspects of various types of wetlands, including his favorite *dabries* (the micro-minor tank irrigation system). I owe a special debt to him. I had received help from R.N. Sharma (retired Additional Director), Subject Matter specialist, soil and water conservation wing and C.L. Jain (retired Director), the then Assistant Director, Department Agriculture, M.P/CG. My sincere thanks to them.

2 S.C. Mukherjee, Director Extension Services, Indira Gandhi Agricultural University Raipur, CG, and R.K. Chandrawanshi, Joint Director, Agriculture, Department of Agriculture, GOCG, coordinated data collection for this study. SCOs, Deputy/Assistant Director Agriculture and KVK scientists collected data pertaining to *dabries* and farmers' socio-economic and agriculture parameters. Without their sincere efforts it would have not been possible to cover a large sample. I am grateful to all of them for their timely help and support.

3 With the formation of Chhattisgarh State, the government of Chhattisgarh launched an intensive programme for the rehabilitation and construction of small farm reservoirs/ponds (with one ha. Water-spread area) on individual farmer's fields with 100 percent subsidy, completion/construction, and strengthening of existing water storage and conveyance structures under the harnessing of surface water resources programmes, implementation of lift irrigation schemes, and use of water-saving devices such as sprinklers, and drip irrigation systems. The SDF also re-introduced *Munda* technology (a traditional technology with small water ponds of 0.30 ha of water-spread area) for fish culture and to provide irrigation to paddy crops. Similarly, the State Department of Rural Development, the Department of Forests, and the Rural Engineering Department have also started an initiative for creating small water bodies for domestic and irrigation purposes after the formation of the state. Most of the water structures were created under the relief measures as the state has been facing frequent droughts. All of these rural water creation activities have become part of MGNREGA (Marothia, 2004b).

4 An earlier study conducted by Marothia et al. (2007) clearly indicates the specialized nature of cropping patterns due to mono-cropped areas, dominated by rice-based farming systems. Farmers grew a range of other crops in rabi season, depending on irrigation availability. The poor irrigation facilities, however, restrict the cultivation of rabi season crops and leave little scope for crop diversification. As a result, even prime agricultural land has been kept fallow in rabi season. Complete dependence of agriculture on rainfall, which is also erratic in nature, inhibits

the farmers from taking chances for diversifying crop patterns. Rice is an essential component in the food basket of people in Chhattisgarh and has been a cultural expectation for farmers to grow rice. Also, open grazing is a serious problem after the rainy season (rabi season) in most of the villages of Chhattisgarh. Even if a few farmers choose to grow rabi crops, it is difficult to save the crops from grazing. The extent of varietal diversification in rice was estimated by the authors applying the Simpson Diversification Index. Coefficient value of the Diversification Index indicates the evidence of varietal diversification in the cultivation of rice crops, which in turn also indicates that varietal diversification in rice varieties is comparatively common under large farms in Bastar, Surguja, and districts of Chhattisgarh's Central Plains. The co-efficient value of the diversity index indicates a reverse trend across the farm size of holdings.

5 This observation is consistent with the findings reported by Siy (1982); Sugden (1984); Ostrom (1992a); Tang (1992); Schlager et al. (1994); Swallow and Bromley (1994) in the cases of complex and simple water systems.

8 Culture Fisheries in Multiuse Wetlands Under Alternative Property Rights Regimes

8.1 Introduction

Fisheries in inland freshwater bodies date back to time immemorial in several countries, including in many Indian states and have been important sources of nutritional and livelihood security for a large number of people. Freshwater bodies have traditionally been managed and controlled for fish production under the common property regime in Chhattisgarh. However, after 1952, most such water bodies (ponds, tanks, and reservoirs) have been transferred to Panchayats or irrigation departments, depending on their water-spread area. Water bodies are being administered and controlled in the state under different institutional hierarchies or property rights regimes for aquaculture. The DWR, the soil and water conservation wing of the State Development of Agriculture, the SDF, the Department of Panchayats Raj Institutions and Rural Development, and private owners have legal ownership over the water bodies. Nevertheless, they are still multipurpose or multiuse and multifunctional in nature, with inherent interdependencies among users.

The issues related to managing multiuse ponds and irrigation tanks by fisheries are complex due to different the different categories into which they fall and the range of characteristics of these water bodies in terms of scale, location, size, and coverage of FCSs, FGs, Individual Fishers (IFs), and multiple agencies involved in governing the water resources. Against this background, this chapter[1] is organized in ten sections to provide an exhaustive analysis of the performance of fisheries in MWs under alternative governance structures. This introductory section outlines the structure of the remaining sections. Section two provides a stock of the water resources available for freshwater aquaculture activities in Chhattisgarh State. Section three describes cooperative governance in aquaculture. Section four provides details about the database and coverage of the study. Section five provides an in-depth analysis of five FCSs and individual fisherman, for which information was collected about types of ponds/tanks used for fisheries, along with ownership, management, and leasing authority structure. In section six and its subsections, the salient features of FCSs, the characteristics of fishermen's households, the physical and technical attributes of ponds and tanks under the control

DOI: 10.4324/9781003456742-8

of a particular FCS, and an analysis of comparative performance of the selected five FCSs are presented. The analysis of decision-making mechanisms adopted by five FCSs and its effects on patterns of collective interaction is presented in section seven. The decision-making mechanisms across the five FCSs differ significantly, despite the fact that they are all registered FCSs. Comparative performance of IFS (leased by Panchayats and privately owned ponds)[2] and FCSs in terms of yield, operating expenses, and net income are presented in section eight. A brief description of indigenous Water Harvesting Technology (*Munda*) for collective livelihood is given in section nine. In the last section, institutional and policy options are suggested to sustain fish culture in multiuse water bodies under different property regimes.

8.2 Water Resources and Fish Culture Activities

Freshwater aquaculture resources (FARs) in the form of *Mundas*, *tar bunds*, village ponds, *Dabries* (ranging in size from 0.40 ha to 40 ha), irrigation and multipurpose tanks, rivers and their tributaries, and reservoirs are extensively distributed among all the villages of the state (Marothia, 2004c, 2006b, 2007, 2012a). This section takes stock of the water resources available for freshwater aquaculture activities in the state (Table 8.1). Total area under fish culture developed in the state is 180,000 ha (about 90%) against the total water resources available 190,000 ha. Fisheries in the state are mostly culture-based, in which major carp fish groups play a leading role (approximately 75% major carps, 15% minor carps, and 10% exotic carps). FARs cover 88,671 village ponds and 1,770 irrigation tanks with 108,300 ha and 82,600 ha water-spread area respectively. Of the total water-spread area (190,900 ha) available in the state, 93 percent is used for fish culture in 78,275 village ponds, and 97 percent is used in 1,635 irrigation tanks (Table 8.1). The state has a network of several rivers, rivulets, streams, and streamlets running 3,573 km. This provides an immense opportunity to expand the scope of riverine, sport, and recreational fishing. With current production at 531,000 tonnes (90.79% from ponds, 4.19% from reservoirs, and 1.02% from rivers), the state is in sixth position in terms of inland fisheries production and in fish-seed production. Annual fish production recorded an average growth of 283% between 2007–08 and 2019–20. Current average fish production in rural water bodies is 3,287 kg/ha/year against the national average of 2,850 kg/ha/year, and the average production of reservoirs is 222 kg/ha/year against the national average of 69 kg/ha/year. The state is self-sufficient in fish production, besides catering to the needs of the fish-farmers of the state, the fish seeds are exported to the farmers of Odisha, MP, Maharashtra, Bihar, and Uttar Pradesh. Present demand for fish seed is placed at 140 crores standard fry, against which 273 crore were produced in 2019–20. The state aimed to produce 285 crores of standard fry during 2020–21. Fishery seeds are being produced by the state's fishermen using the latest technology. With

Table 8.1 Water Resource and Fish Culture Activities

S.No.	*Activities*	*Achievement (2018–19)*
1	Water Resources (in lakh ha)	
	1.1 Available water area for fish culture	1.994
	1.2 Irrigation reservoir water area	
	(i) Number	1770
	(ii) Area	0.826
	1.3 Village tanks ponds	
	(i) Number	88671
	(ii) Area	1.011
	1.4 Riverine water area (154 *Anicut*)	0.157
2	Water area under fish culture (in lakh ha)	1.862
	2.1 Irrigation reservoirs	0.801
	2.2 Village ponds	1.006
	2.3 Riverine water area (138 *Anicut*)	0.123
3	Resource available for fish-seed production (No./ha)	1263/639.29
	3.1 Government (No./ha)	677/336.51
	3.2 Private (No./ha)	586/303.28
4	Fisheries seed production (standard fry) (crore)	265
5	Fish production (in lakh metric tonnes)	5.31
6	Fish productivity (kg/ha)	
	6.1 Village ponds	3287
	6.2 irrigation reservoirs	222
7	Construction of ponds on private land (no./ha)	4722/4374
8	Registered FCS (no./members)	1606/55363
9	Women's FCS (no./members)	77/1848 (4.79/3.34)
10	Men's cooperative societies (no. of society/ members)	1514/49367 (94.27/89.17)
	10.1 General (no. of society/members)	1040/35127 (64.76/63.45)
	10.2 Scheduled tribe (no. of society/members)	352/11190 (21.92/20.21)
	10.3 Scheduled caste (no. of society/members)	122/3050 (7.60/5.51)
11	Women's Cooperative Societies (no. of society/ members)	64/1563 (3.99/2.82)
	11.1 General (no. of society/members)	28/676 (1.74/1.22)
	11.2 Scheduled tribe (no. of society/members)	21/594 (1.31/1.07)
	11.3 Scheduled caste (no. of society/members)	15/293 (0.93/0.53)
12	Beneficiaries engaged in fish culture (no./ members)	220000/38815
13	Employment (man-days) (in lakh)	242

Source: Information compiled from Annual Report and Administrative Report, Year 2019–20a, Directorate of Fisheries, Indirawati Bhawan, Atal Nagar, Nawa Raipur

Note: Figures in parenthesis indicate percentages of a particular fish cooperative society and members

the local production of fish seed, farmers get timely and improved fish seed at a low cost. The work of fish-seed production is being done at the private and government level and by the Chhattisgarh Fisheries Federation. Around 220,000 fishermen are engaged in aquaculture and fisheries for their livelihood representing 2.42 million man-days (see Table 8.1). After the formation of Chhattisgarh, culture fisheries had received focused attention of the state government. The Government of Chhattisgarh (GoCG) prepared and implemented the State Fisheries Policy. The major policies adopted by GoCG regarding culture fisheries aim to create more water-spread area on community and private lands, increase seed production, provide financial assistance through central and state schemes to extend or renovate water bodies, diversify fisheries activities, and build the capacity of fisheries communities. Special assistance has also been given to promote fish culture through the traditional *Munda* technique (small water structures in paddy fields are used for fish culture, for which rain and surplus irrigated water is the main source of water – see Marothia, 2004b for more details). Similarly, in other tribal and scheduled-caste-dominated districts, special assistance has been provided to create small ponds on private lands to promote culture fisheries. It is worth mentioning here that Panchayats (village, janpad, and district) earn around Rs. 150 million from leasing out ponds/tanks at the rate of Rs. 1,500 per ha. Chhattisgarh State Fish Cooperative Federation (CSFCF) earns around Rs. 24.899 million from lease rents from reservoirs of above 200 ha Water-spread area. From Dulhara (62.25 ha water-spread area) and Tuppa reservoir (38 ha water-spread area), it earns Rs. 528,000 and Rs. 220,000 respectively (SDF, GoCG, 2019–20b).

8.3 Cooperative Governance in Fish Culture

As mentioned earlier, MWs governance involves many governments, local bodies, and private players for freshwater aquaculture activities. Some evidence is available in Chhattisgarh when many wetlands (particularly rural ponds and irrigation tanks) degraded under an open-access regime, but were brought under a state or private or community management regime through appropriate changes in technical, institutional arrangements, and authority structure (Marothia, 1992a, 1992b, 1993b, 1995b, 1997a, 2002a, 2009b, 2010a, 2012a, 2019). As per draft of National Fisheries Policy, 2020 (https://nfdb.gov.in/PDF/National_Fisheries_Policy_2020.pdf), efforts will be made by states to enhance fish production and productivity through the application of technology and the formation of Farmers Producer Organizations (FPOs) to cater to small pond-holder's needs. The state is already following these policy guidelines. In recent years, fish culture in ponds and tanks has been managed under internal institutional structures of distributed governance or shared management systems in India including in the State of Chhattisgarh.[3] Fishermen communities/resource users groups and the state or local government (Panchayats) shared the responsibility of managing freshwater aquaculture by combining the skills of local resource users/local committees with technical, administrative, and financial resources available within the states. In Chhattisgarh, freshwater aquaculture has been largely managed under a cooperative governance structure – an internal institutional structure model of distributed governance system. The most important component of the policy document is to carry out fish farming in all the village ponds and irrigation tanks through FCSs in order of preference.[4] FCSs have been assigned usufruct rights to use MWs for culture fisheries subject to certain socio-economic and administrative conditions (Marothia, 2004c, 2006b, 2007, 2009b, 2012a, 2019). Under the Fisheries Policy of Government of Chhattisgarh, first preference has been/is being given to FCSs to lease village ponds/village irrigation tanks irrespective of the water-spread area of these water bodies. The village ponds/tanks can only be leased to fishermen's groups and individual fishermen if FCSs do not bid for leasing in the aqua cultural water bodies, or they do not exist in a particular village. Further, in order of priority, the fishermen's community is given first preference to lease out ponds/tanks.

Chhattisgarh state has a well-structured organizational network to manage freshwater aquaculture in village ponds, irrigation tanks, and reservoirs. The state has a two-tier cooperative structure to manage and develop freshwater aquaculture. At the top level, CSFCF is responsible for managing reservoirs, fish farms, and hatcheries for fisheries development. At the second level of the cooperative structure there is a three-tier Panchayat organizational setup (village Panchayat, janpad Panchayat, and district Panchayat) which assigns fishing rights or leases village ponds and irrigation tanks. The administrative and functional jurisdiction of village Panchayat, janpad Panchayat, and district Panchayat are restricted according to the water-spread area of ponds/tanks. Leases of ponds or tanks with a water-spread area of below 10 ha, between 10 ha and 100 ha, and between 100 ha and 200 ha is assigned by village Panchayat, janpad Panchayat and district Panchayat respectively. The SDF grants leases for all ponds/tanks with a water-spread area above 200 ha. The lease amount is used for promoting fisher's activities by Panchayats and the State Department of Fisheries (SDF).

The SDF is engaged in coordinating state and central government schemes, building the capacity of fishermen communities, and assisting cooperative societies in promoting fisheries' activities and fish production. A large number of ponds and tanks are used for fish culture by the FCSs in the state (see Table 8.1). The state has nearly 1,606 FCSs and 55,363 active members (GoCG, 2019–20a). All the FCSs listed under general categories have members from the fisheries community (locally known as Kewat, Dhimar, and Nishad). Scheduled tribe and caste FCSs also, by and large, have members from the same caste i.e. in scheduled-tribe FCSs, all members belong to homogeneous tribal castes. Similarly, all members of scheduled-caste FCSs are from scheduled castes. This categorization of FCSs clearly indicates that all the three types of FCSs have a homogeneous group of fishermen within them.

There are also socio-political reasons behind adopting homogeneous group/caste-based formations of FCSs in the State Fisheries Policy of Chhattisgarh.[5]

For FCS, FG, and individual fishers (IFs), the maximum water-spread area per member/person of village pond and irrigated tank is restricted to 0.50 ha and 4.00 ha respectively. However, ponds with 1 ha water-spread area can be leased to local fishermen. The lease duration for ponds/irrigation tanks is five years. The ponds/tanks can be leased to the same FCS, FG, and IF based on performance. In the case that FCS, FG, and IFs are not interested in leasing the ponds/tanks, they can be leased out to self-help groups from the local village. An FCS can be leased more than one pond/tank with the restricted norms of water-spread area per member. Similarly, based on the size of the water-spread area of ponds and tanks, leases can be assigned to more than one FCS. Village Panchayat, janpad Panchayat and district Panchayat advertise widely in villages and also issue notifications in local newspapers for the allotment of ponds/tanks on lease within their working zones. The representatives of SDF, the Fish Farmers Development Agency (FFDA), and Panchayat scrutinize the application before leasing the ponds/tanks to FCS, FG, and IF.

The lease amount for village ponds and irrigation tanks is currently fixed at Rs. 1,000 per ha and Rs. 240 per ha per year respectively, with provision for a ten percent increase after every two years. The lease amount can be deposited in three installments of 35 percent, 30 percent, 35 percent respectively during the same financial year.[6] The lower lease rent for irrigation tanks is due to restrictions imposed by the DWR on the use of feed and manure in all departmental tanks. The FCS, FG, and IF have to deposit lease money within a stipulated time. An interest rate of 2.5 per cent is charged for the late deposit of lease money. Leases can be cancelled after three months, in cases of non-payments of lease, after serving three notices. For seasonal irrigation tanks, the lease amount is fixed on the availability of water during a year, and fish production levels (maximum fish production level achieved during the last five years is taken into account). The State Fisheries Department has also developed a lease policy for urban water bodies (see New Fisheries Policy, 2019–20b, SDF, GoCG).

8.4 Database and Coverage

This study was confined to village ponds and irrigation tanks situated in Boriya Khurd, Barbanda, Serikhedi, and Kura of Raipur District. Two urban tanks located in Raipur city were also selected. These ponds and tanks have been used for culture fish farming by FCSs and IFs. The study area, in which the selected FCSs and IFs are located, represents the agro-climatic socio-economic conditions of Chhattisgarh's Central Plain (see Chapter 2 for other agro-climatic parameters).

In Chhattisgarh, as mentioned previously, the lease of village ponds/irrigation tanks is assigned under a three-tier structure of Panchayats to registered FCSs, FGs, and IFs in order of priority. The functional jurisdiction of a particular FCS is within an eight km radius. Furthermore, in FCSs, members should be from the fisheries communities (Kewat, Dhimar, and Nishad) as per the Fisheries Leasing Policy of GoCG, except in the STs and SCs FCSs. STs and SCs FCSs are functioning in the STs and SCs populated villages. As a result, FGs, IFs, and self-help groups (SHGs) have not effectively become part and parcel of culture fisheries in Chhattisgarh. Furthermore, their numbers are abysmally low. This is particularly true in the case of FGs and SHGs.

In view of these facts, this study was carried out to understand the performance of FCSs, and IFs in culture fish farming in village ponds and tanks, which are being administered and controlled by different agencies in the state of Chhattisgarh. After having a series of discussions with the SDF, water resources, agriculture, rural development, and Panchayats, to collect some baseline information, and interacting with cooperative societies' office holders and individual fishermen, a set of four questionnaires was designed and pre-tested to collect required information for FCSs, fishermen's households, village profiles, and the profile of tanks and ponds covered by respective FCSs and IFs. Five FCSs were selected for in-depth analysis. Out of five FCSs, four were located in the four villages (Boriya Khurd, Barbanda, Serikhedi, and Kura) of Dharsiwa Block of Raipur District and one is located in Raipur city (see Map 8.1). All four FCSs that operate in rural areas have usufruct rights over village water bodies, whereas the FCS located in Raipur city is engaged in leasing urban water bodies (see Maps 8.1–8.6 for the locations of water

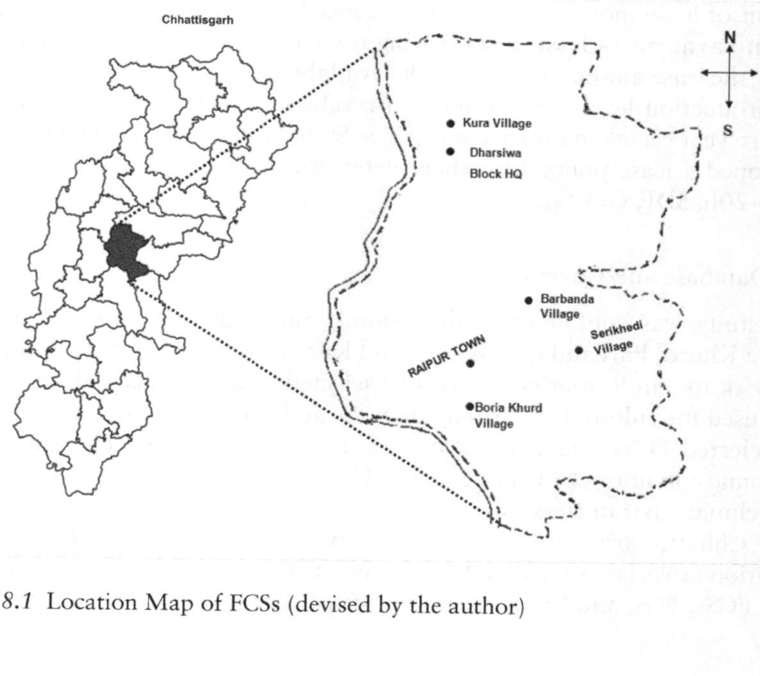

Map 8.1 Location Map of FCSs (devised by the author)

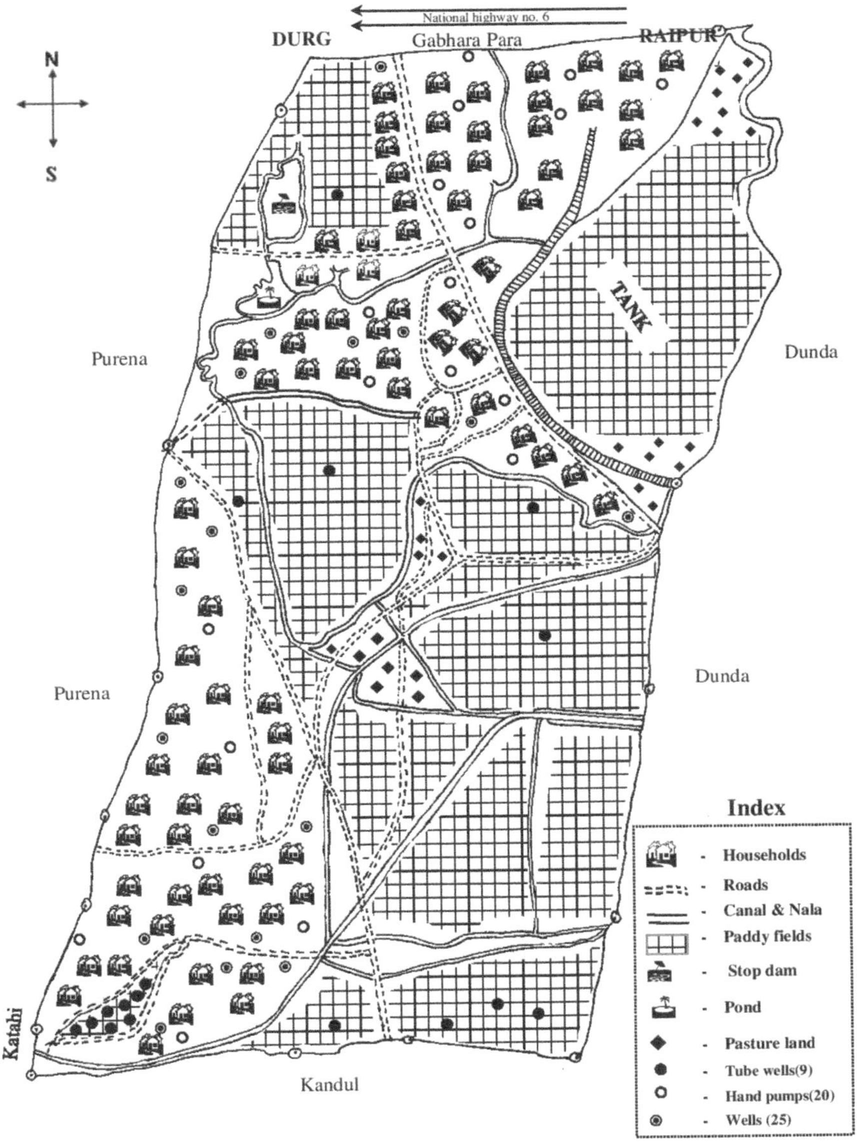

Map 8.2 Location Map of Tank Covered by Boriya Khurd FCSs (devised by the author)

bodies covered by the five FCSs, IFs, and SHGs). The FCSs and IFs selected cover different categories of ponds and tanks administered under different property rights regimes. We could not include FGs for our analysis, as there are none in the study area. Fisheries SHGs are no longer in the development

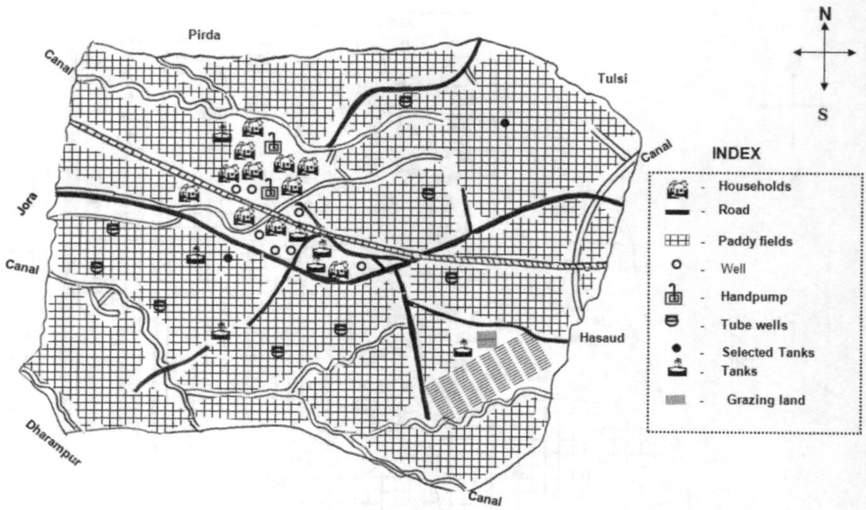

Map 8.3 Location Map of Ponds/Tanks Covered by Serikhedi FCSs (devised by the author)

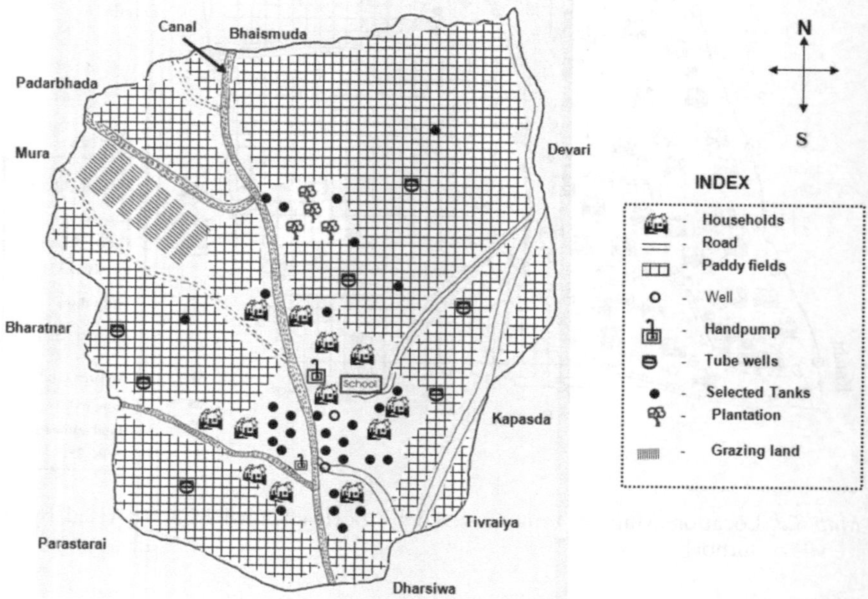

Map 8.4 Location Map of Ponds/Tanks Covered by Kura FCSs (devised by the author)

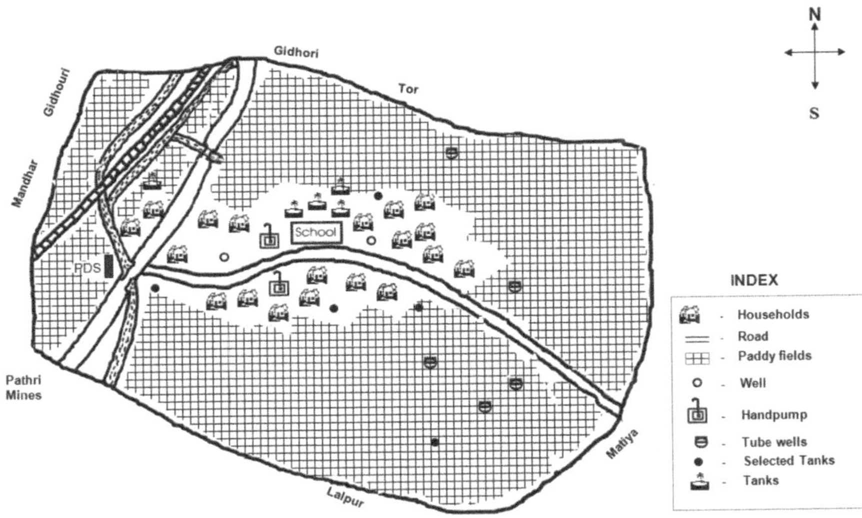

Map 8.5 Location Map of Ponds/Tanks Covered by Barbanda FCSs (devised by the author)

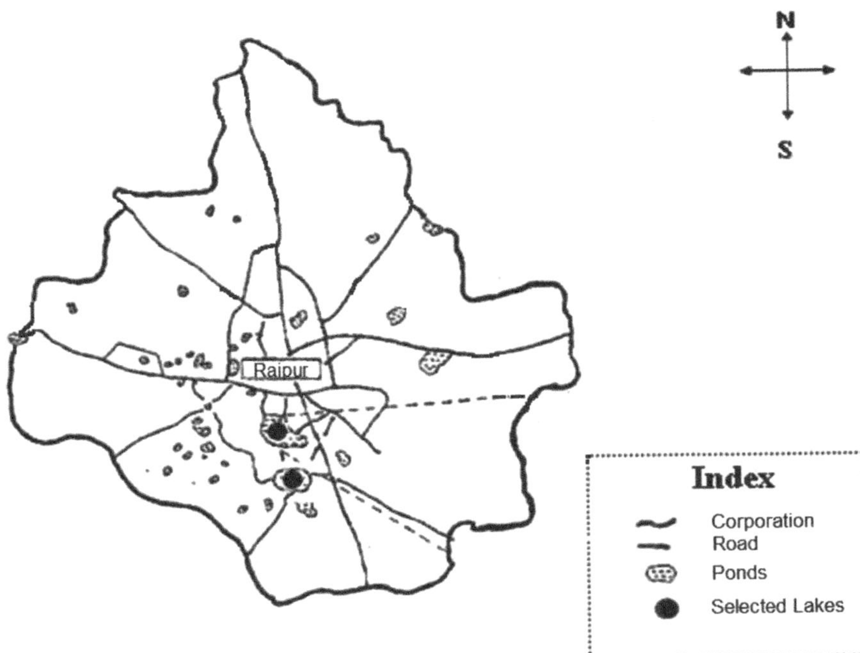

Map 8.6 Location Map of the Tank Covered by Mahamaipara FCSs (devised by the author)

portfolio of the Department of Panchayats and Rural Development, and the Department of Fisheries. However, four community ponds which were leased by individual fishermen from village Panchayats, and six privately owned ponds either leased out to fishermen or self-used for fish culture were included for comparative analysis.

All the tanks/ponds covered by the five FCSs and IFs are multiuse and multifunctional in nature, involving multiple stakeholders. The four primary FCSs are known by their respective village names. The fifth FCS covered two urban tanks. Information regarding the profile of FCSs and IFs (including salient features of management structures, decision-making arrangements, output, and distributing gains), physical and technical attributes of ponds and tanks, and general characteristics of fishermen's households was collected from respective FCSs and IFs. Information about fish yields and input use, disposal patterns, and distributive gains were collected for three years from the records of the FCSs. However, this information was only available for one year for pond/tanks leased by IFs from village Panchayats or private owners or self-managed ponds. The information required to analyse the performance of FSCs and IFs, and was gathered from all five FCSs and IFs during June-August 2004. Audit reports of the FCSs were the main source of information on yields, income, input use, disposal patterns, and distributive mechanisms. Data from audit reports were cross checked with information gathered during focused discussion sessions with members of the FCSs. The basic information regarding the physical and technical characteristics of irrigation tanks was collected from the DWR and soil and water conservation wing of the SDA. Data regarding ponds/tanks constructed during ex-zamindars (landlord) tenure were collected from village Panchayats. Information for urban tanks was collected from Raipur Municipal Corporation Office. To analyse the strengths and weaknesses of cooperative governance structures primarily adopted in the state of Chhattisgarh for the sustainable use of tanks/ponds for fish culture, an institutional framework, developed in Chapter 2, was applied. A brief description of types of ponds/tanks, management, ownership, and leasing authority structure of the selected water bodies is provided.

8.5 Categories of MWs and Management Systems

The details of types of ponds/tanks used for culture fisheries along with ownership, management, and leasing authority structure are given in Table 8.2. Although this table is self-explanatory, a brief description of the ponds/tanks leased out by village, janpad, and district Panchayats to FCSs and IFs and by private owners to fishermen or used by themselves is given in the following paragraphs.

Table 8.2 Types of Ponds/Tanks, Management, Ownership, and Leasing Authority Structure of the Sample Multiuse Wetlands

Name of Tank	*Type of Tank*	*Source of Water*	*Availability of Water*	*Major Uses of Water in Order of Priority*	*Local Institutions Involved in Water Uses*	*Ownership*	*Leasing Authority*	*Lease Rent (Rs./HWSA)*	*Restriction/ Conditions to Use Water for Fish Culture*
Village ponds/tanks leased in by FCS									
Boriya Khurd (Boriya Khurd FCS)	Perennial (65 ha.)	Canal-fed	Adequate, year round	Irrigation, domestic use, fish culture	FCS, WUAs, Panchayat	DWR	Janpad Panchayat	240.00	Feed and manure can't be used by FCS, other uses can't be excluded from using water
Bundha Tank (Serikhedi FCS)	Perennial (200 ha.)	Canal-fed	Adequate, year round	Irrigation, domestic use, fish culture	FCS, WUAs, Panchayat	DWR	District Panchayat	243.00	Feed and manure can't be used by FCS, other uses can't be excluded from using water
Raipuriha Pond (Serikhedi FCS)	Perennial (0.60 ha.)	Rainfed	Up to Feb.-March	Domestic and fish culture	Village Panchayat	Village Panchayat	Village Panchayat	2000.00	No restriction on use of fish feed
Kura Ponds (Kura FCS)	Perennial (8)* (1.15 ha.)	Canal-fed	Throughout the year	Irrigation, domestic use, fish culture	Krishi Samiti of Panchayat, Panchayat, FCS	Village Panchayat	Village Panchayat	252.00	No restriction on use of fish feed
Kura Ponds (Kura FCS)	Seasonal (9)* (1.15 ha.)	Canal-fed	Up to Feb.-March	Irrigation, domestic use, fish culture	Krishi Samiti of Panchayat, Panchayat, FCS	Village Panchayat	Village Panchayat	252.00	No restriction on use of fish feed
Barbanda Tank-1 (Barbanda FCS)	Perennial (2.88 ha.)	Canal-fed	Up to Feb.-March	Irrigation, fish culture, domestic use	Panchayat	Panchayat	Village Panchayat	857.00	Restriction on feed and manure
Barbanda Tank-2 (Barbanda FCS)	Perennial (2.30 ha.)	Canal-fed	Up to Feb-March	Irrigation, fish culture, domestic use	Panchayat, Krishi Samiti	Panchayat	Village Panchayat	857.00	Restriction on feed and manure

(*Continued*)

Table 8.2 (Continued)

Name of tank	*Type of tank*	*Source of water*	*Availability of water*	*Major uses of water in order of priority*	*Local Institutions involved in water uses*	*Ownership*	*Leasing authority*	*Lease rent (Rs./HWSA)*	*Restriction/ Conditions to use water for fish culture*
Barbanda Tank-3 (Barbanda FCS)	Perennial (17 ha.)	Canal-fed	Up to Feb.-March	Irrigation, fish culture, domestic use	Panchayat	Panchayat	Janpad Panchayat	857.00	Restriction on feed and manure
Matwa Talab (Barbanda FCS)	Perennial (0.40 ha.)	Rainfed	Up to December	Domestic use, fish culture, Irrigation,	Village Panchayat, FCS	Village Panchayat	Village Panchayat	2600.00	Limited restrictions
Dev Talab (Barbanda FCS)	Perennial (0.90 ha.)	Canal-fed	Adequate	Domestic use, fish culture, Irrigation	Village Panchayat, FCS	Village Panchayat	Village Panchayat	2600.00	Limited restrictions
Urban tanks leased in by FCS									
Budha Talab	Perennial (28.75 ha.)	Rainfed	Adequate through the year	Domestic and fish culture	Municipal Corporation, FCS	Raipur, Municipal Corporation	Raipur, Municipal Corporation	2504.00	No restriction on fish feed
Maharajbandh Talab	Perennial (21.45 ha.)	Rainfed	Adequate through the year	Domestic and fish culture	Municipal Corporation, FCS	Raipur, Municipal Corporation	Raipur, Municipal Corporation	2504.00	No restriction on fish feed
Village Owned Ponds Leased in by IF									
Boriya Khurd Pond (Boriya Khurd village)	Rainfed (0.60 ha.)	Tubewell	Up to March	Domestic use and fish culture	Panchayat	Panchayat & individual fishermen	Panchayat	10000.00	No restriction on fish feed and manure application
Chhapar Pond (Barbanda village)	Rainfed (3.50 ha.)	Canal	Up to December	Irrigation, fisheries, and domestic use	Panchayat, Krishi Samitee, FCS	Panchayat	Panchayat	4928.00	No restriction

Jogi Bandh (Barbanda village)	Seasonal (0.40 ha)	Canal, Rain	Up to December	Fish culture, irrigation and domestic	Panchayat Krishi Samitee	Panchayat	Panchayat	2600.00	No restriction
Ledara pond (Barbanda village)	Perennial (4.50 ha)	Canal, Rain	Up to Feb. March	Fish culture, irrigation and domestic	Panchayat Krishi Samitee	Panchayat	Panchayat	2600.00	No restriction
Privately Owned Ponds (POP)									
Mandal Pond (Serikhedi village)	Perennial (1.60 ha.)	Rainfed	Adequate water for fish culture up to February	Fish production	Private	Private	Private (Lease out to fishermen)	5625.00	No restriction
Mandir Pond (Serikhedi village)	Perennial (2.40 ha.)	Rainfed	Up to February.	Fish and domestic use	Private	Private	Self-use	Self-owned	No restriction
Marar Pond (Serikhedi village)	Perennial (1.80 ha.)	Rainfed	Up to December	Fish and domestic use	Private	Private	Self-use	Self-owned	No restriction
Dhobani Pond (Barbanda village)	Perennial (0.70 ha.)	Rainfed	Up to December	Fish and family use	Private	Private	Self-use	Self-owned	No restriction
Gadai Pond (Barbanda village)	Perennial (0.90 ha.)	Rainfed	Up to December	Fish and family use	Private	Private	Self-use	Self-owned	No restriction

* Kura FCS leased 29 ponds (12 perennial and 17 seasonal); eight perennial and nine seasonal were used for fish production. Two seasonal ponds were used for nursery hatching. Figures in brackets indicate the water-spread area of respective tanks. HWSA – water-spread area in hectares. (Author's analysis)

8.5.1 Ponds and Irrigation Tanks Used by FCSs

A brief mention of five FCSs is made here along with the types of ponds and irrigation tanks leased by them from different levels of Phayats.

(i) **Boriya Khurd FCS** has leased one canal-fed, perennial irrigation tank of 65 ha water-spread area. The tank is situated at the outer periphery of Boriya Khurd village. The ownership of the tank is with the DWR. The irrigation tank is the major source of water for farmers for irrigating crops in kharif and rabi seasons and fish culture, as well as catering to the domestic needs of the village population. Usufruct rights have been assigned to the village FCS under the three-tier Panchayat structure for leasing out village water bodies with the understanding that other users cannot be excluded, and the use of fish for feed and manures is prohibited. The irrigation tank has been leased out by the Dharsiwa Janpad Panchayat to the FCS of Boriya Khurd village. The FCS has been engaged since 1966 in culture fisheries in the Boriya Khurd tank under the usufruct rights arrangement with DWR.

(ii) **Serikhedi FCS** covers one perennial, canal-fed irrigation tank and one perennial, rainfed pond. The ownership of the canal-fed irrigation tank (200 ha water-spread area) is with DWR while the perennial rainfed tank (0.60 ha water-spread area), constructed by ex-zamindar (landlord), is being managed by the Panchayat. District and village Panchayats are the leasing authorities for the perennial, canal-fed irrigation tank and perennial, rainfed pond respectively. The FCS has been actively engaged in both water bodies for fish culture since 1996. In the case of the canal-fed irrigation tank, the use of fish feed and manure is restricted. However, such restrictions are not imposed in the case of Panchayat pond (Raipuriha pond).

(iii) **Kura FCS** covers 29 village ponds with different water-spread areas. These ponds were constructed by ex-zamindars and are now managed by the Panchayat. Out of 29 ponds, eight perennial and nine seasonal ponds (from these, two are used for nursery hatching) are used for fish culture. The water-spread area of these ponds varies between 0.15 and 5.50 ha. The FCS has been working since 1958. The village Panchayat is the leasing authority. There is no restriction on using feed and manure in the ponds.

(iv) **Barbanda FCS** covers three canal-fed micro-minor irrigation tanks (MMITs), which were constructed by the soil and water conservation wing of the SDA and two ponds constructed by ex-zamindars. These tanks are being managed and leased out by village Panchayats. In the case of MMITs, there is a restriction on the use of feed and manure, whereas limited restriction is imposed on the other two ponds. The water-spread area ranges between 0.4 ha. to 17 ha. The FCS has been involved in fish culture since 1983–84.

(v) **Mahamaipara FCS** covers two perennial urban tanks located in Raipur city. The water-spread area of the two tanks is 28.75 ha (Lake Vivekanada/Budha Talab) and 21.45 ha (Maharajbandh Talab) respectively. The FCS has been working since 1947. The Raipur Municipal Corporation owns these tanks as well as the leasing authority. There is no restriction on the use of fish feed and manure in the tanks.

8.5.2 Village Panchayat-Owned Ponds Used by IFs

Village Panchayats have leased out one community pond in Boriya Khurd and three in Barbanda to IFs. Usufruct rights of community ponds have been assigned by the village Panchayats to individual fishermen for fish culture. There is no restriction on the use of fish feed, medicine, and manure in these ponds. Boriya Khurd village Panchayat provides water through tube wells to the lessee, in case the level of the pond's water drops below the agreed level for fish culture. In the case of Barbanda village, IFs leased community tanks with the complete knowledge that these tanks are seasonal in nature but receive water from the canal. While Boriya Khurd common pond is used for limited domestic purposes as well, all three ponds of Barbanda are being used for fish culture, domestic use, and one occasional kharif irrigation (refer Table 8.2).

8.5.3 Privately Owned Ponds (POP) Used by IFs and Owner

In the study area, a few ponds were owned by private landowners. In Serikhedi village, out of three POP, one pond was leased out to local fishermen and the other two were used by the owners themselves. In Barbanda, all the three ponds were owned by a single landowner and leased out to one local fisherman.

Fish production from all three ponds were shared in equal proportion. Input cost is borne by lessees of ponds. To this end, we discuss the comparative performance of FCSs and IFs within their respective categories.

8.6 Comparative Performance of FCSs

In this section, the salient features of FCSs, the characteristics of fishermen's households, the physical and technical attributes of ponds and tanks under the control of a particular FCS, and an analysis of comparative performance of selected five FCSs is presented.

8.6.1 A Profile of FCSs and Attributes of Fisher Households

The salient features and management structure of the selected FCSs is shown in Table 8.3. Mahamaipara FCS is the oldest cooperative society with the largest number of members, followed by Kura cooperative society.

Table 8.3 A Profile of Selected FCSs

Particulars	*Boriya Khurd*	*Serikhedi*	*Kura*	*Barbanda*	*Mahamaipara*
Name of the society	Prathmik Matsya Sahakari Samiti Maryadit, Boriya Khurd	Primary FCS, Serikhedi	Matsya Prathmik Sahakari Samiti Kura	Machhua Sahakari Samiti, Barbanda	Primary FCS, Mahamaipara, Raipur
Year of establishment	1966	1996	1958	1983–84	1947
Primary objectives of the society	To enhance livelihood through fish culture	To enhance livelihood through fish culture	To generate employment through fish culture for the member of FCS	To generate employment for members by fish production	To generate employment for members by fish production
Leasing Authority	Three-tier Panchayat institutions systems	Three-tier Panchayat institutions systems	Three-tier Panchayat institutions systems	Three-tier Panchayat institutions systems	Raipur Municipal Corporation
No. of members	33	12	70	15	464
Social structure of members	Fisher community (Kewat, Dhimar, Nishad)	Fisher community (Kewat, Dhimar)	Fisher community (Kewat, Dhimar)	Fisher community (Kewat, Dhimar, Nishad)	Fisher community (Nishad, Dhimar)
Membership fees at the time of joining	Rs. 11.00 per member at the time of formation of society	Rs. 51.00 per member at the time of formation of society	Rs. 20.00 per member at the time of formation of society	Rs. 51.00 per member at the time of formation of society	Rs. 11.00 per member at the time of formation of society Currently Rs. 51.00
Membership criteria	Only the fisher community, above 18 years, transferred to family members after the death	Only the fisher community, above 18 years	Only the fisher above 18 years	Only the fisher community, above 18 years, below the poverty line	Only the fisher community, above 18 years

Working capital (Rs.)					
(a) Members share capital	1300.00	1320	2025	-	17877
(b) Borrowed capital					
(i) From members and money lenders (for purchase of seed etc.)	19216	-	8548	7000	-
(ii) From District Central Cooperative Bank	50000 (for purchase of nets, boats, ropes, repair of tanks etc.)		28428	-	4900 (for purchase of nets, sis etc.)
(c) Saving account deposit	2530	10509	27495	3000	203170
Total	73046	11829	66596	10000	225947
Assets					
(a) Boat	1	-	-	-	-
(b) Net	4	Small-6	Big = 11; Small = 6	Small-2	5
(c) Others (Rope)	40 kg.	5 kg.	-	20 kg.	-
(d) Land	-	-	-	-	0.5 acre
(e) Office building	-	-	1	-	
management committee (a) President/treasurer					
(b) Vice-president	1	1	1	1	1
(c) Secretary	1	1	1	1	1
(d) Executive members	5	2	6	4	5
(e) Treasurer	-	-	1	-	-
No. of ponds/tank leased for fish culture	1	2	29	5	2
Duration of lease (yr)	5	5	5	5	5
Lease rent per year ha^{-1}. (Rs.) with 10% increase every year on the basic lease amount	240	242.72 for Bandha talab and Rs. 2000 for Raipuriha talab	251.42	856.55	2503.95

* Sub-castes of fisher community in Chhattisgarh. Kewat sub-caste dominates the fishing profession. (Author's analysis)

The Barbandha cooperative society came into existence during the FFDA Programme. Boriya Khurd and Serikhedi societies started culture fish practices in 1996. The basic objective of all the FCSs is to enhance livelihoods and generate year-round employment. In terms of the social structure of members, membership criteria, management system, and external institutional support, all the five FCSs have a more or less similar pattern. All the members of the society belong to the fisher community, with predominant sub-castes according to their population in a particular village. For example, in rural FCSs, Kewat and Dhimar sub-castes dominate membership, whereas in urban FCS a large number of members are from Nishad sub-caste. Members from these sub-castes, based on their numbers in a particular FCS, are predominant in the management and executive committees. Sometimes, small, hidden groups are formed on the basis of these sub-castes to control the decisions of FCS. The president, vice-president, secretary, and executive members are elected by the members. Except in Kura FCS, the president also works as treasurer. All FCSs are male-dominated. The details regarding membership fees, working capital, assets, lease periods, and rents are given in Table 8.3.

A visible difference can be observed with respect to these parameters and the difference is basically associated with ownership, leasing authority, size, and water-spread area of the ponds and tanks leased by a particular FCS. The lease rents for common rural and urban ponds are many times higher than irrigation tanks. This is due to the fact that the application of fish feed, manure, and medicine is prohibited in irrigation tanks (see also Table 8.2 for related features of FCSs).

The general characteristics of fishermen's households (FHs) covered under the five FCSs are given in Table 8.4. As mentioned previously, according to the GoCG Fisheries Policy, members of the FCSs should be from the fishermen community and, due to this policy, all members in a FCS belong to the fishermen community. This policy decision had minimized or almost eliminated inter-caste conflict in the States of MP and Chhattisgarh. Agriculture and wage earnings are the other sources of FHs' income. The majority of the FHs are illiterate and fall into the age group of 18–50 years old. A large proportion of the economically active/adult fishermen are engaged in freshwater activities in the target areas of respective FCSs. Most of the fisheries' activities are carried out by male members of FHs.

However, in the case of Serikhedi and Mahamaipara FCSs, female members of fishermen households were also involved in stocking, netting, and marketing activities. Except in Mahamaipara FCS, members do not have any affiliation with political parties. This is due to the interest of some politically strong leaders in lease matters concerning the large urban water bodies. This is particularly true in the case of Budha talab and Maharajbandh talab. A few members of FCSs have leased village ponds on an individual basis from village Panchayats. We will discuss this aspect further in the section on the performance of IFs using individual ponds for fish culture.

Table 8.4 General Characteristics of Fisher Households in FCSs

Particulars	*Boriya Khurd*	*Serikhedi*	*Kura*	*Barbanda*	*Mahamaipara*
1. Number of fisher households in village	45	15	205	45	588
2. Number of fisher households in FCs	18	7	66	14	374
3. Occupation					
a. Main: Fisheries	100	100	100	100	100
b. Supplementary					
i. Agriculture*	33.33 (1.80)	-	57.14 (9.60)	14.29 (1.20)	1.87 (2.60)
ii. Wage earner	66.67	100	37.14	85.71	97.59
iii. Service	-	-	-	-	0.53
4. Yearly income per member (Rs.)	11153	13593	6491	7807	10241
a. Farm income (%)	19.37	-	50.42	23.71	1.48
b. Income from agricultural wages (%)	-	33.1	18.21	49.40	-
c. Service (%)	-	-		-	3.13
d. Other income (%)	28.69	-		-	45.74
e. Fisheries	51.94	66.89	31.37	26.89	49.64
5. Educational background (%)					
a. Illiterate	66.67	100	75.76	78.57	80.48
b. Up to 5th class (up to primary level)	33.33	-	6.06	14.29	9.09
c. 5–8th class (up to middle)	-	-	12.12	7.14	6.95
d. 8–12th class (up to H.S.S.C)	-	-	6.06	-	2.4
e. Up to college level	-	-	-	-	1.07
6. Age (% of total members)					
a. Above 18–30 years	60.60	41.67	18.57	35.71	56.68
b. 30–50 years	30.3	58.33	61.43	64.29	36.21
c. 50–70 9.09	-	20	7.14	7.21	
7. Fisher population (in village)	315	83	1643	203	2244
a. Male (%)	60.95	54.22	51.49	66.5	66.67
b. Female (%)	39.05	45.78	48.52	33.5	33.33

(Continued)

Table 8.4 (Continued)

8. Family size	9	7	10	6	7
i. Males	3	2	3	2	2
ii. Females	2	2	3	1	2
iii. Children	4	3	4	3	4
9. Number of economically active members in the village	152	37	820	45	947
a. Male (%)	70.39	81.08	75	100	76.77
b. Female (%)	29.61	18.92	25	-	23.23
10. % of Adult family members working full time in village (male)	28.57	36.14	24.95	22.17	16.67
11. Active fishermen (no. per/family)	2	2	1	1	2
12. Male-dominated fisheries' activities with allocation of time (days)	Pond preparation 1d; stocking 4d; manuring 4d; netting & selling 79d	Stocking 2d; manuring 2d; netting & selling 78d	Stocking 4d; manuring 4d; netting & selling 64d	-	Pond preparation 1d; stocking 6d; netting & selling 72d
13. Female-dominated activities with time allocation (days)	-	Stocking 2d; manuring 2d; netting & selling 19d	-	-	Pond preparation 1d; stocking 6d; netting 17d
14. Recipient of training from SDF (duration) – 2 weeks to 10 months	-		66 households, four times in Dharsiwa block		
15. % of license holders for selling fish seed	5.55	- -	- -	-	
16. Political affiliation	-	-	-	- -	
a. % politically affiliated	-	- -	- -	- -	11.5
b. % non-politically affiliated	-	- -	- -	- -	88.5

Source: Author's analysis based on field work

8.6.2 Physical and Technical Attributes of MWs

The physical and technical characteristics of ponds and irrigation tanks covered by FCSs are presented in Tables 8.5–8.11. Some other features related to categories of irrigation tanks, village common ponds, the duration of water availability, multiple institutions involved in water use and management, ownership and leasing authority structure, variations in lease rents,

Table 8.5 Physical and Technical Attributes of Boriya Khurd Irrigation Tanks Managed by DWR

S.N.	*Attributes*	*Particulars*
1.	Year of construction	1980
2.	Construction cost (Rs. in lakhs)	16.4
3.	Source of water	Canal-fed
4.	Gross-storage capacity (mcft)	34.34
5.	Catchment area (sq. mile)	2
6.	Total command area (ha.)	385.2
7.	Total cultivable area (ha.)	296
8.	Total area under cultivation	
	a). Kharif (ha.)	266
	b). Rabi (ha.)	43.2
9.	Net cropped area (ha.)	309.2
10.	Irrigated area (ha.)	250.4
11.	Mean monsoon yield (mcft)	84.34
12.	Dead-storage capacity (mcft)	2.81
13.	Live-storage capacity (mcft)	31.53
14.	Full tank level (FTL) (mtr.)	97
15.	Maximum tank level (MTL) (mtr.)	98
16.	Top bank level (TBL) (mtr.)	103
17.	Lowest silt level (LSL) (mtr.)	85
18.	Water-spread area	
	a). Full tank level (FTL) (ha.)	65
	b). Maximum tank level (MTL) (ha.)	93.83
19.	Length of dam (earth (ft.)	5247
20.	Length of main canal chain (mtr.)	133
21.	Head discharge (cusec)	18
22.	No. of villages under command area	4
23.	No. of sluice gates	1
24.	Soil type of the tank	Loamy clay
25.	Soil type of command area	Sandy loam
26.	External rate of silting (%)	2
27.	Estimated life of tank (yrs.)	100
28.	No. of waterings	
	a). Kharif	3
	b). Rabi	1
29.	Irrigation charges (Rs./ha.)	
	a). Kharif	230
	b). Rabi	600
30.	Leasing authority	Janpad Panchayat

Table 8.6 Physical and Technical Attributes of Serikhedi Irrigation Tank Managed by DWR

S.N.	*Attributes*	*Particulars*
1.	Year of construction	1919
2.	Construction cost (Rs. in lakhs)	0.43
3.	Source of water	Canal-fed perennial
4.	Gross-storage capacity (mcft)	45.93
5.	Catchment area (sq.mile)	0.53
6.	Total command area (ha.)	480
7.	Total cultivable area (ha.)	524.8
8.	Total area under cultivation	
	a). Kharif (ha.)	480
	b). Rabi (ha.)	44.8
9.	Net cropped area (ha.)	480
10.	Irrigated area (ha.)	480
11.	Mean monsoon yield (mcft)	NA
12.	Dead-storage capacity (mcft)	0.3
13.	Live-storage capacity (mcft)	45.93
14.	Full tank level (FTL) (mtr.)	977.5
15.	Maximum tank level (MTL) (mtr.)	978
16.	Top bank level (TBL) (mtr.)	982
17.	Lowest silt level (LSL) (mtr.)	964
18.	Water-spread area	
	a). Full tank level (FTL) (ha.)	200
	b). Maximum tank level (MTL) (ha.)	240
19.	Length of dam (earth (ft.)	3500
20.	Length of main canal chain (mtr.)	234
21.	Head discharge (cusec)	16
22.	No. of villages under command area	3
23.	No. of sluice gates	1
24.	Soil type of the tank	Loamy clay
25.	Soil type of command area	Sandy loam
26.	External rate of silting (%)	3
27.	Estimated life of tank (yrs.)	100
28.	No. of waterings	
	a). Kharif	3
	b). Rabi	1
29.	Irrigation charges (Rs./ha.)	
	a). Kharif	227
	b). Rabi	600
30.	Leasing authority	District Panchayat

number of ponds to cater to the domestic needs of the villagers. For nursery hatching, two seasonal tanks have been used. The remaining 12 tanks have not been used for fish production due to inadequate water availability as these tanks are also being used for irrigation and other activities (Table 8.8). Kura FCS have been granted usufruct rights for fish culture in these common water bodies. The water is mainly shared between farmers and fishermen under well-defined water-sharing arrangements. Barbanda FCS have leased

Table 8.7 Physical and Technical Attributes of Raipuriha Pond of Serikhedi Village Managed by Village Panchayat

Attributes	*Particulars*
Location	2.5 km. from Panchayat office
Area (ha.)	0.69
Source of water	Rainfed perennial
Water-spread area (ha.)	0.60
Constructed by	Ex-Zamindar
Original ownership	Common
Current ownership	Village Panchayat
Management system	Panchayat
Current status of tank	Perennial
Multiple-uses of tank	Fish production, bathing, washing clothes, and cattle tending
Fish production (qtl./year)	7.20
Number of fishermen households	7
Social structure of beneficiaries	Dhimar, Kewat
Leasing authority	Panchayat

Table 8.8 Physical and Technical Attributes of Ponds of Kura Village Managed by Village Panchayat

Attributes	*Particulars*
Original property rights regime	Private (Zamindars)
Current property rights regime	Common property owned by village Panchayat
Purpose of construction of tanks	For domestic use & drinking water for humans & cattle, irrigation
Maintenance of tank	By village Panchayat & FCS
Management system of tanks	By Krishi Samiti of Panchayat & FCS
Source of water	Rain & canal
Current status of tanks	
(i) Water quality	No aquatic weeds, clean, but undrinkable
(ii) Maintenance	Good
(iii) Bund condition	Good
(iv) Plantation	Local palm, mango, mahua, banyan, pipal, ganga imli & tamarind
Total no. of tanks	29
a. Perennial	12
b. Seasonal	17
c. Percentage of total tanks having multiple water uses	
i) Fishing	59%
ii) Irrigation	55%
iii) Tending cattle	72%
iv) Bathing	45%
v) Washing clothes	45%

(*Continued*)

Table 8.8 (Continued)

Attributes	*Particulars*
Leased in tanks (No.)	29
No. of tanks used for fish Production	
a. Perennial	8
b. Seasonal	7
c. Total	15
No. of tanks used for nursery hatching	
a. Perennial	-
b. Seasonal	2
c. Total	2
No. of tanks not using for fish production	
a. Perennial	6
b. Seasonal	6
c. Total	12
Average size of tanks (ha.)	1.87
Average water-spread area (ha.)	1.15
Distribution of tanks according to size	
a. up to 0.5 ha.	4
b. 0.5 to 2 ha.	17
c. above 2 ha.	8
Average tank area per fishermen household (ha.)	0.77
Number of fishermen household	70
Social structure of beneficiaries	Fisherfolk community
Average fish production per tank (kg)	21
Leasing authority	Panchayat
Fees for water use (irrigation fees Rs.ha^{-1})	230
Penalty for non-payment of fees	Rs. 230 with interest

Table 8.9 Physical and Technical Attributes of Soil and Water Conservation Tanks of Barbanda Village Managed by Village Panchayat

Attributes	*Tank-1*	*Tank-2*	*Tank-3*
Year of construction	1980	1981	1983
Construction cost (Rs. Lakh)	1.56	0.53	1.42
Source of water	Canal	Canal	Canal
Water-spread area (ha)	2.8	2.3	17
Gross-storage capacity (Mcft)	1	0.9	6.1
Catchment area (Sq. mile)	0.36	0.1	0.3
Total command area (ha)	48	14	40
Total cultivable area (ha)	192	56	160
Total area under cultivation (ha)			
a. Kharif	169	49	140
b. Rabi	23	7	20
Net cropped area (ha)	169	49.00	140
Irrigated area (ha) in 2002–03	48	11.64	40
Dead-storage capacity (Mcft)	0.16	0.15	1.01
Live-storage capacity (Mcft)	0.83	0.75	5.09
Number of villages under command area	1	1	1

Attributes	*Tank-1*	*Tank-2*	*Tank-3*
Number of sluice gates	1	1	1
Soil type of the tank	Kanhar	Kanhar	Kanhar
Soil type of command area	Kanhar & Matasi	Kanhar & Matasi	Kanhar & Matasi
External rate of silting	2	2	2
Estimated life of tank	100	100	100
Number of waterings			
a. Kharif	2	2	2
b. Rabi	1	1	1
Irrigation charges (Rs./ha)			
a. Kharif	230	230	230
b. Rabi	600	600	600
No. of cultivators			
a. SC	11	01	06
b. ST	01	-	-
c. Others	44	20	44
d. Total	56	21	50
Irrigated area (ha) in			
1979–80	-	-	-
1980–81	-	-	-
1981–82	40	9.60	-
1982–83	40	12.40	39.20
1983–84	40	13.20	40.80
Leasing authority	Panchayat	Panchayat	Janpad Panchayat

Table 8.10 Physical and Technical Attributes of Ponds of Barbanda Village Managed by Village Panchayat

Attributes	Matwa Talab	Dev Talab
Area (ha.)	0.903	1.522
Water-spread area (ha.)	0.400	0.900
Constructed by	Ex-Zamindar	Ex-Zamindar
Original ownership	Private	Private
Current ownership	Panchayat	Panchayat
Catchment area (ha)	3.20	4.00
Command area (ha)	8.00	8.00
Management system	Panchayat	Panchayat
Current status of tank	Perennial	Perennial
Source of water	Rainfed	Canal fed
Number of fillings	-	1
Multiple uses of tank	Bathing, washing clothes, cattle tending, fish production, and irrigation	Bathing, washing clothes, cattle tending, fish production, and irrigation
Fish production (qtl. per year)	1.83	4.12

(*Continued*)

Table 8.10 (Continued)

Attributes	Matwa Talab	Dev Talab
Number of fisher households	14	14
Social structure of beneficiaries	Homogeneous	Homogeneous
Rules and regulation for water use	First preference to irrigation	First preference to irrigation
Fees for water use (Rs. per ha)	230; only for kharif	Rs. 230; only for kharif
Penalty for non-payment of fees	Irrigation fees with interest	Irrigation fees with interest
Leasing authority	Panchayat	Panchayat

Table 8.11 Physical and Technical Attributes of Ponds of Tanks of Mahamaipara, Raipur Town Managed by FCS

Attributes	*Budha Talab/LV*	*Maharajbandh talab*
Original property right	Private (King)	Private (King)
Current property right	State Government	State Government
Source of water	Rainfed	Rainfed
Current status of tank	Perennial	Perennial
Pond area (ha)	30.250	23.050
Water-spread area (ha)	28.750	21.450
Water-holding capacity (m^3)	8.900	5.610
Multiple uses of tank	Fish production, bathing washing clothes, park, and water recreation	Fish production, bathing, washing clothes, cattle tending
Number of fisher households	374	374
Social structure of beneficiaries	Homogeneous	Homogeneous
Management system	Municipal Corporation, Raipur	Municipal Corporation, Raipur
Fish production (x 100 kg) in 2002–03	887.56 (30.87 per ha.)	697.37 (32.51 per ha.)
Year of construction	1401 BC	1850
Income to government (Rs. per year)	90,000	54,000
Lake pollution monitoring authority	Chhattisgarh Pollution Control Board	Chhattisgarh Pollution Control Board
Users of pond (no. per day)		
a. Fisheries	140	150
b. Bathing	1500	1200
c. Washing (washerman)	25	35

Attributes	*Budha Talab/LV*	*Maharajbandh talab*
d. Cattle tending	-	40 (Buffaloes)
e. Recreational (excluding rainy days)		
i. Boating	50	-
ii. Park visit	150	-
iii. Swimming	25	-
Leasing authority	Municipal Corporation, Raipur	Municipal Corporation, Raipur

and restrictions on the use of feed, manure, and medicines to cure or prevent fish diseases are also provided in Table 8.3. Thus, information presented in Tables 8.5–8.11 needs to be viewed in relation to the details given in Table 8.3 in order to understand the importance of varIous attributes.

The information regarding year of construction, construction cost, gross-storage capacity with dead-storage capacity and live-storage capacity, catchment area, command area, cultivable and irrigated areas, water-spread area, length of dam earth, length of main canal, head discharge, number of villages under command area, and irrigation fees/charges indicate the physical and technical parameters of the Boriya Khurd Irrigation Tank (BKIT) and the Serikhedi Irrigation Tank (SIT) (see Tables 8.5 and 8.6). The institutional parameters of BKIT and SIT are given in Table 8.3. The DWR owns BKIT and SIT and the fishing rights under the three-tier Panchayat system has been assigned to Boriya Khrud and Serikhedi primary FCSs respectively. Janpad and district Panchayats are the leasing authorities for BKIT and SIT respectively. The lease rents for these tanks are much lower than other Panchayat-owned ponds in the same villages. Lower lease rents are due to restrictions on the use of fish feed. Both of these tanks are canal-fed and the minimum water level required for fish culture is available throughout the year (see Tables 8.3, 8.5–8.6). However, there is a restriction on adding manure and fertilizer as feed input for the fish culture. Both these tanks also provide irrigation water to adjoining villages. Water from these tanks is provided for irrigation of kharif paddy and one is used to irrigate rabi crops. However, the recovery rate of irrigation fees is quite low. The distribution of irrigation water is managed under participatory irrigation management by the WUAs. Serikhedi FCS also leased a village common pond constructed by ex-zamindar, and now the village Panchayat owns this pond (Table 8.7). The Panchayat is the leasing authority, and lease rent for this pond is almost four times higher than SIT as there is no restriction on using feed, manure, and medicine in the pond (see Table 8.3). The pond is perennial in nature and monsoon rains are the main water source; the pond is also near to a perennial water stream. In these ponds, there is no restriction on the use of feed and manure as most of them are located at the outer periphery of the village and there are a good three tanks and two common ponds for fish culture. Three tanks were constructed by the soil and water conservation wing of the SDA, and, after construction, handed responsibility for managing them over to the Panchayat. These three tanks, along with two other common ponds constructed by ex-zamindars, have been leased out by the Panchayat to Barbanda FCS (see Tables 8.9–8.10). The lease rents between these two categories of water bodies varies significantly due to restriction/limited restriction imposed by local village Panchayats (see Table 8.3). Water from three tanks has priority for kharif irrigation. However, adequate water is available up until February-march for fish culture. For the other two ponds, domestic and fish culture activities are important (Table 8.3).

Tanks constructed by SWC receive water from the canal and are perennial in nature (Table 8.9). The other two village tanks are also perennial in nature. The Matwa talab is rainfed and Dev talab is canal fed. All five tanks have multiple uses (Table 8.10).

Mahamaipara FCS is essentially an urban FCS and has leased two tanks (Lake Vivekanada and Maharajband Talab). These tanks are located in the heart of Raipur city. Municipal Corporation Raipur owns these tanks as well the leasing authority. Both the tanks are multipurpose and multifunctional in nature (Table 8.11). There is adequate water throughout the year and there is no restriction on the application of feed, manure, or medicine. These tanks are not used for irrigation (see Tables 8.3 and 8.11).

The extent of water availability for multiple uses (non-excluding component), and the restriction on feed, manure, and medicine are the two most important factors which affect the FCSs' performance. For example, water is adequately available in BKIT and SIT for the complete growth cycle of fish, but due to the restriction of feed and manure, productivity of these ponds is quite low in comparison to urban tanks leased by Mahamaipara FCS, where there is no restriction on the use of external feed materials. More elaborative discussion on this aspect is provided in the sections on decision making and outcomes and impact.

8.7 Decision-Making Arrangements

The analysis of decision-making mechanisms adopted by five FCSs and their effects on patterns of collective interaction is presented in Tables 8.12–8.16. Despite the fact that they are all registered FCSs, the decision-making mechanisms used by each differ significantly in regard to harvesting fish, thinning fish populations of both local and improved fish varieties to encourage growth for the remaining fish, fixing prices, modes of sale, receiving of financial and technical assistance, and influencing leasing authorities, Panchayats, and other organizations responsible for maintaining water levels in the tanks and ponds for fish culture. For example, in the case of Boriya Khurd FCS, most harvesting of fish is carried out in the months of March to June as fishes gain maximum weight during these months. Further, the decision to intermediate fish catch during this period helps fishermen to sustain their interest and to supplement their food requirement. The members collectively decided to sell fish to a wholesaler at the tank site in presence of as many members as possible. The FCS also has been paying higher wages to members for netting activities. Members of FCSs, through collective action, can compel DWR and Panchayats to maintain the minimum water required for fish culture, in case the water level drops below normal required levels. But in the case of Serikhedi FCS, members harvest fish throughout the year and sell almost 90 percent of their catch to retailers.

As a result, members receive 80% of the total profit from fish production. The FCSs have also appointed a guard to watch tank sites to restrict fish poaching. In the case of Boriya Khurd, members collectively guard the tank. In the case of lower fish yields as a result of less water in the common village pond leased in by Serikhedi FCS, members are collectively exempted from paying lease rent to the Panchayat. Kura FCS has a long tradition of cooperative governance and has been sharing common pond water with a large number of farmers without much conflict. This FCS coordinates rationally with krishi samitee and the Panchayat to receive adequate water and to use feed material. Fishing operations are generally organized on market day during the year to sell produce to village merchants and wholesalers. Kura FCS has appointed a secretary to take care of financial matters. Members guard against poaching and killing of fish. In the case of Barbanda FCS, harvesting fish is sometimes undertaken by contractors. The same contractors buy fish and make payments on the spot at mutually agreed prices. The Panchayat also exempt members from paying lease rent if water-stress conditions affect fish yield. Mahamaipara FCS has been intensively involved in regular harvesting and direct sale to wholesalers and it is politically strong enough to influence the state pollution board to minimize water pollution. Two guards continuously protect the tanks against poaching. Members get fish for consumption and sale at lower prices. There are a few sub-groups among the members based on their political linkage. This often creates conflict over even small issues.

Here, we examine the impact of decision-making mechanisms on outcomes in terms of fish production, income, and employment, rational disposable patterns, and the equitable distribution of benefits in the case of each of the five FCSs.

8.7.1 Outcomes

Effective interrelationships among ponds and tanks, the characteristics of fishermen community sub-castes, decision-making arrangements, and patterns of interaction can yield mutually beneficial economic outcomes.

The performance of five selected FCSs in relation to members' implicit goal of enhancing livelihoods and generating employment through fish culture is evaluated in terms of fish yield per ha of water-spread area, net

Table 8.12 Decision-Making Arrangements and Patterns of Interaction in Boriya Khurd FCS

Attributes	*Description*
A. Decision-making arrangements	
1. Legal and administrative relationship with state	Registered FCS under the Cooperative Act and managed under three-tier Panchayat institution system
2. Harvesting period of fish in a year	Throughout the year, maximum during March to June
3. Days of intermediate fishing catches	Twice a week
4. Harvesting method	Collective operations
5. Reasons for harvesting fish	During March to June, water level is relatively low and by this time good growth of fish is achieved
6. Arrangement for netting, marketing, and distribution	By all members of the fish cooperative society with support from the SDF.
7. Price fixing	At village level, FCS fix the price according to market, and at mandi level it depends upon market forces
8. Mode of sale of fish	Village level (members/villagers/village merchant) and mandi (wholesaler). Large quantities sold to wholesalers at tank sites.
9. Time and mode of payment	Spot payment in cash
10. Ability to raise funds	1. By selling fish, contributions of members can take loans from banks/ moneylenders 2. Financial assistance from SDF up to Rs. 25,000 for three years (share capital 8%, lease rent 22%, purchase of seed and stocking 20%, nylon rope and boat 50%)
11. Ability of society to influence other Panchayat/government activities	Members of FCS can collectively compel the DWR through Janpad Panchayat to maintain minimum water levels in the tank for fish culture
12. Technical assistance from external organizations	State Fisheries Department and FFDA assist in supply of fish seeds, rearing of fish, and supplying nets, boats, and nylon ropes

13. Competition and conflict over water use and conflict-resolving mechanism	Due to the protective nature of the irrigation system and field-to-field method of irrigation, there were some conflicts observed among head, middle, and tail-end farmers during low-rainfall years. Since DWR and Panchayats ensure a minimum level of water required for fish culture, there is no conflict between the fishermen community and farmers. In cases where water levels in the tank drop to a level that may affect fish culture, FCS collectively forces the village and Janpad Panchayat, and DWR, to release additional water. With the additional release of water, fish growth period continues until June. The FCS can't use fish feed and manure in the tank, and the Panchayat resolves conflicts, if any, between domestic users of tank water and FCS
B. Patterns of interaction	
1. Membership, legitimacy, and outsiders	Only fishermen are members of society; the president, vice-president, secretary, and executive members are elected by members in the presence of one officer from the Registrar Office. Outsiders are not allowed to interfere in FCS working. Members stake and control is quite high
2. Reciprocal interactions	Collective action by the members of FCS through collective interaction and decision, if differences of opinion on any issue, senior members along with village head and official members of cooperative assist to resolve the issue The society's performance, in the view of members, is satisfactory in relation to members' implicit goal to achieve higher employment and income from fish sales
3. Rules of protection of fish and tank by members	Members watch against poaching and killing of fish. Offenders are fined by the village Panchayat. If the offender is one of the members themselves, the executive body heavily punishes that member or excludes him from the fisheries cooperative
4. Use regulation of fishes	Members and non-members can buy fish for consumption. Members must participate in all activities of the FCS. For netting, members receive wages (Rs. 40 per day).

Table 8.13 Decision-Making Arrangement and Patterns of Interaction in Serikhedi FCS

Particulars	*Fish Cooperative*
A. Decision-making arrangements	
1. Legal and administrative relationship with state	Registered FCS under the Cooperative Act and managed under three-tier Panchayat institution system.
2. Harvesting period of fish in a year	Throughout the year
3. Days of intermediate fishing catches	Twice a week
4. Harvesting method	Collective
5. Reasons for harvesting fish	Growth of fish and based on need
6. Arrangement for netting, marketing, and distribution	By all members of the fish cooperative society
7. Price fixing	According to market forces or prices
8. Mode of sale of fish	Direct to consumer and retailers
9. Time and mode of payment	Spot payment in cash
10. Ability to raise funds	Selling fish, contributions from members, can take loan from moneylenders
11. Ability of society to influence other Panchayat/government activities	Members of fish cooperative society can collectively compel the DWR/ Panchayat to maintain minimum water level in the tank for fish culture. If Fish die due to low water levels, Panchayat exempt the lease amount. In the case of irrigation tanks, water availability remains adequate throughout the year
12. Technical assistance from external organizations	State Fisheries Department assists in supply of fish seeds, rearing of fish, and technical guidance, periodically arranged training
13. Competition and conflict over water use and conflict-resolving mechanism	Users of irrigation tanks do not have any conflict over water use as the tank gets enough water throughout the year. District Panchayat ensures adequate availability of water until June for fish culture and harvesting. The restriction on feed and manure use has been a concern of members. In the case of Raipuriha pond, no conflict has been observed between FCS and domestic users as the pond located quite far from the village. Panchayats resolve conflicts, if any.
B. Patterns of interaction	
1. Membership, legitimacy, and outsiders	Only fishermen are members of society. They take collective decisions for their fish farming. Outsiders are not allowed to interfere in FCS's work. Members have a high stake and control in FCS

Particulars	*Fish Cooperative*
2. Reciprocal interactions	Collective action by the members of the FCS through collective interaction
3. Rules for protection of fish and tank by members	One guard appointed at Rs. 1200/- per month to watch tank sites and fisheries. Offenders are fined by members of the FCS. There have been three cases of poaching in the irrigation tank and cases were filed in the local police station. Poaching cases have been observed in the case of irrigation tanks
4. Use regulation of fishes	Members can get a quantity of fish free of cost for their consumption (2–3 Kg.) each catching time. Non-members can buy fish for consumption. Members must participate in all activities of FCS, for netting members get wages (Rs. 30 per day).

Table 8.14 Decision-Making Arrangement and Patterns of Interaction in Kura FCS

Particulars	*Fish Cooperative*
A. Decision-making arrangements	
1. Legal and administrative relationship with state	Registered FCS under the Cooperative Act and managed under three-tier pPanchayat institution system
2. Harvesting period of fish in a year	Throughout the year
3. Days of intermediate fishing catches	Twice a week
4. Harvesting method	Collective
5. Reasons for harvesting fish	Netting is fixed on Wednesday, being market day, and it is mandatory to mulch the pond for good growth of fish
6. Arrangement for netting, marketing, and distribution	Largely, all members of the fish cooperative society collectively make the net and decide the marketing of fish with support from the SDF.
7. Price fixing	At village level, village Panchayat fixes the price according to market, and at mandi level depends upon market forces – prices vary from day to day
8. Mode of sale of fish	Village level (members/villagers/village merchant) and mandi (wholesaler)
9. Time and mode of payment	Spot payment in cash at cooperative office site/pond site

(*Continued*)

Table 8.14 (Continued)

10. Ability to raise funds	1. By selling fish, contributions of members can take loans from banks/moneylenders 2. Financial assistance from the SDF up to Rs. 25,000 3 Years (share capital 8%, lease rent 22%, purchase of seed and stocking 20%, nylon rope and boats 50%)
11. Ability of society to influence other Panchayat/government activities	Members of the cooperative society requested that the Panchayat maintain a minimum water level for fish culture. For cooperative office building, FCS has acquired common wasteland from village Panchayat. Panchayat puts restrictions on farmers using ponds water to maintain a minimum level of water
12. Technical assistance from external organization	State Fisheries Department and FFDA assist in supplying fish seeds, rearing of fish, and arrangements for nets, boats, and nylon rope
13. Competition and conflict over water use and conflict-resolving mechanism	Kura village has a long tradition of sharing pond water without conflict. The Panchayat ensures the minimum level of water required for fish culture and coordinates between Krishi Samitee (responsible for maintaining irrigation water use) and the FCS. With lease money, the Panchayat repairs bunds of the ponds to stop outflow of fish. In cases of water-stress conditions in rainfed ponds, the FCS transfers fish to perennial ponds.
B. Patterns of interaction	
1. Membership, legitimacy, and outsiders	Only fishermen are members of society, president, vice-president, secretary, and executive members are elected by members. Secretary gets Rs. 500 per month to maintain records of the society. Outsiders are not involved in any activities of FCS. Extremely high stake and control of the member over FCS
2. Reciprocal interactions	Collective action by the members of FCS through collective interaction
3. Rules for protection of fish and tank by members	Members watch against poaching and killing of fish. FCS announced an award of Rs. 2,000 for information on poaching of fish. Offenders are fined by the village Panchayat. If the offender is from the members themselves, the executive body heavily punishes that member or excludes him from the fisheries cooperative. Every year, two or three cases of poaching have occurred, with the FCS punishing the offenders.
4. Use regulation of fishes	Members and non-members can buy fish for consumption. Members must participate in all activities of FCS, for netting members get wages (Rs. 30/day).

Table 8.15 Decision-Making Arrangement and Patterns of Interaction in Barbanda FCS

Particulars	*Fish Cooperative*
A. Decision-making arrangements	
1. Legal and administrative relationship with state	Registered FCS under the Cooperative Act and managed under three-tier Panchayat institution system
2. Harvesting period of fish in a year	Throughout the year, maximum during April to June
3. Days of intermediate fishing catches	Once a week
4. Harvesting method	Collective, and sometimes through contract
5. Reasons for harvesting fish	During April to June, water level is low and growth of fish is good
6. Arrangement for netting, marketing, and distribution	By all members of the FCS with support from the SDF.
7. Price fixing	At village level, village Panchayat fixes the price according to market
8. Mode of sale of fish	Village level (members/villagers)
9. Time and mode of payment	Spot payment in cash in case of contract harvesting
10. Ability to raise funds	Selling fish, contributions from members, can take loans from bank/moneylenders
11. Ability of society to influence other Panchayat/government activities	Members of FCS can collectively compel the Panchayat to maintain minimum water level in the tank for fish culture. If fish die due to low water level (in the case of rainfed tanks), the Panchayat will suspend lease payments.
12. Technical assistance from external organization	State Fisheries Department and FFDA assist in supply of fish seeds, rearing of fish, and supplying nets, boats, and nylon rope
13. Competition and conflict over water use and conflict-resolving mechanism	In the case of water scarcity, farmers put pressure on the Panchayat for using ponds water (Barbanda I, II & III) to irrigate crops. If fish growth is affected due to inadequate water and the Panchayat releases water due to pressure from farmers, the lease amount is exempted. In the case of domestic ponds, conflicts can get serious if FCS attempts to add fish feed or manure.
B. Patterns of interaction	
1. Membership, legitimacy, and outsiders	Only fishermen are members of society, president, vice-president, secretary, and executive members are elected by members. Outsiders are not allowed to interfere in FCS work. Members have a high stake and control over the functioning of FCS

(*Continued*)

Table 8.15 (Continued)

Particulars	*Fish Cooperative*
2. Reciprocal interactions	Collective action by the members of FCS through collective interaction
3. Rules of protection of fish and tank by members	Members watch against poaching and killing of fish. Offenders are fined by the village Panchayat. If the offender is from the members themselves, they are expelled from the FCS
4. Use regulation of fishes	Members and non-members can buy fish for consumption. Members must participate in all activities of FCS, for netting members get wages (Rs. 30 per day).

Table 8.16 Decision-Making Arrangement and Patterns of Interaction in Mahamaipara FCS

Particulars	*Fish Cooperative*
A. Decision-makingarrangements	
1. Legal and administrative relationship with state	Registered FCS under the Cooperative Act
2. Harvesting period of fish in a year	October, November, December, March, and April.
3. Days of intermediate fishing catches	Two or three times in a week
4. Harvesting method	Collective
5. Reasons of harvesting fish	Growth and based on need
6. Arrangement for netting, marketing, and distribution	By all members of the FCS
7. Price fixing	According to market prices
8. Mode of sale of fish	Direct to wholesaler
9. Time and mode of payment	Spot payment in cash
10. Ability to raise funds	Members share capital, sale of fish among members and outsiders, and can take loans from District Cooperative Bank.
11. Ability of society to influence other Panchayat/government activities	Influence CGPCB to regulate water quality and compel Raipur Municipal Corporation to eradicate water hyacinth
12. Technical assistance from external organizations	FFDA provides information about modern technology and composite fish culture, and the State Fisheries Department assists in supply of fish seeds and rearing of fish.
13. Competition and conflict over water use and conflict-resolving mechanism	There is no conflict over water use. Adequate water for fish culture throughout the year. However, internal conflicts among the members is quite visible due to political interface and affiliations of members with political parties. Rival members are often involved in fish poaching and fish killing. FCS attempts to resolve the issue. Sometimes the issue remains unresolved and further exacerbates the problem

Particulars	*Fish Cooperative*
B. Patterns of interaction	
1. Membership, legitimacy, and outsiders	Election every five years under the Cooperative Act, president, vice-president, secretary, and executive members are elected by members. Outsiders are not allowed to interfere in FCS work, but hidden political forces often do interfere. Only fishermen are members of society.
2. Reciprocal interactions	Collective action by the members of FCS through collective interaction only in the case of the fish production process
3. Rules for protection of fish and tank by members	Two guards appointed at Rs. 1500/- per month. Offenders are fined by the management committee of FCS on condition basis. There are no cases of poaching. However, fish killing cases were reported to the police.
4. Use regulation of fishes	Members can buy fish for their consumption at Rs. 15 per kg at every catching time (2–3Kg.). Non-members cannot get fish from the pond site. Members can get the fish for sale at Rs. 20 per kg. Members must participate in all activities of FCS; for netting, members get wages (Rs. 30 per day).

income, and man-days employment per member per year. The other parameters for evaluating performance include patterns of distribution of benefits, and mechanisms of sale of fish produced (Tables 8.17–8.21). In terms of yield and operating expenses, Mahamaipara FCS ranks first, followed by Kura, Barbanda, Boriya Khurd, and Serikhedi respectively. This may be attributed largely to the use of fish feed and medicine in the first three FCSs. However, in terms of net returns, Barbanda is at the bottom level, Boriya Khurd and Kura are at equal level. Mahamaipara is the highest earner followed by Serikhedi FCS. The variation in net income is mainly attributed to prices (Tables 8.17–8.21). In terms of employment generation, Mahamaipara, Boriya Khurd, and Barbanda are almost equal, followed by Serikhedi and Kura. The pattern of distribution of benefits varies extensively across the five FCS in terms of wages paid to members, funds kept aside for next year's investment, and asset creation. The disposal pattern of fish produced across the FCSs centred around members, non-members, retailers, and wholesalers with visible price differentiation at different levels. For members and non-members of the same village, prices are generally not too different (Tables 8.17–8.21).

Table 8.17 Outcomes and Impacts in Boriya Khurd FCS

Particulars	*2000–01*	*2001–02*	*2002–03*
Economic gains			
i. Yield (x 100 kg ha^{-1})	2.57	2.71	2.86
a. Rohu, Katla, Mrigal	2.06	2.17	2.29
b. Local	0.51	0.54	0.57
ii. Price Rs. per 100 kg)			
(a) Rohu, Katla, Mrigal	2500	2500	2500
(b) Local	1000	1000	1000
(c) Gross return (Rs./ha)	5657	5971	6286
(d) Total operating cost (RsIa)	4057	4131	4205
(e) Net return (Rs./ha)	1600	1840	2081
(f) Net income (Rs./kg)	6.22	6.78	7.28
(g) Input-output ratio	1:1.39	1:1.45	1:1.49
(h) Employment generated for members (man-days)	79	79	79
Distributive gains			
(a) Pattern of distribution of benefits (percent)*			
(b) Wages paid to members	32	31	26
(c) Share of benefits kept aside for next year's expenditure on fish rearing	18	20	21
(d) Remaining amount for creating assets (fishing boats/nets/community building, recltion, etc.)	45	49	53
(e) Sustainable development of fish and tank resources	5	-	-
(f) Users are fully satisfied	Yes	Yes	Yes
Disposal pattern (% to total produce)			
At village level			
Member	10 (@Rs. 20/kg)	10 (@Rs. 20/kg)	10 (@Rs. 20/kg)
Non-member	10 (@Rs. 30/kg)	10 (@Rs. 30/kg)	10 (@Rs. 20/kg)
Village merchant	-	-	-
Wholesaler	80 (@Rs. 25/kg)	80 (@Rs. 25/kg)	80 (@Rs. 20/kg)
Retailer	-	-	-
Others	-	-	-
Net price to the fishermen (Rs./kg.)	22	22	22
Wholesaler's purchase price (Rs./kg.)	25	25	25
Wholesaler's sale price (Rs./kg.)	30	30	30
Retailer's purchase price (Rs./kg.)	30	30	30
Retailer's sale price (Rs./kg.)	35	35	35

* Based on net income

Table 8.18 Outcomes and Impacts in Serikhedi FCS

Particulars	*2000–01*	*2001–02*	*2002–03*
Economic gains/ha			
(a) Yield (qtl./ha)	2.45	2.08	2.72
i. Rohu, Katla, Mrigal	1.96	1.66	2.18
ii. Local	0.49	0.42	0.54
(b) Price Rs./qtl.)			
i. Rohu, Katla, Mrigal	3000	3000	3000
ii. Local	1500	1500	1500
(c) Gross return (Rs./ha)	6615	5610	7350
(d) Total Operating cost (Rs.)	3063	2470	3537
(e) Net return (Rs./ha)	3552	3140	3813
(f) Net income/kg (Rs.)	14.50	15.10	14.02
(g) Input-output ratio	1:2.16	1:2.27	1:2.08
(h) Employment generated for members (man-days)	85	56	78
Distributive gains			
Pattern of distribution of benefits (per cent) *			
(a) Wages paid to members	80	80	79
(b) Share of benefits kept aside for next year's expenditure on fish rearing	10	13	13
(c) Remaining amount for creating assets (fishing boats/nets/ community building, recreation, etc.)	10	07	08
(d) Sustainable development of fish and tank resources	-	-	-
(e) Users are fully satisfied	No (deepening pond)	No (deepening pond)	No (deepening pond)
Disposal pattern (% to total produce)			
At village level			
Member	-	-	-
Non-member	10 (@Rs. 30/kg)	10 (@Rs. 30/kg)	-
Village merchant	-	-	-
Wholesaler	-	-	-
Retailer	90 (@Rs. 30/kg)	90 (@Rs. 30/kg)	100 (@Rs. 30/kg)
Others	-	-	-

(*Continued*)

Table 8.18 (Continued)

Particulars	*2000–01*	*2001–02*	*2002–03*
Net price to the fishermen (Rs./kg.)	(equal distribution of benefits)	(equal distribution of benefits)	(equal distribution of benefits)
Wholesaler's purchase price (Rs./kg.)	-	-	-
Wholesaler's sale price (Rs./kg.)	-	-	-
Retailer's purchase price (Rs./kg.)	30 (improved) 15 (local)	30 (improved) 15 (local)	30 (improved) 15 (local)
Retailer's sale price (Rs./kg.)	35 (improved) 20 (local)	35 (improved) 20 (local)	35 (improved) 20 (local)

* Based on net income

Table 8.19 Outcomes and Impacts in Kura FCS

Particulars	*2000–01*	*2001–02*	*2002–03*
Economic gains/ha			
(a) Yield (qtl./ha)	7.16	3.35	6.12
i. Rohu, Katla, Mrigal	5.72	2.68	4.89
ii. Local	1.44	0.67	1.23
(b) Price Rs./qtl.)			
i. Rohu, Katla, Mrigal	2500	2500	2500
ii. Ial	1000	1000	1000
(c) Gross return (Rs./ha)	15740	7376	13452
(d) Total operating cost (Rs.)	11779	8569	8887
(e) Net return (Rs./ha)	3961	-1193	4565
(f) Net income/kg (Rs.)	5.53	-3.56	7.46
(g) Input-output ratio	1:1.34	1:0.86	1:1.51
(h) Employment generated for members (man-days)	84	50	64
Distributive gains			
Pattern of distribution of benefits (percent) *			
(a) Wages paid to members	52	-	41
(b) Share of benefits kept aside for next yI's expenditure on fish rearing	14	-	12
(c) Remaining amount for creating assets (fishing boats/nets/community building, recreation, etc.)	34	-	47
(a) Sustainable development of fish and tank resources	-	-	-
(b) Users are fully satisfied	Yes	No	Yes

Particulars	*2000–01*	*2001–02*	*2002–03*
Disposal pattern (% to total produce)			
At village level			
Member	10 (@Rs. 15/kg)	10 (@Rs. 15/kg)	10 (@Rs. 15/kg)
Non-member	10 (@Rs. 30/kg)	10 (@Rs. 30/kg)	10 (@Rs. 30/kg)
Village merchant	20 (@Rs. 25/kg)	20 (@Rs. 25/kg)	20 (@Rs. 25/kg)
Wholesaler	50 (@Rs. 25/kg)	50 (@Rs. 25/kg)	50 (@Rs. 25/kg)
Retailer	10 (@Rs. 30/kg)	10 (@Rs. 30/kg)	10 (@Rs. 30/kg)
Others	-	-	-
Net price to the fishermen (Rs./kg.)	22	22	22
Wholesaler's purchase price (Rs./kg.)	25	25	25
Wholesaler's sale price (Rs./kg.)	30	30	30
Retailer's purchase price (Rs./kg.)	30	30	30
Retailer's sale price (Rs./kg.)	35	35	35

* Based on net income

Table 8.20 Outcomes and Impacts in Barbanda FCS

Particulars	*2000–01*	*2001–02*	*2002–03*
Economic gains/ha			
(a) Yield (qtl/ha)	4.61	4.9495	4.48
i. Rohu, Katla, Mrigal	3.69	0.99	3.58
ii. Local	0.92		8.0.90
(b) Price Rs./qtl.)			
i. Rohu, Katla, Iigal	1500	1500	2000
ii. Local	500	500	600
(c) Gross return (Rs./ha)	5995	6421	7701
(d) Total operating cost (Rs.)	5168	5836	6638
(e) Net return (Rs./ha)	827	585	1063
(f) Net income/kg (Rs.)	1.79	1.18	2.37
(g) Input-output ratio	1:1.16	1:1.10	1:1.16
(h) Employment generated for members (man-days)	78	78	78
Distributive gains			
Pattern of distribution of benefits (percent) *			
(a) Wages paid to members	-	-	-
(b) Share of benefits kept aside for nelyear expenditure on fish rearing	100	100	100

(*Continued*)

Table 8.20 (Continued)

Particulars	*2000–01*	*2001–02*	*2002–03*
(c) Remaining amount for creating assets (fishing boats/nets/community building, recreation, etc.)	-	-	-
(d) Sustainable development of fish and tank resources	-	-	-
(e) Users are fully satisfied	No (price is very low of fish, fixed by the sarpanch)	No (price is very low of fish, fixed by the sarpanch)	No (price is very low of fish, fixed by the sarpanch)
Disposal pattern (% to total produce)			
At village level			
Member	3	3	3
Non-member	97	97	97
Village merchant	-	-	-
Wholesaler	-	-	-
Retailer	-	-	-
Net price to the fishermen (Rs./kg.)	Rs. 13/kg.	Rs. 13/kg.	Rs. 17/kg.
Wholesaler's purchase price (Rs./kg.)	-	-	-
Wholesaler's sale price (Rs./kg.)	-	-	-
Retailer's purchase price (Rs./kg.)	-	-	-
Retailer's sale price (Rs./kg.)	-	-	-

* Based on net income

Table 8.21 Outcomes and Impacts in Mahamaipara FCS

Particulars	*2000–01*	*2001–02*	*2002–03*
Economic gains/ha			
(a.) Yield (qtl./ha)	30.92	28.97	31.57
i. Rohu, Katla, Mrigal	23.2	23.28	24.18
ii. Local	7.903	5.69	7.389
(b.) Price Rs./qtl.)			
i. Rohu, Katla, Mrigal	2200	2200	2200
ii. Local	800	800	800
(c) Gross return (Rs./ha)	56969	55768	59114
(d) Total operating cost (Rs.)	53789	43262	43624
(e) Net return (Rs./ha)	3180	12506	15490
(f) Net income/kg (Rs.)	1.028	4.32	4.906
(g) Input-output ratio	1:1.059	1:1.29	1:1.35
(h) Employment generated for members (man-days)	79	79	79

Particulars	*2000–01*	*2001–02*	*2002–03*
Distributive gains			
Pattern of distribution of benefits (percent) *			
a) Wages paid to members	-	69	69 (kept for next year)
b) Share of benefits kept aside for next year's expenditure on fish rearing	60	11	13
c) Remaining amount for creating assets (fishing boats/nets/community building, recreation, etc.)	40	20	18
d) Sustainable development of fish and tank resources	-	-	-
e) Users are fully satisfied	-	-	-
Disposal pattern (% to total produce)			
At village level			
Member	10 (@Rs. 20/kg)	10 (@Rs. 20/kg)	10 (@Rs. 20/kg)
Non-member	-	-	-
Village merchant	-	-	-
Wholesaler	90 (@Rs. 25/kg)	90 (@Rs. 25/kg)	90 (@Rs. 25/kg)
Retailer	-	-	-
Other	-	-	-
Net price to the fishermen (Rs./kg.)	15 improved 6 local	15 improved 6 local	15 improved 6 local
Wholesaler's purchase price (Rs./kg.)	22 improved 6 local	22 improved 6 local	22 improved 6 local
Wholesaler's sale price (Rs./kg.)	30 improved 15 local	30 improved 15 local	30 improved 15 local
Retailer's purchase price (Rs./kg.)	30 improved 15 local	30 improved 15 local	30 improved 15 local
Retailer's sale price (Rs./kg.)	35 improved 20 local	35 improved 20 local	35 improved 20 local

* Based on net income

8.8 Comparative Performance of IFs and FCSs

Common ponds are leased out by village Panchayats to individual fishermen belonging to the fishermen community in Boriya Khurd and Barbanda villages. These individual fishermen are also members of the FCSs working in the respective villages. In fact, the president of Boriya Khurd FCS has leased the pond. The lease rents of all these ponds are several times higher than for other water bodies existing in these villages. Since there is no restriction on the use of feed, manures, and medicines, and higher prices are fixed by individual fisher menand the availability of minimum required water is assured, the yield levels, net income, and employment generated from Boriya Khurd and Chhapar ponds are much higher than collectively managed tanks by FCSs in these villages. Even in less productive fish producing ponds (Ledara) employment opportunities were higher in comparison to Barbanda FCS (see Table 8.22).

Few privately owned ponds are also used for fish culture, either by the owners themselves, or leased out to an individual local fishermen or fish production sharing an equal proportion without sharing input costs (see Table 8.23). In terms of yield, net income, and man-days employment generation levels, the performance of all the three ponds of Serikhedi was much higher than FCS levels of respective parameters in this village. Although these ponds are basically rainfed, due to higher investment in fish feed and manure, private individual fishermen can achieve higher yields, income, and employment levels. Such levels, however, were not obtained in other privately owned ponds as these ponds are under sharing arrangements (Table 8.24).

Comparative performance of FCSs and IFs in terms of yield, operating expenses, and net income are presented in Table 8.24. It is clear from Table 8.24 that, with adequate water availability and the application of feed, manure, and proper and timely use of medicine to prevent fish motility, higher yields, incomes, and employment can be achieved (see yield, income, and employment levels of tanks leased by Mahamaipara FCS and Boriya Khurd common pond). It is also clearly evident in the case of the IFs who leased ponds. Further, higher performance can also be achieved even in the case of rainfed ponds without imposing restrictions on the use of growth-promoting inputs, as seen in case of Mandir, Marar, and Mandal.

Table 8.22 Fish Culture in Panchayat-Owned Ponds

Particulars/Village	*Boriya Khurd*	*Barbanda*		
Name of pond	Boriya Khurd	Chhapar	Ledara	Jogibandh
Leasing authority	Panchayat	Panchayat	Panchayat	Panchayat
Leased in by	Fisherfolk *	Fisherfolk *	Fisherfolk *	Fisherfolk *
Water-spread area (ha.)	0.60	3.45	4.50	0.40
Leased rent (Rs./ha.)	10,000	4,928	2,600	2,600
Source of water	Tubewell – 1	Canal	Canal, Rain	Canal, Rain
Net Income per ha.	22650	18430	2033	425
Employment days				
(i) Male	106	132–	82–	66–
(ii) Female	37			
Conflict with village community	Nil	Nil	Nil	Nil

Source: Author's analysis based on field work

* Lessee belongs to the fisher community

Table 8.23 Fish Culture in Privately Owned Tanks

Particulars/Village	*Serikhedi*			*Barbanda*	
Name of pond	Mandir	Marar	Mandal	Dhobani	Gadai
Leasing authority	Private	Private	Private	Private	Private
Leased in by	Self-use	Self-use	Fishermen	Fishermen	Fishermen
Water-spread area (ha.)	2.40	1.80	1.60	0.70	0.90
Leased arrangement (Rs.)	Self-owned	Self-owned	Rs. 5625	Sharing *	Sharing *
Source of water	Rainfed	Rainfed	Rainfed	Rainfed	Rainfed
Production (in quintal/ha.)	22.01	22.01	22.01	0.71	0.55
Netcome per ha.	23025	23120	23270	786	575
Employment days					
(i) Male	109–	111–	113–	45–	42–
(ii) Female					
Conflict with village community	Nil	Nil	Nil	Nil	Nil

Source: Author's analysis based on field work

* Fish production is shared in equal proportion between owner and leases

Table 8.24 Fish Yield and Income From the Ponds Leased by FCS or Owned by Individual Fishermen (average of three years)

Name of ponds	*Fish yield kg/ha*	*Operating Cost, Rs./ha*	*Net return Rs./ha*	*Employment Man-days*
Boriya Khurd FCS	271	4131	1840	79
Serikhedi FCS	242	3023	3502	73
Kura FCS	554	9745	2444	66
Barbanda FCS	457	5881	825	78
Mahamaipara FCS				
a. Budha Talab	3087	43624	15490	79
b. Maharajbandh	3251	56789	53789	82
Boriya Khurd Pond	1855	23725	22650	143
Mandal Pond	2201	31755	23270	113
Chhapar Pond	15	23245	14255	102
Mandir Pond	2201	32000	23025	109
Marar Pond	2201	31905	23120	111
Dhobani Pond	71	1000	786	45
Gadai Pond	55	800	575	42
Jogi Bandh Pond	125	2700	425	66
Ledara Pond	333	6292	2033	82

8.9 *Munda*: Indigenous Water-Harvesting Technology for Collective Livelihood

Munda are small water-storage structures that have a played vital role since the Chhandak dynasty (1023 BC to 1313 BC) to Kakatiya dynasty (1313 BC to 1358 BC) in sustaining livelihoods and water security in the tribal villages of Bastar region of Chhattisgarh.[7] During the Kakatiya dynasty, Dalput sagar and Barsur tanks were constructed (Jagdalpuri, 1994, pp. 5–15). *Munda* are a small water-harvesting structures constructed in a slightly low-lying location on farmland owned by four or five farmers. All participating farmers willingly part with a small piece of land for the construction of a *Munda* (common structure size: 40m x 50m to 555m to 65m). The depth of *Munda* varies from two to six metres. The location for *Munda* construction is the collective decision of a participatory group of farmers. *Munda* water is shared by a group of participatory farmers for the irrigation of paddy and vegetables, and the cultivation of small local fisheries. *Munda* are CPRs constructed on pooled private land used by farmers under common property regime with soft, workable institutional arrangements and a collective system. The average life of *Munda* is 25 to 30 years but can be extended if regular de-siltation is carried out. Desilted soils are used for backyard farming and the embankment of *Munda*. Water from *Munda* is lifted through traditional methods (*tenda pathi*) and equal inflow of water is maintained to supply water to the fields. Mainly local fish are cultivated, namely, *Turu, Bami, Khokhasi, Baliya, Kotari, Mongari, Rohu, Katla, Mrigal* (these fish are part of the consumption basket of farmers). The normal weight of these fish varies from 100 gm to 1.5 kg. Approximately one to three quintals of fish are produced in a one-year cycle. Fish production is equally shared by farmers. *Munda* techniques were promoted, even after splitting from MP, by the State Fisheries Department in tribal areas under special assistance programmes. The number of *mundas* has recently been reducing due the withdrawal of financial support and division of land between family members of farmers. However, in some cases, *Munda* has been transferred from one generation to another. The emerging trend of tube wells and water tanks, coupled with free electricity under subsidized schemes, is also an important reason for the decline of *Munda* in tribal areas near towns (Marothia, 1994, 2012a).

8.10 Institutional and Policy Choices

From the point of view of managing fisheries in common water bodies, this study provides some meaningful findings. Despite the fact that they are all registered FCSs, the decision-making mechanisms used by each differ significantly in regard to harvesting fish, thinning fish populations of both local and improved fish varieties to encourage growth for the remaining fish, fixing prices, modes of sale, receiving of financial and technical assistance, and influencing leasing authorities, Panchayats, and other organizations responsible for maintaining water levels in the tanks and ponds for fish culture.

The extent of water availability, multiple-use (non-excluding component), and restrictions on the use of feed, manure, and medicine are the two most important factors which affect the FCSs' performance. For example, water is adequately available in BKIT and SIT for the complete growth cycle of fish but due to restrictions on feed and manure, the productivity of these ponds is quite low in comparison to urban tanks leased by Mahamaipara FCS, in which there is no restriction on the use of external feed materials. It is also clearly evident in the case of ponds leased to IFs. Further, higher performance can also be achieved with rainfed ponds, without imposing restrictions on the use of growth-promoting inputs, as seen in case of Mandir, Marar, and Mandal.

In the villages of Chhattisgarh, community ponds and irrigation tanks have traditionally been allotted by local Panchayats for different uses. In almost every village, ponds were separately allotted for female groups. This can still be seen in many villages. A few ponds are exclusively used for SC communities under the social caste hierarchy. Taking a leaf from the traditional allocation of ponds, a few tanks may be left out for domestic use and social rituals in a village, and the rest of the ponds can be exclusively used for culture fisheries without any restrictions on the use of growth-promoting inputs to raise potential yields. Fortunately, in every village, a minimum of seven to ten community ponds still exist. Close-in ponds can be reserved for common use and more distant ponds for fish culture. Such simple political and administrative decisions can substantially reduce instances of intercommunity conflict. Similarly, irrigation ponds/tanks can be exclusively used for irrigation and fish culture, if some ponds can be kept aside for catering to the needs of villagers. Further, a feasible solution can be worked out between FCSs and the DWR for de- silting the tanks, as one of the major concerns of the DWR is increasing silt load and reducing the water-intake capacity of a tank. Due to the use of fish feed and manure, such institutional arrangements can increase fish yields and the total productivity of common water bodies by many times, besides minimizing social conflicts. It has been observed in the study area that FCSs have tendency to lease all the common ponds/irrigation tanks within eight km (a norm prescribed in leasing policy) of a particular village to strategically eliminate FGs and IFs leasing these common water resources. Further, in some cases, not all the leased ponds/tanks have been used for fish culture. For example, none of the 29 ponds leased in by Kura FCS are being used for fish culture. In the case of a village pond leased by individual fishermen, it has also been observed that the lessee is either the office bearer of a local FCS or a member of any sub-caste of the fishermen community. The SDF may seriously consider reviving the *Munda* technology to benefit a large number of tribal farmers. In view of these suggestions, the fishery policy of the state needs to be changed in consultation with the DWR, the SDA, the Department Panchayat and Rural Development, and fisheries organizations. The findings of this study confirmed that the cooperative form of internal institutional arrangements of distributed governance is a better option for managing MWs sustainably. Similar observations were made in the case of local fisheries management under distributed governance in other countries.[8] Our study has the potential for cross learning lessons.

Notes

1 This chapter draws heavily on the updated and modified version of the study "*Performance Evaluation of Culture Fisheries under Alternative Property Rights Regimes in Chhattisgarh*" carried out by the author with support from Sir Ratan Tata Trust under IWMI-TATA Water Policy Programme during 2004. The author is grateful for financial support to Sir Ratan Tata Trust under IWMI-TATA Water Policy Programme. An earlier version of this chapter appeared in the *International Journal of Ecology and Environmental Sciences*, 38(4): 163–207. I am grateful to the journal for permission to reproduce a large part of it here. Post script additions have been incorporated to update information about the fisheries sector in the state.
2 Except in SHGs, the lessee has to be assigned to the fisher community for Panchayat-owned ponds.
3 See endnote 4 of Chapter 1.
4 See endnote 5 of Chapter 1.
5 Privatizing fishing rights of community ponds has created social and economic disequilibrium in the several villages in Chhattisgarh during 1983 to 1992 due to the policy of selecting leaseholders from ST and SC communities, excluding fisher communities that otherwise would have had access as village members to their communities' ponds. After visible social unrest, lease policy in MP and Chhattisgarh was revised to assign lease of community ponds on a priority basis to members of the fishing community (see Marothia, 1997a).
6 During the postscript of this chapter, lease rents were revised in 2016–17. It is important to note that ponds and tanks included in this chapter are not affected with current lease rents. New lease rents for village ponds less than 10 ha is Rs. 1,500 and more than 10 ha it is Rs. 3,000, and for irrigation reservoirs it ranges from Rs. 500 (10–50 ha) to Rs.1000 (200–1000 ha). State government, according to modified state fishery policy, barrages, anicuts and small dams can be leased out to the corporate sector for fish production.
7 Santosh Nag, Senior Scientist and Head, KVK, Jagdalpur, provided information regarding *Munda* technology. Thanks to him.
8 See Jentoft (1989); Pinkerton (1989); Ruddle (1989); Townsend (1990, 1995); Townsend and Pooley (1995); Etiegni et al. (2016).

9 Emerging Policy, Governance and Research Issues

In recent years, the role of crafting institutional arrangements in policymaking for governing different categories of wetlands has gained prominence at the Ramsar Conference of Parties (COPs), country, and state levels. The goal of this chapter is first to critically examine the current Rules of Wetland Conservation and Management (MOEFCC, GoI) in the context of the Ramsar Convention's priorities. Thereafter, the chapter will discuss in brief the management performance of MWs under alternative governance structures. Finally, a policy, governance, and research agenda is proposed for getting a national and international importance tag for Chhattisgarh's wetlands that fall under different categories. The proposed agenda may also be applicable to other states and neighbouring countries for sustaining their inland wetland ecosystems.

9.1 Policy, Law, and Rules of Wetland Conservation in India

In the past 50 years, the Ramsar Convention had introduced many resolutions for sustaining different types of wetlands in the changing socio-ecological systems. However, despite Ramsar's incredible achievements, the threat to the health and survival of many wetlands across the globe has not diminished since 1971. The majority of wetlands in human-dominated areas continue to be threatened due to urbanization (Bowman, 2002; MEA, 2005; Smardon, 1983, 2015; Hettiarachchi et al., 2015; Marothia, 1995a, 1997c, 2001a, 2004d, 2014, 2015, 2019). The trend is still continuing despite urban wetlands being formally recognized as a concern in the Ramsar discourse. Similar trends have been observed in other categories of wetlands. Even after joining the Ramsar Convention, India has no wetland-specific policy or law (Panini, 1998; Gopal, 2018). Of course, the country has taken a few important decisions to protect lakes and rivers within the Convention's institutional framework. These include the Environment (Protection) Act, 1986 (29 of 1986), the National Wetland Conservation Programme (1985) under state central coordination, the National Lake Conservation Plan (1993), and National River Conservation Plan (1995). The significance of wetlands has legally been

DOI: 10.4324/9781003456742-9

enshrined within National Environmental Policy (2006), which recognized the ecosystem services of wetlands and mentions the need for maintaining their ecological character. The National Water Policy (2012) also recognized the importance of wetlands. The Wetland (Conservation and Management) Rules 2010 (GoI, 2010) were enacted by the Ministry of Environment Forest and Climate Change (MoEF&CC) under the Environment Protection Act, 1986, to regulate various activities affecting wetlands adversely. These rules adopted an extended version of Ramsar's definition of wetland (see Chapter 1 for Ramsar's definition of wetlands). The definition is also covered under the notification of the Government of India in the Ministry of Environment and Forest, S.O. No. 114(E), dated 19 February, 1991, published in the *Gazette of India*, Extraordinary, Part II, Section 3, subsection (ii) dated 20 February, 1991 (provision of 2010).

The Wetlands (Conservation and Management) Rules, 2010 (GoI, 2010), state that "wetlands in areas that are ecologically sensitive and important, such as, national parks, marine parks, sanctuaries, reserved forests, wildlife habitats, mangroves, corals, coral reefs, areas of *outstanding natural beauty or historical or heritage areas and the areas rich in genetic diversity*" (2(ii) of Rules 2010) (emphasis added by the author).

The Ministry of Environment, Forest and Climate Change has notified Wetlands (Conservation and Management) Rules, 2017 (GoI, 2017) (hereafter "Wetlands Rules"), under the provisions of the Environment (Protection) Act, 1986, as the regulatory framework for conservation and management of wetlands in India. The Wetland (Conservation and Management) Rules, 2017, has replaced the earlier Rules of 2010, which, together with some judgments of the National Green Tribunal in cases involving wetlands, have cast a long shadow over the future of wetlands in India. These recent developments highlight the inefficiency of the policies and the inherent weakness of laws which are not driven by science (cited from Gopal, 2018).

However, the 2010 and 2017 Rules emphasized that the ecological character of wetlands ought to be maintained for their conservation. The 2017 Rules, however, came under severe criticism from environmental experts for seriously weakening existing regulations and omitting important wetland types.[1] For instance, under Rule 2017, "wetland means an area of marsh, fen, peat land or water; whether natural or artificial, permanent or temporary, with water that is static or flowing, fresh, brackish or salt, including areas of marine water the depth of which at low tide does not exceed six metres, but does not include river channels, paddy fields, human-made water bodies/tanks specifically constructed for drinking water purposes and structures specifically constructed for aquaculture, salt production, recreation and irrigation purposes" (Rule 2 (g) of Wetland Rules 2017). Excluding "human-made water bodies/tanks specifically constructed for drinking water purposes and structures **specifically** constructed for aquaculture, salt production, recreation and irrigation purposes" (Rule 2. (part of g) of Wetland Rules 2017). These wetlands are excluded from notification under these Rules (Rule 2 (g) and Rule 3 of Wetlands Rules). It is interesting to note that excluded categories are very much part of wetland definition and classification of Ramsar Convention Secretariat (2016 page 9–10).

Thus, the future of wetlands in India appears to be very bleak because they cannot claim any protection unless notified after due process by the state wetland authorities under the Wetland Rules 2017. This trait goes a step beyond, because the Wetland Rules 2017 do not take into account the more than half a million wetlands smaller than 2.25 ha recognized in the National Wetland Inventory. It is quickly forgotten that these small wetlands provide an important and valuable ecosystem services for the overall well-being of the people living around them in rural areas, and that these wetlands are crucial to the water, food, and livelihood security of the people and in meeting the challenges of climate change (Cited from Gopal, 2018; see also Gopal and Marothia, 2017; Marothia, 2015, 2017).

It is important to recognize and understand that wetlands under the "excluded category" are multiuse and multifunctional. For example, in Chhattisgarh and in many eastern states, wetlands (talabs, tanks, and complexes) were constructed centuries ago by kings, rulers, and *Malgujars* for meeting the multiuse requirements of communities (see Chapters 3, 4, 5, 6 and 8). Many wetlands and complexes were constructed using principles of ancient Indian water science and hydrology (see Chapter 4). These unique wetlands were constructed during a period when many developed nations were not even born, and some were only learning the basics of water resource science and management. These wetlands were rarely constructed "specifically" for aquaculture, recreation, and irrigation purposes. These centuries-old wetlands still exist and serve societies living around them. Ecosystem services provided by these MWs and complexes (see Table 3.21 of Chapter 3, and Chapters 4, 5, 6, 8 of this book) are consistent with the opening Para of Notification: "wetlands, vital parts of the hydrological cycle, are highly productive ecosystems which support rich biodiversity and provide a wide range of ecosystem services such as water storage, water purification, flood mitigation, erosion control, aquifer recharge, micro-climate regulation, aesthetic enhancement of landscapes while simultaneously supporting many significant recreational, social and cultural activities, being part of our rich cultural heritage" (G.S.R.1203(E) of Notification of Rules 2017).

With the passage of time and growing urbanization in the last five decades, these wetlands are now categorized as urban, peri-urban, and village wetlands and wetland complexes. It is extremely important to understand that these MWs are the arteries of village, peri-urban and urban landscape. Under Rule 2017 these wetlands may not be considered for notification as nationally important wetland, but ignoring their wise use will certainly lead to ecological disaster. Although Rule 2017 also defined *"wetlands complexes" means two or more ecologically and hydrologically contiguous wetlands and may include their connecting channels/ducts" (Rule 2(g) of wetland 2017)*. However, failed to recognize that most of the wetland complexes are human-made in southern, eastern and north-eastern states. For example, Ratanpur and Chandrakhuri have century old human-made wetland complexes and still sustaining ecosystem services (reported in Chapter 3 and 4). This is a major confusion in the Wetland Rules 2017. In this context, it is further important to note that a city with wetland complexes (natural or man-made) is regarded as a potential city for Ramsar Wetland City Accreditation.

Further, the Wetland Rules 2017 explain that, in an urban lake-type of wetland, intervention like concretization of shoreline for beautification will increase the aesthetic value and tourism benefits, but will lead to decrease in the "ability to accommodate monsoon flows" and thus may not be a "wise use" (para 43, page 12 of Guidelines for Implementing Wetlands (Conservation and Management) Rules, 2017). Again, a large number of urban wetlands in Chhattisgarh, eastern and many other states, are human-made, and these were constructed beyond "specific" uses.

Even during the British regime, few common water bodies were initially constructed to meet the water requirements of the community. Even today, in these water bodies, although largely used for irrigation, other users are not excluded. Since 1951, these community resources have been administered and managed by the DWR and therefore are categorized as irrigation tanks. Such tanks are also used for freshwater aquaculture, tending cattle, and meeting villagers' household needs (see Chapter 6 and 8).

The relevance of including tanks and ponds as important types of wetland was duly recognized in the Wetland Rules 2010. Exclusion of human-made MW and the dilution of rules under the Wetland Rules 2017 will have serious implications for achieving sustainable development goals and the wise use of excluded wetlands. Such rules also contradict other resolutions of the Ramsar Convention (for example, Wetland and Urbanization, and Ramsar City Accreditation Resolution). Also, these rules create a disconnect between federal and state governments and wetland researchers in building resilient thinking for wetland restoration and arresting their loss. Further, such provisions raise questions about the Recommendations of the Committee from civil societies and environmental experts who have spent considerable time throughout their careers in wetland-operations research. Finally, without scientific logic, the exclusion or inclusion of any types of wetlands exasperates long-term research investigations and policy recommendations that emerge from such research efforts.

The Wetland Rules 2017 (MoEF&CC) provide for the setting up of a SWA for the purpose of regulating activities in and around wetlands and taking action against activities which adversely affect wetlands – an important provision towards encouraging decentralized governance of wetlands. The Wetland Rules 2017 provide that, "All State Governments shall set up a State Level Wetland Authority entrusted with affairs related to wetland conservation, regulation and management under the relevant State bye-laws". As the central government seeks to empower the states in such matters, it is necessary that a SWA is established soon. To oversee the work carried out by states, the rules stipulate the setting up of the National Wetlands Committee. The Constitution of State and National Wetland Committees, their powers and functions, are given in the Wetland Rules 2017 (see Gopal, 2018 for a critique of the Wetlands Rules, see also Paul et al., 2011). On 2nd February 2021 to mark the 50th anniversary of the Ramsar Convention, India established the Centre for Wetland Conservation & Management, which is the first in the country. It is set up under the Ministry of Environment, Forest and Climate Change (MoEF&CC), at the National Centre for Sustainable Coastal Management (NCSCM) in Chennai.

9.2 Ramsar Convention Resolutions and National Wetland Rules: Implications for Indian States

As mentioned in Chapter 1, India has no specific policy for wetland conservation and management. Many states, including Chhattisgarh, have not been able to implement various critical provisions of the Ramsar Convention Resolutions and National Wetland Rules (Conservation and Management) Rules 2010 and 2017 to promote the sustainable development of natural and human-made urban, peri-urban, and rural wetlands and wetland complexes. This is also partly due to inconsistency and frequent changes in wetland conservation and management rules framed by MOEFCC. For example, despite the fact that many wetlands in Chhattisgarh, both outside and inside forests, have inimitable attributes of socio-cultural-economic-spiritual and religious, historical, and ecological significance, the state does not have a single Ramsar site under any category. Sadly, until today the state did not even have a nationally important wetland. Although Chhattisgarh is exceptionally rich in wetlands, there has hardly been any effort to document or investigate the state's wetland wealth (Gopal and Marothia, 2017). To this end, a brief discussion of different Ramsar Resolutions is needed to understand how far wetland restoration efforts in Chhattisgarh are aligned with the Provisions of different Ramsar Resolutions. Such an analysis is also required in other states.

9.2.1 Ramsar Resolution X.27: Wetlands and Urbanization

Urban wetlands were formally recognized as a concern in the Ramsar discourse in 2008 (Ramsar Convention, 2008) (Res X.27), an offshoot of the Curitiba Declaration on Cities and Biodiversity (2007). The resolution highlights that "urban wetlands" and "peri-urban wetlands" provide many important ecosystem services. The resolution recognized that "many wetlands in urban and peri-urban environments are becoming degraded through encroachment of surrounding populations, pollution, poorly managed waste and infilling or other developments, and that these activities have diminished both the ecosystem services that urban wetlands can provide and the recognition of their value." The resolution emphasized the "safeguarding important ecosystems against in appropriate urban encroachment . . . recreational, spiritual, and aesthetic dimensions of wetlands, as a means of supporting the health and well-being of people by facilitating access to such wetlands". The resolution also encourages "to connect urban wetlands with Convention on Biodiversity Diversity initiatives". Further, the resolution encourages "involving municipalities in their planning processes and operational actions on wetland conservation and wise use. Local governments, including municipalities and communities can play a crucial role in ensuring conservation and wise use of these wetlands" (see Resolution X.27 for detailed provisions).

A more comprehensive declaration on urban wetlands was made in Resolution XI.11 (Ramsar Convention Secretariat, 2012), "Agreement of joint UN HABITAT-Ramsar overarching principles which define the philosophical and conceptual approach to urban planning, development and management and the protection, enhancement and conservation of wetland biodiversity". The resolution adopted four overarching principles of urban wetland governance: "(i) No further degradation or loss of wetlands as a result of urban development or management. (ii). Wetlands should be considered as essential water-management infrastructure. (iii). Wetlands should be restored and created as elements of urban infrastructure in order to optimise ecosystem service delivery. (iv). The wise use of wetlands should be considered as a key component within sustainable human settlements". (See Figure 1 of Ramsar Resolution XI.11, 2012, page 7.) Resolution XI.11 further stated five practical principles to guide management: (i) wetland conservation; (ii) wetland restoration and creation; (iii) understanding the value of wetlands; (iv) stakeholder engagement; and (v) integrated planning (Ramsar Resolution XI.11, 2012, pp. 8–9, see Hettiarachchi et al., 2015 for critical analysis on the role of the Ramsar Convention and urban wetlands conservation; see also Matthews, 1993; Smardon, 2015). A large number of research studies have established that urban wetlands are valuable multiuse and multifunctional ecosystems.[2]

Despite the above two resolutions, adequate evidence is available of substantial loss and degradation of urban and peri-urban wetlands in rapidly expanding cities in various states in India. A large number of wetlands are threatened despite restoration efforts. The trend of urban and peri-urban wetlands loss is quite visible in Chhattisgarh (see Chapters 3, 4, 5, 6 and 8 for the ecological status of different categories of wetland including urban and peri-urban wetlands and wetland complexes).

Raipur and Bilaspur cities in Chhattisgarh[3] were included under the Smart City Mission. Raipur Smart City Limited (RSCL) had developed *An Action Plan to Repair, Renovate, Restore and Rejuvenate Water Bodies* across the Raipur City during 2016. To begin with, a pilot project 'Lake Purification and Restoration at *Telibandha Talab* (lake)' was launched in collaboration with other agencies. Besides improving water quality, the walking tract was developed by reducing the water-spread area of the lake. Several old trees of different species around the boundaries of the lake were uprooted. As a result, biodiversity loss has almost touched threshold level, and, in turn, collectively, space for making nests for domestic and migratory birds has reduced to almost zero. RSCL has developed *Telibandha* lake as a tourist spot and named the surrounding area as 'Marine Drive'. An action plan was also prepared by RSCL to restore 43 wetlands in two phases. In the first phase, 14 wetlands were identified to restore, develop pathways, de-silt and build boundary walls. RSCL has assigned restoration work to the different agencies. All the identified lakes are suffering from degradation, pollution, and encroachment (see Map 9.1 for location of identified wetlands). In a few cases, litigation has been filed by encroachers, for example Raja Talab and Jora Talab (Indira Gandhi ward), are still pending. Similarly, although beautification works in Maharajband wetland started by RMC and RSCL, the matter of alteration is still being scrutinized by the National Green Tribunal (NGT). In another case, RMC under *'Sarover Dharovar Yojana'*, had prepared a plan for development along the lines of 'Marine Drive' of Telibandha talab, through connecting Mahant Talab, Darri Talab, and Dabra Talab adjacent to Shaheed Manmohan Singh Bakshi Ward. But protests by residents started before the plan could be implemented, as, under the proposed scheme,

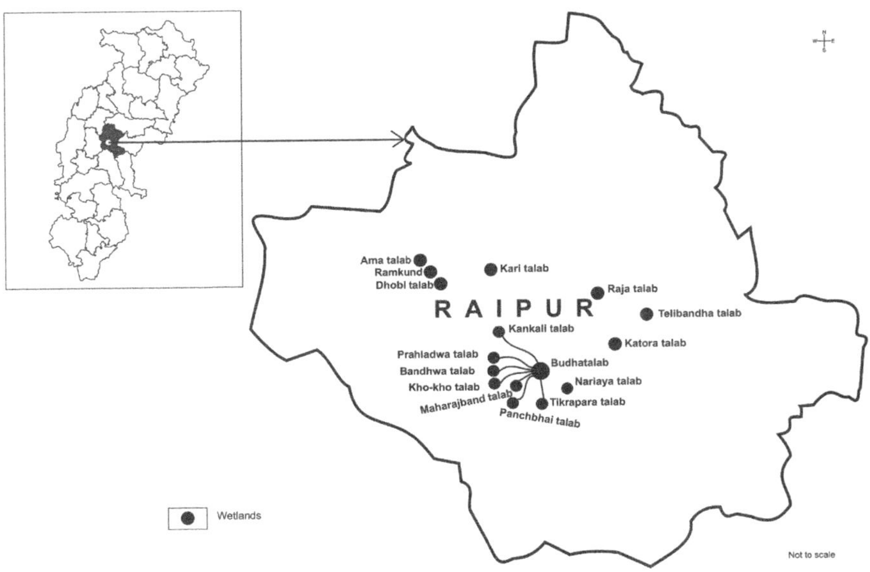

Map 9.1 Lakes Identified (Black Bullets) by Raipur Smart City Limited (2016, p. 15). Hydrological connectivity (added by the author, indicated through dark black arrows in the map) of seven lakes, with LV (Kankali talab, Maharajbandh, and Panchbhai talab, Tikrapara talab, Khokho Group of talab consists of three sets of talab, namely Khokho talab, Pehladwa talab, and Bandhwa talab) (devised by the author)

nearly 1,300 families were asked to move to other localities. The majority of people encroached the surrounding three talabs and illegally constructed houses (these talabs are not shown in Map 9.1). The biggest challenge for RMC and RSCL is to get rid of encroachment and illegal construction from the identified lakes.

As mentioned in Chapter 3, the process of alteraing and destroying urban and peri-urban wetlands had an adverse and cumulative impact on wetland wealth, and on the quality of civic life in major cities and towns of Chhattisgarh. But serious damage to wetlands can be seen in Raipur, Bilaspur, Ratanpur, Ambikapur, and Jagdalpur. For example, in Raipur city, a sizable area of Maharajband historical lake, built in 1770, has been reduced through encroachment by surrounding populations, pollution, and infilling for the construction of a 60-foott wide road along the north-east boundary (approximately 700 metres long) of the lake. On an encroached area of nearly 300 metres along the north-east boundary, houses and huts were constructed and the illegal cultivation of crops and vegetables is in operation. Also, people living along the edge of the lake are using encroached areas for backyard cultivation and it looks like part of the back of their houses. Pathways along the lake boundary were completed, but washed away during the first rains. Similar stories can be seen in the cases of Raja talab, Bandwa, Kori, Narriya, Tarriaya, Kho-kho, Jora Talab (under Indira Gandhi Ward) and many more (see the location of lakes in Map 9.1). Interestingly, RMC had spent millions of rupees on renovating many talabs, even before RSCL identified some of them for renovation and beautification under Smart City Mission[4] (lakes identified by RSCL for beautification are marked in Map 9.1).

Boriya Khurd, a highly productive peri-urban MW, has recently become a sufferer of Smart City Mission. RMC and RSCL have initiated beautification work and taken possession of this multifunctional body to make it a part of the upcoming Kamal Vihar Township for recreation. It is important to note that nearly 432 households, 245 cultivators, and 45 fishers have been deprived of water resources forever. This water body covered a total 385 ha of command area covering four villages. The majority of villagers will be pushed into irreversible food and nutritional insecurity traps. In the process of urbanization, inter-generational externalities will also be created. (See Chapter 8 for the extent of perpetual gains from Boriya Khurd to the village community.) Further, Boriya Khurd is connected with a canal irrigation system to provide irrigation. Now the canal is dismantled and the main source will be rains. If so, the water body will be dry during January and/or February. Water supply through the village common tube well to the water body will certainly create water insecurity for the village community. Another peri-urban MW – the community talab of Jora village (near an agricultural university campus) – was encroached for the illegal construction of houses, and in the remaining in-filled area, a weekly market is being organized.

Urban wetlands in Bilaspur city, under Smart City Mission, are equally in bad ecological state. To demonstrate the extent of degradation, three urban wetlands were visited by the author. Bandhwapara talab, Bilaspur, was renovated sometime during 2005. Under renovation and beautification, households that had illegally occupied a sizable area of the lake for many years were moved to a new settlement. A bridge was constructed across the lake with water-recreation facilities. After two to three years, the lake again suffered with pollution and degradation. Now Bilaspur Municipal Corporation (BMC) entered an agreement to manage the lake with a private agency under a Private-Public Partnership (PPP) mode for 20 years, for an amount of Rs. five crores. In Tarbahar talab, the boundary was fenced, but the buffer zone was encroached. The lake is famous for naturally grown lotus. The third urban wetland (Jora talab, Sarkanda, Bilaspur) is seriously degraded.

Historical talabs of Ambikapur City are degraded and encroached by all kinds of sections of society. A sizable number of wetlands disappeared in recent decades.

Similarly, a sizable area of medieval Ganga Sagar and Dalpat Sagar of Jagdalpur city has been encroached and urbanized (Map 9.2). It is worth mentioning that the concept of rainwater harvesting was initiated by the Kakatiya dynasty 300 years ago in the Bastar kingdom. A large number of water-storage structures were constructed during their rules. Among these structures, Dalpat Sagar and Ganga water bodies were the most important ones which still exist in Jagdalpur city. Both the urban wetlands are multiuse

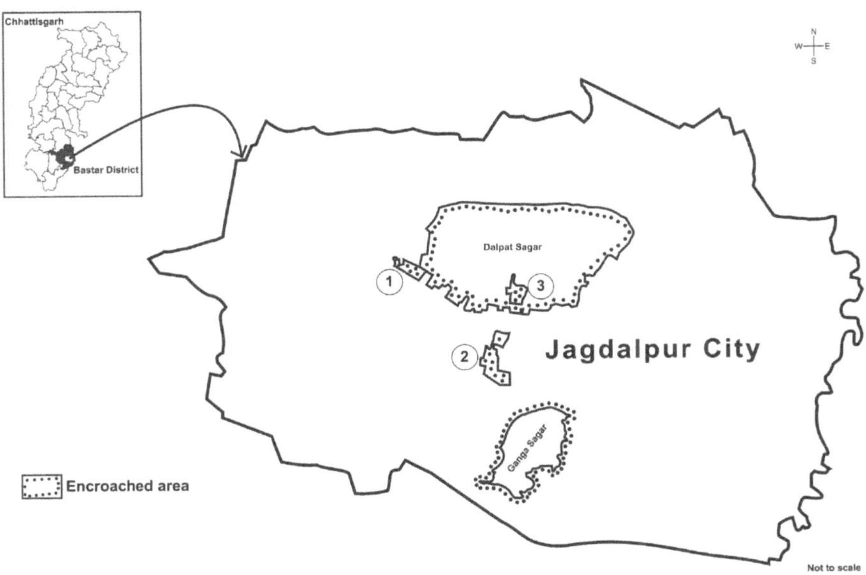

Map 9.2 Encroachment and Ecological Alteration Process (From 2006 to 2020) in and Around Dalpat Sagar and Ganga Sagar, Jagdalpur (devised by the author with technical input from Aadikant Pradhan)

and multifunctional and provide all ecosystem services. The slope direction of Jagdalpur city is south to north, which promotes the flow of rainwater into the Indrāvati river, which was used for constructing storage structures, to store rainwater in sequence. In between, three bodies of water and a few small ponds were constructed to collect base flow; one pond remained intact and the rest of the ponds disappeared. Rainwater first rests in Ganga Sagar, and the overflow moves to Dalpat Sagar. After water is retained in Dalpat Sagar, further surplus rainwater moves to the Indrāvati River. Nowadays, encroachment holds back the flow of rainwater due to the construction of colonies and cultivated lands. Further, due to encroachment, water enters residential areas. As a result, the area always faces a water stagnation problem. Also, the construction of retaining bunds around these water bodies restricts rainwater from entering the water bodies (see Pradhan and Patil, 2017; Pradhan et al., 2019, for effects of catchment and hydrology of Dalpat Sagar Reservoir on water resources availability and flow).

The trend of wetland extinction, degradation, or destruction, continues in spite of the Wetland Rules 2010 and the current Rules 2017, and the Ramsar Resolution of Wetland and Urbanization. Irrespective of the tenure of mayors and village heads, the wetland wealth of the state has disappeared to large extent. Besides, alteration of the associated ecosystem, especially green canopy cover, showed a sharp decline in most of the city's wetlands. This has altered the micro-climate of the town with higher temperatures during summer and flooding during the monsoon. During the last two decades, streets and residential colonies of Raipur city have been flooded just after the first rains. Drained water enters into the houses, residential, and commercial complexes – a common sight that can be seen every year during the rainy season. This situation is worse in the areas of the city where wetlands once existed but were destroyed for the construction of houses, colonies, and urban infrastructure. In such areas, a wide spread of dengue, malaria, and jaundice has been reported every year and many human lives lost due to consumption of contaminated water. Furthermore, many people have lost their property due to choking of drainage systems. Earlier in such areas, wetlands played an important role in storing rainwater, filtering pollutants from the water, monitoring the water cycle, the nutrient cycle, and the carbon cycle, and channelling the water flow to prevent water-logging and flooding. Urban flooding risks have also increased as a result of increasing concretized areas and wetland encroachment. Municipal authorities and the state government have largely failed to control urban floods. Unrelenting flooding during the rainy season can be contained if urban and peri-urban wetlands are restored with active participation of the local community. Catchment areas, if encroached, also need to be dispossessed to revive dead drainage systems.

Due to the conversion of wetlands for urban development, the availability of water for domestic use has reduced significantly. The market for water through tankers has grown exponentially in the last three decades in Raipur city. This is mainly because groundwater level dipped below threshold levels in many localities of Raipur city. Thus, wetlands served as a kidney and lifeline for these areas and Raipur city prior to the loss of wetlands. The water bodies also acted as a refuge for migratory birds and supported an abundance of animals and plants (Marothia, 2014, 2015, 2019). The LV restoration approach has clearly indicated that Smart Cities need a polycentric governance approach more than investment on heavy infrastructure (see Chapter 5).

The Smart Cities Project is faced with multiple challenges and indistinctness in terms of its understanding about wetland hydrology, ecology, the application of green infrastructure, urban socio-economic development and planning, governance coordination skill, and up-scaling good practices.[5] The consequences of ecological alteration proved, beyond doubt, that smart cities need nature's friendly governance bodies to protect the lake's unique biodiversity in the emerging urban area. RSCL has done a similar ecological alteration in the influence zone of Lake Vivekananda (Chapter 5).

9.2.2 Ramsar Resolution Vi.5 (1996): Inclusion of Subterranean Karst Wetlands

Under Resolution Vi (Inclusion of Subterranean Karst Wetlands as a Wetland Type Under The Ramsar Classification System, Convention on Wetlands (Ramsar, Iran, 1971), 1996), the Convention recognizes the "fundamental ecological functions of wetlands as regulators of water regimes and as habitats supporting a characteristic flora and fauna". The resolution recognized that "some subterranean cave and Karst systems provide natural underground wetlands and constitute a resource of ecological, cultural, scientific, aesthetic and recreational value, providing an environment for specialist vertebrate and invertebrate species and in many cases being the groundwater source for otherwise arid areas". The Conference of the Contracting Parties (COP) decided that subterranean Karst and cave hydrological systems shall be added to the Ramsar Wetland Classification System, and urged Contracting Parties to assess the significance of Karst and cave wetland systems within their territories and to consider their designation for the list. In Chhattisgarh, several pre-historical Karst wetlands exist (see Chapter 3) which fit well within the framework of Resolution Vi. Many Karsts are located in the national parks and provide valuable ecosystem services (see Chapter 3 for the distribution of Karst/caves in Chhattisgarh).

9.2.3 Ramsar Resolution XII.10: Wetland City Accreditation

Recognizing the important role of cities in protecting and enhancing the benefits from urban wetlands, the Convention introduced a Wetland City Accreditation Scheme (Ramsar Resolution XII.10 of 2015) (Ramsar

Convention, 2015). This voluntary scheme provides an opportunity for cities that value their natural or human-made wetlands to gain international recognition and positive publicity for their efforts. The Wetland City Accreditation Scheme encourages cities to highlight and strengthen their positive relationship with their wetlands, with an emphasis on creating public awareness of wetland benefits, incorporating wetland protection into municipal planning and decision-making, and promoting wetland benefits for local people. This scheme provides positive branding opportunities for cities that demonstrate strong and positive relationships with their wetlands (see annex 3 for detailed provisions of the Resolution XII.10 of 2015, Natori, 1999).

During COP13, the Ramsar Convention recognized 18 cities, located in France (four cities), China (six cities), Republic of Korea (four cities), Hungary (one city), Madagascar (one city), Sri Lanka (one city) and Tunisia (one city). These countries have taken extraordinary steps to safeguard their urban wetlands. These pioneering cities can serve as examples and inspire authorities in other countries to plan sustainable urbanization. The secretariat has launched a call for applications to the second edition (2019–2021 trienniums) of the Wetland City Accreditation (Wetland City Accreditation of Ramsar Convention-Guidance Note for Administrative Authorities, 25th July 2019).

As mentioned in Chapter 4, the group of wetlands or wetland complexes in Ratanpur can be identified for Wetland City Accreditation for their historical, socio-cultural, religious, and/or economic importance, as well as for biodiversity and hydrological ecosystem services. I along with late Prof. Brij Gopal and other colleagues, initiated steps towards the conservation of Ratanpur wetlands with the objective of obtaining their designation under the Ramsar Convention (together with Ratanpur as a Wetland City) in view of its uniqueness, water-centric and diversified livelihoods it supports, and its very high stake of local community (Chapter 4 deals with Ratanpur wetland complex).

9.2.4 *Ramsar Convention Resolution VIII in 2002: Interactions Between Agriculture, Wetlands, and Water Resource*

The Ramsar Convention adopted Resolution VIII in 2002 to enhance the interactions between agriculture, wetlands, and water resource management, and subsequently in support of UN International Year of Family Farming, Ramsar's theme for World Water Day 2014 was Wetland and Agriculture. Article VIII.34 of the Ramsar Convention Resolution VIII adopted in 2002 called for setting guidelines to enhance the interaction between agriculture, wetland, and water.[6] Resolution VIII.34 requested the Conference of Contracting Parties (COP) to establish a framework for identifying, documenting, and disseminating good agriculture-related practices. For developing guidelines, a scoping document on Agriculture-Wetland Interactions was produced and published (Wood and Halsema 2008, see Zingstra, 2009 for further elaboration) to explore the nature of WAIs through the application of the drivers, impacts, pressures, state changes, impacts and responses (DPSIR) framework (see Chapter 3 for water-wetland-agriculture interactions).

9.2.5 COP 14: Wetlands Identified for Priority Restoration

The Fourteenth Session of the Conference of the Parties to the UN Convention to Combat Desertification (UNCCD COP 14) took place in New Delhi, India, between 2–13 September 2019 and underlined the importance of wetlands not only as a source of water but also as ecosystems that prevent land degradation and desertification. The Government of India has identified 130 wetlands for priority restoration in the next five years and asked states to submit their respective integrated management plans by October 15, 2019. The highest number of such identified wetlands are in Uttar Pradesh (16) followed by MP (13), Jammu & Kashmir (12), Gujarat (eight), Karnataka (seven), and West Bengal (six). Each of these wetlands will be restored under a comprehensive scheme of the National Plan for Conservation of Aquatic Ecosystems (NPCA) to conserve and restore wetlands and lakes. Not a single wetland from Chhattisgarh figured in 130 wetlands identified for restoration. The pro-active engagement of the government of Chhattisgarh might have added at least a few wetlands to the list.

In view of the above implications of the Ramsar Convention Resolutions and India's efforts of conservation of wetland ecosystems, a comparative performance of MWs under alternative governance is presented.

9.3 Comparative Performance of MWs Under Alternative Governance

An effort is made in this book to analyse the performance of diverse institutional arrangements designed for managing MWs under different property right regimes or governance structures. A synthesis of institutional frameworks (see Figure 2.1, Chapter 2) forms the key methodology, applied in the case studies reported in this book. All the attributes of the framework (Figure 2.1) are now used for the assessment of different governance structures in managing MWs. Sub-attributes of the main nine attributes are listed in Tables 9.1 to 9.3. I also discuss briefly herein the process by which institutional arrangements affect MWs governance, and in turn the transformation of WAIs.

The description of performance indicators presented in the three tables is basically extracted from the findings of Chapters 4 to 8. Findings from Chapters 4 to 8 show that the success or failure of wetland policies essentially depend on the intensity of adoption of institutional arrangements and their potentiality to induce collective action among users. Further, results of Chapters 4, 5 and 6 clearly indicate that ineffective and non-functional governance systems have resulted in poor adoption of institutional arrangements, and in turn failed to define choice-sets of individual resource users and groups of resource users under different governance or property regimes. This ultimately led to wetland degradation and affected their ability to provide key ecosystem services. However, a case of self-organized and self-governed *dabri* system (Chapter 7) indicated that even soft or informal institutional arrangements supported by a vigilant local governing body (village Panchayats) can sustain wetlands over long periods of time. The policy problem is alarmed with recovering weak authority systems (e.g. Panchayats, municipalities, and concerned departments engaged in managing MWs) for making institutional arrangements work under different categories of governance. These results have obvious policy implications for the governance of MWs and the welfare of poor people, but also suggest a general approach for designing policies for MW management. Institutional arrangements (both external and internal) are important attributes in MW management under any property rights regime.

Institutional arrangements in a dynamic framework have the potential to redefine the property rights structures. Transition of institutional arrangements can become a major source of ecosystem degradation in many cases due to poor understanding on the part of policymakers about the social-ecological linkages, users' participation in the programs, and patterns of interaction. For example, ineffective authority systems resulted in poor implementation of external and internal institutional arrangements in cases of multiple tank-irrigation systems. Similarly, in the case of the wetland complex, ineffective and the poor political legitimacy of the municipality was responsible for degradation of many ponds within the wetland complex. On the other hand, in the case of fisheries and paddy cultivation in community wetlands and the *dabries* irrigation system, the internal institutional arrangements of distributed governance systems have effectively managed the WAIs. Similarly, small group farmers have effectively managed riverbed cultivation and fox nut cultivation with informal and soft institutional arrangements (see Table 3.21) Under the Smart City Mission, for example, RSCL had developed *An Action Plan to Repair, Renovate, Restore and Rejuvenate Water Bodies* across the Raipur City during 2016–2020 without understanding the complex peri-urban and urban wetland ecosystems and spatial and temporal socio-ecological-cultural externalities and possible irreversible damage. Similar trends can be observed in other emerging urban cities of the country.

9.4 Policy and Governance Issues

Previous chapters of this book have documented that all MWs provide use and non-use values and represent sources of ecosystem services and functions. It has also demonstrated that MWs have been lost and degraded due to the ineffective implementation of institutional structures which are the core moving attributes constructed in different wetland policy and governance. The conservation of wetlands with the objective of protecting their

Table 9.1 Attributes of MWs Governance

Attributes of MWs	*Wetland Complex Ratanpur*		*Urban Wetland (Lake Vivekananda)*	*Tanks and Tank Irrigation*	*Dabri Irrigation System (Micro-minor Irrigation Tanks)*	*Fisheries in MWs*
	Ponds of the Wetland Complex	*Dulhara Lake (Tank)*				
Governance regime						
i. Construction period	Medieval	Medieval	Medieval	Prc and post-independence	2000–01 to 2019–20	Pre and post-independence
ii. Original ecological status	Land with fine silty to loamy skeletal soils	Land with fine silty, to loamy skeletal soils	Land with fine silty, to loamy skeletal soils	During pre-independence constructed on fertile land of farmers, degraded common land during post-independence	Degraded common and revenue land	During pre-independence constructed on fertile land of farmers, degraded common land during post-independence
iii. Original resource regimes	Kalachuri rulers	Kalachuri rulers	Kalachuri rulers	British rulers, State of Chhattisgarh (erstwhile MP)	GoCG	British rulers, State of Chhattisgarh
iv. Current property rights/resource regime	*Usufructuary* rights to FCSs. FCSs managing ponds for fish culture and aquatic crops (lotus, water chestnut) under common property regime (CP Regime)	Under state property regime irrigation is provided to farmers of four villages in the command area of the tank, *usufructuary* rights to FCSs under (CP Regime)	Raipur Municipal Corporation (RMC) and RSCL	State property regime, irrigation to farmers through seasonal (rainfed) and perennial (connected with canal irrigation network) tanks, *usufructuary* rights to FCSs under (CP Regime)	Panchayats directly managing *Dabries* and *Dabries* irrigation system, water users committee of irrigator farmers representing head, middle, tail enders, farmer's group under (CP Regime)	State property regime, to provide irrigation to farmers, *Usufructuary* rights to FCSs by DWR, GoCG, Panchayats operating at three-tier level FCSs managing ponds for fish culture under (CP Regime)

(*Continued*)

Table 9.1 (Continued)

Attributes of MWs	*Wetland Complex Ratanpur*		*Urban Wetland (Lake Vivekananda)*	*Tanks and Tank Irrigation*	*Dabri Irrigation System (Micro-minor Irrigation Tanks)*	*Fisheries in MWs*
	Ponds of the Wetland Complex	*Dulhara Lake (Tank)*				
v. Authority over MWs	Ratanpur Nagar Palika	DWR, GoCG	RMC	DWR, GoCG	Village Panchayat is the legal unit of three- tier Panchayats institutions (village, janpad, and district level Panchayats)	DWR, GoCG. Village Panchayat is the legal unit of three-tier Panchayats institutions (village, janpad and district level Panchayats)
vi. Number of MWs included	118 ponds of the wetland complex with 195.906 ha water-spread area	One wetland with 30.756 ha water-spread area	One wetland with water-spread area (28.619 ha)	Four perennial and four seasonal tanks	95 *dabries* from 85 villages of 11 blocks of 11 districts spread over three ACZs of Chhattisgarh state	Five FCSs (individual fishermen operating under private resource regime were also included for analysis)

vii.	Organizations engaged in water-sharing control mechanism of MU	SDF	DWR, GoCG and SDF	Transferred lake management to a private company. RMC and RSCL have exclusive control on the lake and buffer zone of the lake. Water and parks inside the lake are used for outdoor recreation, other uses are excluded. Prior to RSCL control, the fisheries' activities were carried out by FCS	DWR, GoCG and SDF and Panchayats (village, janpad and district for leasing out water bodies for fish culture)	Panchayats, water users' committee constituted by Panchayats democratically, and farmers groups	DWR, GoCG, Village Panchayats and SDF and FCSs
viii.	Multiple objectives/ uses of MW management (also includes Wetland-agriculture interactions)	To ensure adequate water in ponds for fish and aquatic-crop cultivation (lotus, water chestnut) and other ecosystem services	Irrigation, fisheries, lotus cultivation, socio-cultural, spiritual, and religious activities	Currently being used for water-centric recreation activities	To sustain rainfed farming system, ensure minimum water level for potential growth of fisheries	Primarily to provide irrigation for paddy. A good number of *Dabries* are also being used for fisheries & domestic activities	Irrigation, fisheries, domestic socio-cultural, and religious activities

Table 9.2 Current Biophysical and Technical Attributes, Social Structures of Users and Non-Users, and Decision-Making Arrangements

Attributes of MWs	*Wetland Complex Ratanpur*		*Urban Wetland* **(Lake Vivekananda)**	*Tanks and Tank Irrigation*	*Dabri Irrigation System (Micro-minor Irrigation Tanks)*	*Fisheries in MWs*
	Ponds of the Wetland Complex	*Dulhara Lake (Tank)*				
A. Current biophysical and technical attributes	Suffering with pollution, encroachment, and siltation – affecting water-retention capacity of water bodies	Catchment area drainage network has been distorted over the years; waste disposal and encroachment. Periphery of tank's bunds are illegally encroached, tank bed is silted, infected with weeds and pollution	Altered ecological characteristics resulted in irreversible damage to biodiversity of buffer zone	Water-carrying channels are damaged due to infected weeds and siltation water outlets are damaged. Encroachment of catchment area impeding water flow	Physical condition of *dabries* and water delivery structure of 90 % *dabries* are in very well-maintained condition	Biophysical and technical attributes of a large number of ponds leased by FCSs are in fairly good condition. However, condition of urban ponds leased in by FCSs are polluted and encroached
B. Attributes of users and non-users	Homogeneous users (fish and aquatic-crop growing community), heterogeneous non-users (tourist, recreationists, bird watchers, non- users perform socio-cultural-religious activities)	Heterogeneous (irrigators, fishers, local residents, pilgrims, socio-cultural-religious uses)	Largely urban users, non-users number reduced sustainably as indirect values almost vanished	Heterogeneous social class: Farmers and fishers are main water users, households of villages use water to meet domestic requirements and tending cattle	Homogeneous	Heterogeneous (irrigators, fishers, local residents)

C. Decision-making institutional arrangements	Internal institutional arrangements are effective to induce collective decisions among FCSs members	Non-existence of WUA resulted in absence of coordination with DCs and PCs. Hence, poor coordination between internal and external institutional arrangements resulted in continuous degradation of tanks and irrigation structure	Authoritative and bureaucratic control without participation of citizens and population living on the periphery	WUA has not been functioning. All tanks and tank-irrigation systems are managed by sub-engineer irrigation and Amin (field staff) including financial allocation Complete absence of workable internal institutional arrangements	Farmer's group collectively make decisions and comply with internal institutional arrangements	FCS work collectively to take decisions the Panchayat resolves conflicts, if any, between domestic users of tank water and FCS
D. Action situation	Positive relations among users, poor between RNP and users	Almost negative action situation between users and DWR	Negative action situation due to poor coordination between administration and society	Almost negative action situation between users and DWR	Positive action situation among users and Panchayats	Positive relation among members of FCSs, Panchayat, and irrigation and fisheries department
E. Patterns of interaction	Conducive to collective behaviour	Not contributing to induce collective	Even within RMC, RSCL, Department of Urban Development and other agencies responsible for monitoring water bodies have poor coordination and convergence	Not contributing to induce collective behaviour, competition to extract maximum water due to absence of rules and authority	Individual co-irrigators make reasonable efforts to avoid over use of water and negative externality	Collective action by the members of FCS produced collective interaction

Table 9.3 Performance of MWs Governance

Outcome and Assessment						
i. Efficiency-productivity and income levels	In terms of productivity, income criteria fish and aquatic-crop cultivation activities are economically efficient without affecting other ecosystem services and values	Lower productivity levels of crop, fish and lotus resulted in lower income than potential levels	In terms of private gains to a private company efficient, largely higher income group is enjoying urban amenities created.	Average productivity levels of paddy are almost close to the state average. This is due to competitive water extracting tendency of farmers to maintain maximum possible level of water to avoid drought condition, Uneven income and yield levels across head, middle and tail ends farmers	Outcomes of workable institutional arrangements in terms of shift from direct/broadcasting of seed to improved sowing method of paddy, increase in cropping intensity, increase in yield of main crops, average fish production and diversification. Rational use of available water promoted diversification.	Productivity levels of fish varies across FCSs depending on extent of restrictions on uses of feed and manure for fish production, also while using water for fish other users can't be excluded from using water for domestic and other activities. Within a FCS there is no variability in yield and income levels. Across FCSs variation in yield and income are there.
ii. Effectiveness	Effective internal institutional arrangements of FCS, but poor external arrangements to manage ponds of wetland complex. RNP needs to look beyond earning from lease and maximizing revenue, human welfare, e.g. aesthetic, existence, bequest, option and quasi-option values are	Due to non-functional WUA the basic institutional arrangement of PIM are not effectively implemented, collective irrigation management collapsed due to poor patterns of interaction and decision-making arrangements mechanism	Ineffective, as ecological alteration reached to the irreversible point. Lake biodiversity affected seriously	WUAs under PIM System-A Decentralized governance regime (although currently default).	To enforce use regulation Panchayats, water users committee and farmers groups have played critical role to minimize water subtractive behaviour of irrigators through engaging irrigators as limited amount of flow is available for limited months in *dabries*	The DWR, GoCG, and Panchayat ensures the minimum level of water required for fish culture there is no conflict between fishermen community and farmers. As specified in leasing policy FCS can't use fish feed and manure in the tank, and the Panchayat resolves conflicts, if any, between domestic users of tank water and FCS.

iii. Equitable gains	Among members of FCSs distributive gains are equally shared	In-equitable gains in terms of crop production and income. Equitable gains in case of FCS and lotus cultivation.	Restoration of lake targeted to create urban amenities, constant irritant for the population living at the periphery of the lake	In-equitable gains in terms of crop production and income. Equitable gains in case of FCS and lotus cultivation.	Panchayats protect *dabries* and restrict free ridding behaviour with irrigators support. This has resulted in-equitable gains from available water resource	The pattern of distribution of benefits varies extensively across the five FCSs in terms of wages paid to members, funds kept aside for next year investment and asset creation. The disposal pattern of fish produced across the FCSs centred around members.
iv. Political Legitimacy	Institutional arrangements to regulate resource users participation in the decision-making process, basically linked to political legitimacy criteria. Poor performance of Ratanpur Nagar Palika (weak authority) is largely related to it's inability to craft rules to make decisions to protect and conserve ponds of the wetland	Non-functional of WUA and poor power-sharing arrangement between WUA DWR, GoCG resulted in failure of PIM institutional arrangement. Visible consequences are continuous degradation of the lake ecosystem and water-carrying canals connecting to lake.	Poor coordination among various agencies responsible for managing the lake. Violation of the urban wetland resolution of the Ramsar Convention	Non-functional of WUA and poor power-sharing arrangement between WUA DWR, GoCG resulted in failure of PIM institutional arrangements. Visible consequences are continuous degradation of tanks and water-carrying canal connecting to tanks.	Due to well-defined institutional provisions in terms of operational rules and their effective implementation by Panchayat and irrigator groups, free riding behaviour could not breed among irrigators. The effective inter-relationships of *dabri* and irrigators community attributes, decision-making arrangements and patterns of interactions provided economic	Managed under cooperative governance structure – an internal institutional structure of distributed governance system. Collective action by the members of FCS through collective interaction and decision, performance of FCSs, in view of members, is satisfactory in relation to members' implicit goal to achieve higher employment, income and food security from fisheries activities.

(*Continued*)

Table 9.3 (Continued)

	complex. Also Disconnect between urban development department and RNP results in weak political legitimacy				and distributive outcomes for irrigators in terms of efficiency (crop or fish output and income) and equity (distributive gains in terms of equitable distribution of water among irrigators)	
v. Sustainability of MW ecosystem	Sustainable as wetland complex survived over centuries despite degradation of physical and technological parameters	Continuous degradation of/ lake tank and irrigation network resulted in unsustainable use of MWs	Un-sustainability created due to alteration of lake ecosystem, uproots trees and biodiversity associated with lake and buffer zone	Continuous degradation of tanks and irrigation network resulted in unsustainable use of tanks and tank irrigation	Sustainable dabries system	Internal institutional governance can sustain different activities even under some constraint due to collective action and internal decentralization
vi. Degree of conformance of rules by multiusers in interdependence situation	Internal institutional arrangements of FCSs are Conducive to collective action	Water distribution rules are non-operative. Hence, conflict over water use among farmers at head, middle and tail ends has become common. Also conflicts among farmers of four villages is common.	Non-existence – poor participation of citizens	Water distribution rules are non-operative. Hence, conflict over water use among farmers at head, middle and tail ends has become common.	Individual co-irrigators make reasonable efforts to follow rules to avoid over use of water and negative externality to co-users and system	Internal institutional arrangements of FCSs are Conducive to collective action even in complex interdependent situations due to multiusers

vii. Degree of external and internal institutional arrangements workability/ effectiveness	Effective between FCS and SDF, but poor between RNP and FCS	Very poor due to non-functional of WUAs, Effective in case of FCS	Total disconnect even within government agencies, RSCL and RMC	Very poor due to non-functional of WUAs, Effective in case of FCS	Soil Conservation officer, Rural extension officer and KVK provide technical guidance to farmers	Effective workability between FCS and SDF, Panchayats, and water resource department
viii. Flexibility of redesigning the external and internal Institutional arrangements	Very high within FCS, effectiveness of external arrangements is re-assessed by SDF with a interval of 3 to 5 years	There is enough flexibility built in PIM to continuously improve capacity of irrigators and executives of WUAs. But WUAs are defunct and DWR, GoCG has to initiate process of reviving WUAs with new and effective external and internal institutional arrangements.	Due to privatization of community and later of State ownership has little hope for re-crafting rules to conserve the lake	There is enough flexibility built in PIM to continuously improve capacity of irrigators and executives of WUAs. But WUAs are defunct and DWR, GoCG has to initiate process of reviving WUAs with new and effective external and internal institutional arrangements.	quite flexible in making even midterm adjustments in water-sharing mechanism through re-crafting decision-making rules	Very high within FCS, effectiveness of external arrangements is re-assessed by SDF with a interval of 3 to 5 years
ix. Effectiveness in sharing authority systems	within FCS fairly effective, but poor between FCS and RNP	within FCS fairly effective, but poor between DWR and users of MWs	Very Poor, as control by a private company on behalf of RMC and RSCL. Other agencies responsible for looking after urban wetland have no participation in managing the lake. Society's participation also have been neglected	within FCS fairly effective, but poor between DWR, WUAs and farmers	Fairly effective under all three management system: governed by Panchayats, group of farmers, and water users committee	Within FCS fairly effective, between DWR, Panchayats and users of MWs

(*Continued*)

Table 9.3 (Continued)

x.Total economic values (use and non-use values/ benefits	Very high-use and non-use values	Lower use and non-use values due to degradation, pollution, and poor authority system to manage MWs.	After restoration non-use values in terms of outdoor recreation increased, other values negatively affected	Lower use and non-use values due to degradation of tanks and water supply network	Very high-use and non-use values	Very high-use and non-use values
xi. Ecosystems services derived from MWs	Providing all ecosystem services simultaneously (provisioning, regulating, cultural and supporting	Despite poor ecological health providing all the ecosystem services, off course at lower level then potentiality.	At the most regulating and cultural ecosystem services (recreation) are provided	Due to degradation only provisioning and regulatory services are provided at lower level.	Providing all ecosystem services simultaneously (prominently provisioning and regulating, to some extent cultural and supporting services are also provided)	Providing all ecosystem services simultaneously (prominently provisioning and regulating, to some extent cultural and supporting services are also provided
xii. Contribution towards Achieving sustainable development goals	SDGs 1,2,5,6,8, 11,13 and 15	SDGs 1,2,5,6,8, 11,13 and 15	Poor contributes to sustainable development goals (SDGs 11, 13 and 15) due to ecological alteration of lake ecosystem	SDGs 1,2,5,6,8, 11,13 and 15	SDGs 1,2,5,6,8, 11,13 and 15	SDGs 1,2,5,6,8, 11,13 and 15
Signals from feedback loops *	Degradation (Physical and Technical condition) and illegal encroachment in the buffer zone of ponds of historical wetland complex due to weak authority of RNP resulting in lower fish and aquatic	Degradation of historical lake due to inefficient management of DWR to protect physical and technical properties of multiuse wetland. lower than potential yields of fish and aquatic crops	Lack of coordination among various agencies, namely, CGPCB, CGTB, RMC, RSCL, State Ministry of Urban Development, State Ministry of Forest, Environment and Climate Change, and the SWA resultant in degradation of LV ecosystem and efforts of restoration.	Degradation of tanks constructed during pre and post-independence era due to inefficient management of DWR to protect physical and technical properties of multiuse tank system. Due to non-functional WUAs, farmers are	Due to well-defined institutional provisions in terms of operational rules and their effective implementation by Panchayat and irrigator groups, free riding behaviour could not breed among irrigators. The effective inter-relationships of	Only fisherfolk are members of FCSs Outsiders are not involved in any activities of cooperative societies. Extremely high stake and control of the member over FCS. Collective decision making by the members of FCSs produce collective

	crops production frontiers even under common property regimes. Other ecosystem services and non-use values are also at sub-optimal level due to pollution and encroachments of many ponds of wetland complex. Within FCSs, desirable patterns of interaction among users produced outcome in terms of fish and aquatic-crop yield and income as a result of combine effect of physical and technological attributes and collective decision-making arrangements	were observed even under common property regimes. Other ecosystem services are also affected severely due to pollution and encroachments of thc lake. Users and non-users perceived continuous erosion in ecosystem services. Selfishness among farmers to extract maximum water for his field, poor excludability due to unregulated irrigation supply, water insecurity and uncertainty due to ineffective rules and poor authority systems. Not contributed to induce collective behaviour		to ensure adequate water supply and maintenance of tanks. Since Panchayats have no control over this tanks, farmers are unable to insist upon village Panchayat for the proper maintenance of the tanks. Continuous reduction in number of field workers of water resource department at village and command area level is also a major constraints. Water distribution rules are non-operative, as a result conflict over water use among farmers located at head, middle and tail ends has become common phenomena. Poor operation of sluice gates leads to wastage of water.	attributes, decision-making arrangements and patterns of interactions provided economic and distributive outcomes for irrigators in terms of efficiency (crop or fish output and income) and equity (distributive gains in terms of equitable distribution of water among irrigators).	In turn this results in possible high yield and income even under constraints on application of fish feed and manure.

(*Continued*)

Table 9.3 (Continued)

			Emergence of individual interest of farmers, each irrigator made all effort to maximum water for his field from total water supply/available, poor excludability due to unregulated irrigation supply, water insecurity and uncertainty due to ineffective rules and authority systems. Farmers not contributing to induce collective behaviour.		
Re-Crafting institutional arrangements for restoration/ sustainable use of MWs	**Setting up of state wetland authorities:** Wetlands Rules 2017 (MoEF&CC) provide provision for the setting up of state wetland authorities with a recommended composition with representation of several concerned Ministries/Departments and experts from different disciplines	Urban wetlands were formally recognized as a concern in the Ramsar discourse in (Res X.27, 2008), Resolution emphasized the safeguarding peri-urban and urban wetland ecosystem as they provide recreational, spiritual, and aesthetic dimensions of wetlands, as a means of supporting the health and	PIM is passing through power-sharing conflicts between traditional owners of water resources (DWR) and members of WUAs in terms of financial resources allocation for repair and maintenance work. Unless WUAs are empowered in real sense and supported with	Effective internal institutional arrangement under common property regime can sustain self –organized and self-governed small irrigation system even with heterogeneous users. This study has also documented forcefully that while degradation of many village commons, including dabries,	Taking a leaf from the traditional allocation arrangement of ponds, a few tanks may be left out for domestic uses and social rituals in a village and rest of the ponds can be exclusively used for culture fisheries without any restrictions on application of growth-promoting inputs to achieve

State Wetland Policy: Because wetlands are not only water-storage structures but provide multiple tangible and intangible benefits, it is necessary that a separate State Wetland Policy is formulated and Rules and Regulations are framed for the purpose of regulating various activities. For a long-term conservation and management of wetlands (including rivers) in the State, it is necessary to have a clear and concise Wetland Policy somewhat similar to other policies related to Water, Agriculture and Forests. ***Ratanpur needs a tag of Wetland City Accreditation:*** The group of wetlands or wetland complexes in Ratanpur can be identified for Wetland City Accreditation for their historical, socio-cultural, religious and/or economic importance, besides for biodiversity and hydrological ecosystem services.	well-being of people by facilitating access to such wetlands" Resolution also encourages "to connect urban wetlands with Convention on Biodiversity Diversity initiatives". Further Resolution encourages "involving municipalities in their planning processes and operational actions on wetland conservation and wise use. Local governments, including municipalities and communities can play a crucial role to ensure conservation and wise use of these wetlands" (see Resolution X.27 for detailed Provisions)."	irrigation management experts, the objectives of PIM cannot be fulfilled. This requires re-assessment of PIM to redesigned the external and internal institutional arrangements for participatory irrigation policy.	incorrectly attributed to CPR-management systems, actually originates in the dissolution of village-level working rules whose very purpose is to manage and safeguard rural natural resource-use pattern efficiently, equitably, sustainably, and satisfactorily for the rural community as a whole.	potential yield. Fortunately, in every village minimum 7 to 10 community ponds are still existing. Close-in ponds can be reserved for common use and more distant ponds for fish culture. Such simple political and administrative decisions can substantially reduce the inter-community conflicts. Similarly irrigation ponds/tanks can be exclusively used for irrigation and fish culture, if some ponds can be kept aside for catering the needs of villagers. Further, a feasible solution can be worked out between FCSs and DWR for de- silting the tanks, as one of the major concerns of the DWR is increasing silt load and reducing water-intake capacity of a tank.

(*Continued*)

Table 9.3 (Continued)

	State Wetland Authority needs to draft a urban wetland policy to conserve urban wetlands and the idea of Sponge Cities Mission can be part of wetland policy. Under sponge city municipalities in emerging cities may develop Watershed management and emergency drainage plan, drainage routes. Technologies are available for green infrastructure. This requires re-crafting Institutional arrangements to deal at policy, organizational operational levels.		That FCSs have tendency to lease in all the common ponds/irrigation tanks within eight km periphery (a norm prescribed in Leasing Policy) in a particular village to strategically eliminate FGs and IFs chance to lease in these common water resources. Further, in some cases all the leased out ponds/tanks have not been used for fish culture. The fishery policy of the state needs to be changed in consultation with the Water Resource Department, SDA, State Department of Rural Development and Panchayat, Fisheries organizations, farmers, and village community.

* Refer feedback arrows in Figure 2.1. of Chapter 2.

Note: Information presented in tables 9.1, 9.2 and 9.3 is extracted from details given in Chapters 3, 4, 5, 6, 7, and 8 of this book.

biodiversity, specific biophysical characteristics, and obtaining optimum benefits from them requires a major shift in policy related to land and water use. Setting policy priorities needs to be based around four basic questions:

1. What policy instruments are needed to reverse the trend of wetland degradation or loss, and how can restoration efforts or gains be sustained?
2. What policy actions at different governance hierarchy levels are required to ensure the "wise use" of wetlands without altering ecological characteristics?
3. What policy actions need to be initiated to identify wetlands for getting tag/s of nationally and internationally important wetlands?
4. What types of research input is needed to improve understanding of policymakers and wetland managers to conserve and sustain MWs in the changing social-ecological-political environment?

Indian states, including Chhattisgarh, may initiate the following actions to address the above and related questions.

9.4.1 Constitution of State Wetland Authority (SWA)

The provision of constitution of SWAs under the Wetland Rules 2017 is a welcome step towards the decentralized governance of wetlands. A few states have recently instituted state wetland authorities, including Chhattisgarh, but these authorities are yet to be fully functional. SWAs need to be legally equipped with national and international provisions, along the lines of National Green Tribunal, to formulate law and policies and ensure their implementation in case of rural, peri-urban, and urban human-made and natural wetlands. With legal teeth, SWAs may ensure "wise use and conservation" of all categories of wetlands mentioned in Ramsar Convention documents. Wetlands are recognized by the Supreme Court of India as 'Commons' (Supreme Court of India, 2011).[7] The state may consider nominations of experts from other parts of the country, if necessary, in addition to suggested experts under the Wetland Rules 2017. Under the new rules, power has been given to the state governments so that protection and conservation work can be done at the local level. Central government has mainly retained powers regarding monitoring.

SWAs need further decentralization within a state at zonal and district level to form a Zonal Wetland Authority (ZWA), and a District Wetland Authority (DWA) to conserve and manage urban and peri-urban wetlands. I would further suggest that at village Panchayat level, in the Panchayats governing setup, a committee may be formed as a Village Wetland Authority (VWA).

9.4.2 State Wetland Policy

Wetlands are not only water-storage structures but also provide multiple tangible and intangible benefits. It is therefore necessary that a separate State Wetland Policy is formulated and rules and regulations are framed for the purpose of regulating various activities. For the long-term conservation and management of wetlands (including rivers) in the states, it is indispensable to have a clear and concise wetland policy somewhat similar to other policies related to water, agriculture, and forests. An Inventory of Wetlands will be helpful in categorizing wetlands and identifying the issues to be addressed by the policy. The policy document can help ensure better coordination between multiple stakeholders in the government for effective conservation and sustainable management.

9.4.3 Wetland Management Plans and Guidelines

A detailed, long-term management plan is required to be prepared for each identified wetland. For wetlands located within the forest areas, the Forest Department has to prepare and implement a management plan, keeping also in view the rules and provisions of the Forest Act. However, in many cases, the threats to wetlands may originate from outside the forest through changes in hydrology and water quality, and therefore it will be necessary to coordinate with and involve other departments concerned with the catchments outside the forest. In other cases, wetland management has to be entrusted to a committee or local body.

The preparation of management plans also requires the development of specific guidelines that should be followed for uniformity. The guidelines can be developed by consulting the Ramsar Guidance on the subject and adapting them to local conditions. The preparation of guidelines for developing the management plans should be undertaken as a priority. The task may be assigned to suitable experts or an agency by the state wetland authorities. The preparation of management plans should be undertaken for wetlands which may be identified for designation as Ramsar sites, simultaneously with the compilation of necessary data on wetland characteristics, as per Ramsar Information Sheets. The management plans for other identified wetlands can be prepared after completing the inventory and the guidelines. Management plans need to follow the Wetlands (Conservation and Management) Rules of 2017.

> The Wetlands (Conservation and Management) Rules 2017 further require the state to identify wetlands to be notified as such through the MoEF&CC. The notification of wetlands assures that their land use and status are not changed in future without due process of law. Wetlands are recognized by the Supreme Court of India as 'commons' (Marothia, 2017).

9.4.4 Identification of National and International Importance Wetlands

SWAs should make sincere efforts to identify wetlands of national and international importance including cities with wetland complexes to get international wetland city/cities accreditation. The state of Chhattisgarh has not yet identified a single wetland for the purpose. The state should identify at least three important wetlands – one in each of the three ACZs and major river basins, at the earliest opportunity, without waiting for the inventory of all wetlands to be completed. Some wetlands in Ratanpur (Bilaspur) and Bastar (e.g. Dalpat Sagar, Ganga Sagar) may also be considered as Natural Heritage Sites. Such a designation adds to the pride of the state which ensures greater protection and better management of the wetland for optimal societal benefits.

Kotumsar Cave-Karst system in Kanger National Park and the group of wetlands in Ratanpur (Dist. Bilaspur) were identified as important wetlands for their historical, social-cultural, religious and/or economic importance, besides for biodiversity and hydrological ecosystem services (Gopal and Marothia, 2017; Marothia, 2018).

As already mentioned in Chapter 3, the Bombay Natural History Society has identified four Important Bird Area (IBAs) for conservation: Barnawapara and Gomarda wildlife sanctuaries in Raipur and Raigarh districts respectively, Udanti and Sitanadi sanctuaries in Gariyaband and Dhamtari, and Indrāvati tiger reserve in Bijapur District. A summary of avifaunal surveys carried out in wetlands located in different parts of Chhattisgarh has already been reported in Table 3.3. Good numbers of studies have been conducted in recent years and a comprehensive *Report on Birds of Chhattisgarh* is also available.[8]

Chhattisgarh has a few rare, unique protected areas and wildlife sanctuaries with good density of wetlands that may be potential candidates for Ramsar sites. The SWA may initiate work on proposals. For example, Udanti Sitanadi Tiger Reserve – this in first place – is the only habitat for Asian wild buffalo except Indrāvati tiger reserve which is practically inaccessible presently, and Assam, where the genetic purity of their population is in doubt, although five selected ones from there have been brought for ex-situ breeding, to be released in the wild later. Also, this forest is connected with the forest of the adjoining State of Odisha and may therefore be a very good habitat for tigers as well. This also is a place where the largest bovine and the Indian Bison and Asian wild buffaloes coexist. It has well over 250 avifauna varieties. Achanakmar Tiger Reserve is, again, a very good habitat, perfectly suited for tigers. There are recent records of black panther – a melanistic variation of *Panthera pardus fusca* (a subspecies of leopard that is found in Bharat). In Central India, other than here, only Tadoba Tiger Reserve has black panthers. This place again houses around 275 avifauna varieties. Bhoramdeo Wildlife Sanctuary has the most picturesque landscape of all the protected forests of Chhattisgarh. As it has common boundaries with Kanha National Park, this place has the best possibilities of sustaining a tiger population. This place also has a very good avifaunal diversity. In these three places, tiger is the main species – when we have a safe, apex species, the whole ecosystem is automatically taken care of.

There is the possibility of creating a Green Corridor between Barnawapara Wildlife Sanctuary and Hirakund of Odisha, extending up to Simlipal Tiger Reserve in view of the gradual disappearance of tigers from Barnawapara Wildlife Sanctuary. It is so because this place has been cut off from all adjoining jungles and has become an island of 250 sq km. Of late, this has been a home for a visiting herd of Asian elephants as well. As there is no connectivity, the elephants have to pass through human settlements to find foraging grounds (they can't stay in one place since the vegetation will be totally destroyed) and they constantly are on the move. With this corridor, the human-elephant conflict would be taken care of and the tiger population in this region is likely to be restored.[9]

9.4.5 Restoration of Hydrological Regimes

Both natural and human-made wetlands should be declared as a specific land-use category and their hydrological characteristics (sources and regimes) should be identified. Conversion of wetlands to any other land use or any reduction in their area or alteration in their water regime should be prohibited, except for strategic reasons after exploring other options and providing for compensatory measures. For example, in Chhattisgarh, the SWA may consider restoring the hydrological regime of Lake Vivekananda. This lake was a complex of seven lakes, as shown in Map 9.1, nearly seven decades ago. Similar activity may be taken for Chankhuri (Chapter 3) and Dalpat Sagar (see Map 9.2) to connect their disappeared waterways. If it happens, this will be the first restoration of its kind and will be a lesson for remaining states and South Asian countries who wish replicate such experiments to restore wetland complexes.

9.4.6 Protect Rural, Peri-Urban, and Urban Wetlands

Chhattisgarh should collectively make all efforts to safeguard its rural wetlands. Rural wetlands have been keeping four major ecosystem services (see Chapter 3) in equilibrium. The SWA should make a comprehensive, long-term plan to sustain rural wetlands for present and future generations.

The SWA should ensure the protection of peri-urban and urban wetlands and put a stop forever on the tendency of Municipal Corporations or smart city authorities to annex these water bodies for urbanization. If the trend of converting peri-urban wetlands continues, it will result in complete and irreversible ecological disaster.

9.4.7 Sponge Cities to Minimize Urban Floods

One of the promising ideas to deal with urban floods is "sponge cities". The idea of a sponge city is to make cities more permeable so as to hold and use the rainwater. Sponge cities absorb the rainwater, which is then naturally filtered by the soil and allowed to reach urban aquifers (cited from *www.Insightsonindia.com/2020/11/03/insights-into-editorial-time-for-a-sponge-cities-mission-in-india*). Sponge cities can be delivered effectively through an urban mission along the lines of the Atal Mission for Rejuvenation and Urban Transformation (AMRUT), National Heritage City Development and Augmentation Yojana (HRIDAY), and the Smart Cities Mission. To take forward the idea of the sponge cities, it can be part of wetland policy. Under sponge city municipalities, emerging cities may develop watershed management and emergency drainage plans and drainage routes. Technologies are available for green infrastructure (refer Hussain, 2020; cited from *www.insightsonindia.com/2020/11/03/insights-into-editorial-time-for-a-sponge-cities-mission-in-india*).

9.4.8 Riverine Wetlands/Floodplains Cultivation

Chhattisgarh has extensive and interconnected riverine wetlands. They are found along the edges of rivers, streams, and creeks and include floodplains, marshes, lakes, and billabongs. They are mostly freshwater. Floodplain and riverbed cultivation in the summer months is very common and extensively practiced by poor and underprivileged. Since floodplains/riverbed cultivation is highly profitable compared to regular cropping systems, the SWA should develop a policy to protect this cultivation system. The Department of Agriculture may develop a processing plan for watermelon and other cucurbits to add value. Such an initiative may minimize distress labour migration in the summer and sustain livelihoods at the local level.[10]

All development projects, such as those related to urban or industrial development, or those concerned with storage, diversion, and abstraction of water from any source, should consider all kinds of wetlands that may be affected directly or indirectly, within the project area and far away from them. In the case of interventions on rivers, the hydrological changes may cascade downstream to the entire river course (e.g., on floodplains). These projects should take into account especially the changes in biodiversity and the ecosystem services of wetlands, and their economic valuation should be integrated into the cost-benefit analysis of the project. The assessment of ecosystem services and their economic valuation should particularly address the benefits to the local community and their livelihoods.

9.4.9 Designation of the Ramsar Wetland City

The designation of the wetlands of Ratanpur, and Ratanpur as a wetland city, under the Ramsar Convention requires that: first, wetlands need to be made pollution and encroachment free; second, the lost hydrological connectivity between wetlands is reconnected to revive the wetland complex; third, the wetland complex is notified under the Government of India's Wetland Rules (Government of India and Ministry of Environment, Forest and Climate Change, 2017) with the approval of the Central Wetland Regulatory Authority; fourth, a proposal is made by completing the Ramsar Information Sheet and other documents for the consideration of the Union Government; and fifth, submit proposals for candidate cities consistent with the framework for the Wetland City accreditation of the Ramsar Convention. Chhattisgarh has missed the secretariat's call for applications to the second edition (2019–2021 trienniums) of the Wetland City Accreditation. The SWA should initiate a process to submit the proposal for the next call.

The SWA should initiate steps towards the conservation of the Ratanpur wetlands as a priority, with the objective of getting their designation under the Ramsar Convention (together with Ratanpur as a Wetland City) in view of its unique situation as an urban centre with a large number of wetlands that are managed by the local community for their livelihoods (complete details of the Ratanpur wetland complex are given in Chapter 4; see also details in Ramsar Convention documents, 2010a, 2010b).

9.4.10 Developing Institutional Mechanisms

Appropriate institutional structures and mechanisms should be established for governing wetlands as multiuse systems integrated with the overall management of the state's water resources for ensuring water, food, energy, and livelihood security, without compromising with the ecological integrity of the natural capital of the state, upon which our future depends.

9.5 Researchable Issues

The current state of knowledge of wetlands in most of the states of India is very poor, and at best fragmentary in some respects. Urgent efforts are required for research, training, and capacity-building on wetlands and their integration with the overall management of water resources, in concert with the National Water Framework Law. The following unexplored research areas need collective thinking and action.

9.5.1 Estimation of Different Categories of Wetlands

Estimates for the extent of different categories of wetlands are required for major ACZs and sub-zones. Using a remote sensing database, field studies may be carried out to improve the quality of data. Input from such research efforts can help state wetland authorities to develop wetland databases and design actionable policies within a decentralized governance framework.

9.5.2 Contribution of WAIs

In view of the Ramsar Convention Resolution VIII, research studies on the variety of agricultural systems in different types of wetlands may be undertaken in different ACZs and sub-zones of the country to assess the contribution of WAIs to conservation and the wise use of wetlands, to sustain livelihoods, provide balanced ecosystem services, and meet SDGs. Input from such studies can identify priority research areas in WAIs and provide guidelines to work out payment for ecosystem services generated from WAIs in multiuse wetlands.

9.5.3 Good Agricultural Practices

Very few studies are available, even globally, on good agricultural practices (GAPs) to support sustainable WAIs. To bridge this critical research gap, case studies need to be conducted taking into consideration the ecosystem services that are assigned to a specific wetland. It is also important to investigate whether current WAIs are driven by wetland farmers' needs or consumer demand, and are clearly planned to maintain or restore the ecological conditions of the wetlands under investigation (Zingstra, 2009 investigated such issues in depth).

9.5.4 WAIs In-Situ Interactions

WAIs can either happen in-situ or ex-situ (see Zingstra, 2009 for in-situ interactions). Direct interactions that occur in-situ, in the wetland itself, include, among others, growing rice in rice paddies, fisheries, lotus, chestnuts, and fox nuts, floodplain. The impact of in-situ agriculture differs from the complete transformation of wetlands (lotus or fox nuts or fish, for example, as the sole activity) to grow combinations of aquatic crops (fish + paddy; fish + lotus; fish + chestnuts). Well-informed research on the abiotic and biotic conditions of the specific WAIs is not available (Zingstra, 2009). Such information and analysis is essential to understand the extent of the use of feed and manure/ chemicals as growth-prompting agents and the sustainable use of water of MWs in competing activities (as described in Chapter 8).

9.5.5 WAIs Ex-Situ Interactions

Ex-situ interactions of agriculture mostly take place at a basin or catchment level (if wetlands are situated on agricultural land) outside the wetlands themselves. Indirect interactions of agriculture include non-point sources of pollution in the form of soil erosion and sediments, nutrients, pesticides, phosphors, metals, and bacteria. Ex-situ interactions also interrupt water flow (see the work of Zingstra, 2009 for ex-situ interactions). All these ex-situ factors can create potential spatial and temporal externalities, high transaction costs, and have an irreversible impact on wetland ecosystems, and in turn may have serious implications for WAIs sustainability. Ex-situ or indirect agriculture interactions are equally serious in peri-urban and urban MWs. To address these issues, case studies are needed to understand the complexities of WAIs in MWs.

9.5.6 Estimation of Direct and Indirect Use and Non-Use Values

The fast disappearance of many wetlands in different parts of the country strongly indicates that the wealth of peri-urban and urban wetlands has not been managed sustainably and has irreversible loss of WAIs. Research efforts are required to estimate direct and indirect use and non-use to guide urban and semi-urban development planners and wider society (Marothia, 2001a, 2004d, 2019).

9.5.7 An Interdisciplinary Framework Application

To enhance the research and policy understanding about multiple-use commons in general and MWs in particular, an interdisciplinary framework – for example, one developed in Chapter 2 – is appropriate to develop a synthetic approach that amalgamates concepts from different disciplines.

9.5.8 Establish an Independent Center for Research and Training in Wetlands

States may consider establishing independent centres for research and training in wetlands (including rivers) at the earliest. This may require the services of wetland experts from outside the states. Research on wetlands should be accorded high priority by supporting research scholars from different universities and colleges. This is essential for gathering basin information required for the notification and designation of wetlands. Such centres may also be engaged in creating awareness about wetlands and their benefits, and the importance of their conservation among the common people, school-going children, and students from various disciplines.[11]

Notes

1 For a critique of Wetland Rules 2017 see Gopal (2018) and RGNUL Student Research Review (RSRR): rsrr.in/2019/11/02/legal-responses-to-multiple-challenges-facing-wetland-management/me >Politics >Policy >Environment ministry notifies new wetland rules. See also www.livemint.com/Politics/y6Tr3tkrr3q28AmGKaBFII/Environment-ministry-notifies-new-wetland-rules.html (Live Mint, 28 Sept. 2017).

2 For relevance of urban wetland ecosystems, see Azous and Horner (2001); Marothia (2001a, 2004d); Bulkeley (2012); Ehrenfeld (2000); Ghosh and Sen (1987); Keil (2003); Mahan et al. (2000); Wildfowl and Wetland Trust (WWT) Consulting (2018); Hettiarachchi et al. (2014); Hettiarachchi et al. (2015);

Roy (2009); Smardon (2006, 2009); Wetlands International South Asia (2018); Joshi and Bhandari (2016); Hussain (2020). See also Water, Land and Ecosystems (WLE) IWMI. WETWin, Enhancing Governance in Wetland Management, Factsheet-2. For conservation and wise use of wetland, refer Dugan (1990); Chatterjee et al. (2008); Briggs (2014); Gevers et al. (2018).

3 Saurabh Kumar, Municipal Commissioner, Raipur Municipal Corporation (RMC) provided the documents of RSCL and RMC pertaining to the Action Plan to Repair, Renovate, Restore, and Rejuvenate Water Bodies across the Raipur City, Sudhir Gupta, Executive Engineer, Bilaspur Municipal Corporation for sharing the information and his experience related to status and restoration of urban wetlands in Bilaspur city. Thanks to both of them.

4 Local environmental journalists have frequently been covering reports of misuse of public funds on the pretext of beautification and restoration by multiple urban local bodies. Interestingly, investments have been made twice at different times for beautification and renovation of the same water bodies by different local urban bodies (see endnote 10 of Chapter 5).

5 See Dandekar and Pradhan (2015); Hussain (2020) for ideas of smart city governance. See also www.insightsonindia.com/2020/11/03/insights-into-editorial-time-for-a-sponge-cities-mission-in-india/

6 The origins of the Guidelines on Agriculture and Wetlands Interactions (GAWI) initiative go back at least as far as October 2002. At that time, in two consecutive weeks, in Valencia (Spain), there was the Global Biodiversity Forum 17 (GBF17) workshop "Wetlands and agriculture" and Resolution VIII 34 at the Eighth Meeting of the Conference of the Contracting Parties to the Ramsar Convention on Wetlands of International Importance especially as Waterfowl Habitat (Ramsar COP8) titled "Agriculture, wetlands and water resource management".

7 Supreme Court of India, Bench of Justice Markandey Katju and Gyan Sudha Misra, Judgement of 28 January 2011. Recognizing the importance of Common Pool Resources (CPR) in general and water bodies in particular, the Supreme Court of India noted in a judgment on 28 January 2011, "our ancestors were not fools. They knew that in certain years there may be droughts or water shortages and water was also required for cattle to drink and bathe. Hence, they built a pond attached to every village, a tank to every temple." It directed all state governments to prepare schemes for the eviction of those occupying water bodies and other village commons and to restore them to the community. Unfortunately, there has been little compliance with this directive so far (Marothia, 2017).

8 For bird diversity in Chhattisgarh, see Vishwakarma et al. (2014); Bharos et al. (2016a, 2016b); Bharos et al. (2018); Naidu et al. (2019); Bux and Bharos (2018). Rahmani et al. (2018); Bharos et al. (2019).

9 Input received from Mohit Sahu on these observations is highly appreciated.

10 Kumari et al. (2018); Mishra (2021); Kumari et al. (2020); Ramjan and Chhetri (2018).

11 Brij Gopal has provided useful insights for developing policy and research agenda. I am indebted to him.

Glossary of Hindi and Other Terms

Dabri Micro-minor irrigation tanks

Biasi Distribution of rice seedlings evenly, after ploughing in the field

Utera Sowing of the next crop seeds before harvesting of standing paddy crop

Zamindars/Landlord/Malgujjars Persons nominated to collect land revenue during British rule

Jagirdars Part of the ruling class during the Mogul period

Sar, Sarovar, Taal, Talab Different names of natural and human-made wetlands

Jheel Lake

Kabir Panthis Followers of the philosophy and teachings of saint Kabir

Jawara Planting wheat seeds in small cane baskets

Lakh 0.1 million

Rs Indian currency

Puja Thali A tray on which the entire prayer materials are accumulated and decorated

Tapu Island/atoll

References

Adger, W.N., 2002. "Inequality, Environment, and Planning". *Environment and Planning A*, 34: 1716–1719.
Adger, W.N., K. Brown, J. Fairbrass, A. Jordan, J. Paavola, S. Rosendo and G. Seyfang, 2003. "Governance for Sustainability: Towards A 'Thick' Analysis of Environmental Decision Making". *Environment and Planning A*, 35: 1095–1110.
Agrawal, A., 2001. "Common Property Institutions and Sustainable Governance of Resources". *World Development*, 29: 1649–1672.
Agrawal, A. and S. Narain, 1997. *Dying Wisdom: Rise, Fall and Potential of India's Traditional Water Harvesting System (State of India's Environment, Vol. 4)*. New Delhi: Centre for Science and Environment.
Agrawal, A. and E. Ostrom, 2001. "Collective Action, Property Rights, and Decentralisation in Resource Use in India and Nepal". *Politics and Society*, 29: 485–414.
Alchian, A. and H. Demsetz, 1973. "The Property Rights Paradigm". *Journal of Economic History*, 33: 16–40.
Aligica, P.D., 2006. "Institutional and Stakeholder Mapping: Frameworks for Policy Analysis and Institutional Change". *Public Organization Review*, 6: 79–90.
Alung, S., 2014. *Chhattisgarh Itihaas aur Sanskriti (in Hindi), (Chhattisgarh History and Culture)*. Allahabad: Anamika Prakashan.
Anderson, E., 1993. *Value in Ethics and Economics*. Cambridge, MA: Harvard University Press.
Andersson, K. and E. Ostrom, 2006. *An Analytical Agenda for the Study of Decentralised Resource Regimes*. Paper presented at the Meeting of the International Society of New Institutional Economics, Boulder, CO, 23 September 2006, pp. 1–29.
Andersson, K. and E. Ostrom, 2008. "Analyzing Decentralized Natural Resource Governance from a Polycentric Perspective". *Policy Sciences*, 41(1): 1–23.
Andreo, B., F. Carrasco, J. José Durán and J.W. LaMoreaux (eds.), 2010. *Advances in Research in Karst Media*. Verlag, Berlin and Heidelberg: Springer.
Anek, R.S., 2011. "Early Prehistoric Signatures of Man in Bastar Region, Central India". *Current Science*, 101(9): 1146–1149.
Anithakumari, L.R. and P.K. Aziz, 1989. "A Limnology of a Temple Tank in Thiruvananthapuram, Kerala". *Journal of the Inland Fisheries Society of India*, 21(2): 31–36.
Arnold, J.E.M. and W.C. Stewart, 1991. *Common Property Resource Management in India*. Tropical Forestry Paper No. 24, Department of Plant Sciences, Oxford Forestry Institute, University of Oxford, p. 51.

Ayers, C.E., 1944. *The Theory of Economic Progress*. Chapel Hill: University of North Carolina Press.
Aylward, B. and E. Barbier, 1992. "Valuing Environmental Functions in Developing Countries". *Biodiversity and Conservation*, 1: 34–50.
Azous, A.L. and R.R. Horner (eds.), 2001. *Wetlands and Urbanization: Implications for the Future*. Boca Raton, FL: Lewis Publishers.
Bachmann, M.R., 2016. "Restoration Journey of the Piccaninnie Ponds Karst Wetlands, South Australia". *Ecological Society of Australia and John Wiley & Sons Australia, Ltd, Ecological Management & Restoration*, 17(2): 102–111.
Ball, V., 1877. "On the Geology of the Mahanadi Basin and its Vicinity". *Records of the Geological Survey of India*, 10: 561–567.
Ballabh, V. (ed.), 2008. *Governance of Water: Institutional Alternatives and Political Economy*. New Delhi: Sage Publication, Ch.10, pp. 174–194.
Bandyopadhyay, J., K. Chopra and N. Ghosh (eds.), 2012. *Environmental Governance: Approaches, Imperatives and Methods*. The Indian Society for Ecological Economics, New Delhi: Bloomsbury.
Barbier, E.B., M. Acreman and M. Knowler, 1997. *Economic Valuation of Wetlands – A Guide for Policy Makers and Planners*. Gland, Switzerland: Ramsar Convention Bureau.
Barkley, P.W. and D.W. Seckler, 1972. *Economic Growth and Environmental Decay: The Solution Becomes the Problem*. New York: Harcourt, Brace, Jovanovich.
Bassi, N.M., D. Kumar, A. Sharma, and P. Pardha-Saradhia, 2014. "Status of Wetlands in India: A Review of Extent, Ecosystem Benefits, Threats and Management Strategies". *Journal of Hydrology: Regional Studies*, 2: 1–19.
Beetham, D, 1991. *The Legitimating of Power*. London: Macmillan.
Benson, D. and A. Jordan, 2017. "Environment Governance". In Richardson, D., N. Castree, M.E. Goodchild, A. Kobayashi, W. Liu and R.A. Marston (eds.), *The International Encyclopedia of Geography*. Oxford: John Wiley & Sons, Ltd., pp. 1–9.
Berkes, F., 1992. "Success and Failure in Marine Coastal Fisheries of Turkey". In Bromley, D.W. (ed.), *Making the Commons Work*. San Francisco: ICS Press, pp. 161–182.
Berkes, F., 2002. "Cross-Scale Institutional Linkages for Commons Management: Perspectives from the Bottom Up". In Ostrom, E., T. Dietz, N. Dolsak, P.C. Stern, S. Stonich and E.U. Weber (eds.), *The Drama of the Commons*. Washington, DC: National Academy Press, pp. 293–321.
Berkes, F. and C. Folke, 1992. "A Systems Perspective on the Interrelations Between Natural, Human-Made and Cultural Capital", *Ecological Economics*, 5: 1–8.
Berkes, F. and C. Folke, 1994. "Investing in Cultural Capital for Sustainable Use of Natural Capital". In Jansson, A.M., M. Hammer, C. Folke and R. Costanza (eds.), *Investing in Natural Capital: The Ecological Economics Approach to Sustainability*. Washington, DC: Inland Press.
Berkes, F. and C. Folke, 1998. *Linking Social and Ecological Systems: Management Practices for Social Mechanisms for Building Resilience*. Cambridge: Cambridge University Press.
Bharos, A.M.K., A. Bharos and A. Bharos, 2016a. "Birdlife in Barnawapara Sanctuary Dist. Balodabazar-Bhatapara, Chhattisgarh (Part 1)". *Newsletter for Birdwatchers*, 56(3): 29–34.
Bharos, A.M.K., A. Bharos and A. Bharos, 2016b. "Birdlife in Barnawapara Sanctuary Dist. Balodabazar-Bhatapara, Chhattisgarh (Part 2)". *Newsletter for Birdwatchers*, 56(4): 39–40.

Bharos, A.M.K., F. Bux, K. Basak, M. Ahmed and A. Bharos, 2018. "Avian Diversity and Range Extensions Records from Udanti-Sitanadi Tiger Reserve, Chhattisgarh, India". *International Journal of Fauna and Biological Studies*, 5(1): 214–227.

Bharos, A.M.K., A. Mandavia, R. Naidu and A. Badesha, 2019. "Distribution Range Extension of Grey Headed Lapwing (Vanellus Cinereus) in Chhattisgarh and Eastern Madhya Pradesh, Jharkhand, India". *International Journal of Fauna and Biological Studies*, 6(3): 11–13.

Bhuyan, B. and D.K. Marothia, 1990. "Agro-climatic Zonal Planning and Agricultural Development: A Sub Regional View". *Indian Journal of Agricultural Economics*, 45(3): 284-291.

Binswanger, H.P., V.W. Ruttan, et al., 1978. *Induced Innovation*. Baltimore: The Johns Hopkins University Press.

Bishop, R.C., 1978. "Endangered Species and Uncertainty: The Economics of a Safe Minimum Standard". *American Journal of Agricultural Economics*, 60: 10–18.

Biswas, J., 1992. "Kotumsar Cave Ecosystem: An Interaction Between Geophysical, Chemical and Biological Characteristics". *NSS Bull*, 54: 7–10.

Biswas, J., A.K. Pati and R.K. Pradhan, 1990. "Circadian and Circannual Rhythm in Air-Gulping Behaviour of a Cave Fish". *Journal of Interdisciplinary Cycle Research*, 21: 257–268.

Bodhankar, N. and A.K. Sinha, 2018. *Evaluation of Groundwater Quality and Its Impact on Human Health in Supebeda Village, Deobhog*, Report Submitted to State Planning Commission, Yojana Bhavan, Sector – 19, North Block, Capitol Complex, Naya Raipur, Chhattisgarh – 491602, p. 12.

Bowman, M., 2002. "The Ramsar Convention on Wetlands: Has It Made a Difference?" In Stokke, O.S. (ed.), *Year Book of International Co-Operation on Environment and Development 2002/03*. London: Earthscan Publications, pp. 61–68.

Bregoli, F., A. Crosato, P. Paron, and, M.E. McClain, 2019. Humans Reshape Wetlands: Unveiling the Last 100 Years of Morphological Changes of the Mara Wetland, Tanzania". *Science of the Total Environment*, 691: 896–907. (Access Paper)

Briggs, C., 2014. *Wise Use of Our Wetlands*. Presentation made on 12th November 2014, Ramsar Convention Secretariat, Gland, Switzerland.

Brinkmann, R. and M. Parise, 2012. "Karst Environments: Problems, Management, Human Impacts, and Sustainability – An Introduction to the Special Issue". *Journal of Cave and Karst Studies*, 74(2): 135–136.

Black, R.D., 1958. *The Theory of Committees and Elections*. New York: Cambridge University Press.

Boulding, K.E., 1957. "A New Look at Institutionalism". *American Economic Review*, 47: 1–12.

Boulding, K.E., 1975. "Notes on the Present State of Neoclassical Economics as a Subset of the Orthodox". *Journal of Economic Issues*, 9(June): 223–227.

Bromley, D.W., 1976. "Environmental Quality and Economics: What Are the Issues?" *Economic Issues*, Department of Agricultural Economics, University of Wisconsin, Madison, 4 pages.

Bromley, D.W., 1978. "Property Rules, Liability Rules and Environmental Economics". *Journal of Economic Issues*, 12: 43–60.

Bromley, D.W., 1989a. "Property Relations and Economic Development: The Other Land Reform". *World Development*, 17: 867–877.

Bromley, D.W., 1989b. *Economic Interests and Institutions: The Conceptual Foundations of Public Policy*. Oxford: Basil Blackwell.

Bromley, D.W., 1991. *Environment and Economy: Property Rights and Public Policy*. Oxford: Basil Blackwell.

Bromley, D.W. (ed.), 1992. *Making the Commons Work*. San Francisco: ICS Press.

Bromley, D.W., 1993. "Institutional Economics Revisited: Discussion". *American Journal of Agricultural Economics*, 75(3): 837–839.

Bromley, D.W., 2007. "Environmental Regulations and the Problem of Sustainability: Moving Beyond 'Market Failure'". *Ecological Economics*, 63(4): 676–683.

Bromley, D.W. and M.M. Cernea, 1989. *The Management of Common Property Natural Resources: Some Conceptual and Operational Fallacies*. Discussion Paper 57. Washington, DC: The World Bank.

Bromley, D.W., C. Donald Taylor and E. Donald Parker, 1980. "Water Reform and Economic Development: Institutional Aspects of Water Management in the Developing Countries". *Economic Development and Cultural Change*, 28(2): 365–387.

Bromley, D.W. and J. Paavola, 2002. "Economics, Ethics and Environmental Policy". In Bromley, D.W. and J. Paavola (eds.), *Economics, Ethics, and Environmental Policy: Contested Choices*. Oxford: Blackwell, pp. 261–276.

Buchanan, J.M. and G. Tullock, 1962. *The Calculus of Consent: Logical Foundations of Constitutional Democracy*. Ann Arbor: University of Michigan Press.

Bulkeley, H., 2012. "Governance and the Geography of Authority: Modalities of Authorization and the Transnational Governing of Climate Change". *Environment and Planning A*, 44: 2428–2444.

Bux, F. and A.M.K. Bharos, 2018. "Aquatic Avian Composition, Diversity and Their Food Resources in the Selected Ponds of Naya Raipur, Chhattisgarh, India". *International Journal of Life Sciences Research*, 6(3): 34–48.

Calabresi, G. and A.D. Melamed, 1972. "Property Rules, Liability Rules and Inalienability: One View of the Cathedral". *Harvard Law Review*, 85: 1089–1128.

Castle, E.N., 1978. "Property Rights and the Political Economy of Resource Scarcity". *American Journal of Agricultural Economics*, 60: 1–9.

Centre for Karst (Gospi, Croatia), 2009. "Sustainability of the Karst Environment Dinaric Karst and Other Karst Regions". International Interdisciplinary Scientific Conference (Plitvice Lakes, Croatia, 23–26 September 2009).

Cesario, Jr. F.J., 1969. "Operations Research in Outdoor Recreation". *Journal of Leisure Research*, 1(1): 33–50.

Chambers, R., 1988. *Managing Canal Irrigation: Practical Analysis from South Asia*. New Delhi and Oxford: IBH Publishing Company, Pvt. Ltd.

Chandra, K., S. Kumar Dutta, R. Prasad Gupta and A. Raha, 2015. "Diversity and Conservational Status of Avifauna in Bastar Plateau of Chhattisgarh, India". *Ambient Science*, 2(1): 31–43.

Chatterjee, A., B. Phillips and D.A. Stroud, 2008. *Wetland Management Planning – A Guide for Site Managers*. New Delhi, India: WWF, Wetlands International, IUCN & Ramsar Convention, p. 76.

Chaudhary, J.L., G.K. Das, S.R. Patel, S. Jain, S. Bhelawe, B. Kumar, S.K. Bal and P.V. Kumar, 2018. *Resource Characterization and Crop Statistics Analysis for Strategic Crop Planning in Chhattisgarh State*. Raipur: Technical Bulletin Number DeAM/AICRPAM/2018–19/02, IGKV, p. 70.

Chaudhary, J.L., G.K. Das, D. Unjan, U.K. Diwan, M.A. Sarath Chandran, V.M. Sandeep and P. Vijaya Kumar, 2018. *Studies on Extreme Weather Events in Chhattisgarh State for Strategic Crop Planning in Rice Based Cropping System*. Raipur: Technical Bulletin Number DeAM/AICRPAM/2018–19/01, IGKV, p. 39.

Chaudhary, J.L., S.K. Patil, R. Khavse, P. Santhibhushan Chowdary, N. Manikandan, S.C. Rao and V.U.M. Rao, 2015. *Agroclimatic Atlas of Chhattisgarh*. Raipur: Indira Gandhi Krishi Vishwavidyalaya, p. 128.

Chhattisgarh Council of Science and Technology (CCOST), 2018. *Inventory of Wetlands Using High Resolution Satellite Data (LISS-IV) and Creation of Wetland Database*. Raipur, Chhattisgarh: Chhattisgarh Pollution Control Board, 15 January, p. 34.

Chhattisgarh State Planning Commission, 2018. *Ecological Studies of Microbes, Macrophytes and Physico-Chemical Properties of Water Bodies in Ratanpur Nagar Palika of Bilaspur District in Chhattisgarh State*. Naya Raipur: Preliminary Report Study Conducted by Bilaspur University, Bilaspur, Submitted to Chhattisgarh State Planning Commission, Yojana Bhawan.

Chhattisgarh Tourism Board, Chhattisgarh: Full of Surprises, 2nd Floor, Udyog Bhavan, Ring Road No. 1, Telibandha, Raipur – 492 006 (C.G.) http://tourism.cg.gov.in.

Chopra, K., 2002. "Shared Management of Protected Areas: Possibilities and Constraints in Keoladeo National Park, Bharatpur". In Marothia, D.K. (ed.), *Institutionalizing Common Pool Resources*. New Delhi: Concept Publishing Company, pp. 634–656.

Ciriacy-Wantrup, S.V., 1963. *Resource Conservation: Economics and Policies*. Berkeley: University of California Press (revised edition).

Ciriacy-Wantrup, S.V., 1969. "Natural Resource in Economic Growth: The Role of Institutions and Policy". *American Journal of Agricultural Economics*, 51(5): 1314–1324.

Ciriacy-Wantrup, S.V., 1971. "The Economics of Environmental Policy". *Land Economics*, 47: 36–45.

Ciriacy-Wantrup, S.V. and R.C. Bishop, 1975. "Common Property as a Concept in Natural Resources Policy". *Natural Resources Journal*, 15(4): 713–728.

Clark, J.M., 1957. *Economic Institutions and Human Welfare*. New York: Alfred A. Knopf.

Clawson, M. and J.L. Knetsch, 1978. *Economics of Outdoor Recreation*. Published for Resources for the Future. Inc. Baltimore: The Johns Hopkins University Press.

Cleaver, F., 2000. "Moral Ecological Rationality, Institutions and the Management of Common Property Resources". *Development and Change*, 31: 361–383.

Coase, R.H., 1960. "The Problem of Social Cost". *Journal of Law and Economics*, 3: 1–14.

Coelho, P., 1976. "Rules, Authorities and the Design of Not-for-Profit Firms". *Journal of Economic Issues*, 10: 416–428.

Coleman, J., 1988. "Social Capital in the Creation of Human Capital". *American Journal of Sociology*, 94 (Supplement): S95–S120.

Commons, J.R., 1924. *Legal Foundations of Capitalism*. New York: Macmillan Company.

Commons, J.R., 1931. "Institutional Economics". *American Economic Review*, 21: 649–657.

Commons, J.R., 1934. *Institutional Economics*. New York: Macmillan Company.

Commons, J.R., 1950. *The Economics of Collective Action*. New York: Macmillan Company.

Convention on Wetlands (Ramsar, Iran, 1971), 1996. *Resolution Vi.5: Inclusion of Subterranean Karst Wetlands as a Wetland Type Under the Ramsar Classification*

System. Proceedings of the 6th Meeting of the Conference of the Contracting Parties (Brisbane, Australia, 19–27 March 1996).

Costanza, R., R. d'Arge, R.S. de Groot, S. Farber, M. Grasso, B. Hannon, K. Limburg, S. Naeem, R.V. O'Neill, J. Paruelo, R.G. Raskin, P. Sutton and M. van den Belt, 1997. "The Value of the World's Ecosystem Services and Natural Capital". *Nature*, 387: 253–260.

Costanza, R., S.C. Farber and J. Maxwell, 1989. "Valuation and Management of Wetland Ecosystems". *Ecological Economics*, 1: 335–361.

Cowardin, L.M., V. Carter, F.C. Golet and E.T. LaRoe, 1979. *Classification of Wetlands and Deepwater Habitats of the United States*. Washington, DC: U.S. Department of the Interior, Fish and Wildlife Service, p. 131.

Cunningham, A., 1872. *Report of a Tour in Bundelkhand and Malwa and in the Central Provinces (Vol. 6)*. New Delhi: Archaeological Survey of India.

Cunningham, A., 1881. *Report on Tours in the Central Provinces and Lower Gangetic Doab in 1880–81 (Vol. XVII)*. New Delhi: Archaeological Survey of India.

Curtis, D., 1991. *Beyond Government: Organisations for Common Benefit*. London: Macmillan.

Dandekar, P. and A. Pradhan, 2015. *Smart Cities Need Smart Governance More Than Heavy Infrastructure*, South Asia Network on Dams, Rivers and People (SANDRP), Delhi, June 25, 2015.

Das, D.P., A. Kundu, N. Das, D.R. Dutta, K. Kumaran, S. Ramamurthy, C. Thanavelu and V. Rajaiya, 1992. "Lithostratigraphy and Sedimentation of Chhattisgarh Basin". *Indian Minerals*, 46(3–4): 271–288.

Dasgupta, P., 1982. *The Control of Resources*. New Delhi: Oxford University Press.

Dasgupta, P., 1993. *An Inquiry into Well-Being and Destitution*. Oxford: Clarendon Press.

Dasgupta, P. and K.G. Mäler, 1994. "Poverty, Institutions, and the Environmental – Resource Base". Environmental Paper No. 9, World Bank, Washington, DC.

Davidson, N.C., 2014. "How Much Wetland Has the World Lost? Long Term and Recent Trends in Global Wetland Area". *Marine and Freshwater Research*, 65(10): 934–941.

Davidson, N.C., 2017. "Wetland Losses and the Status of Wetland Dependent Species". In Finlayson, Max C., G. Randy Milton, R. Crawford Prentice and Nick C. Davidson (eds.), *The Wetland Book: Distribution, Description and Conservation*. Dordrecht: Springer, pp. 369–381.

Davis, L. and D. North, 1970. "Institutional Change and American Economic Growth: A First Step Toward a Theory of Institutional Innovation". *Journal of Economic History*, 30: 131–149.

Davis, T.J., 1994. *The Ramsar Convention Manual: A Guide to the Convention on Wetlands of International Importance, Especially as Waterfowl Habitat*. Gland, Switzerland: Ramsar Convention Bureau.

De Groot, R.S., M.A. Wilson and R.M.J. Boumans, 2002. "A Typology for the Classification, Description and Valuation of Ecosystem Functions, Goods and Services". *Ecological Economics*, 41(3): 393–408.

Deb, D. and K.C. Malhotra, 1997. "Interface Between Biodiversity and Tribal Cultural Heritage: An Exploratory Study". *Journal of Human Ecology*, 8(3): 157–163.

Demsetz, H., 1967. "Towards a Theory of Property Rights". *American Economic Review*, 57: 347–373.

Desai, S., P. Vashishtha, and O. Joshi, 2015. *Mahatma Gandhi National Rural Employment Guarantee Act: A Catalyst for Rural Transformation*. New Delhi: National Council of Applied Economics Research.

Dewey, J., 1922. *Human Nature and Conduct*. New York: Henry Holt and Co.

Dey, R.K., 2015. *Ecological Studies of Wetlands of Raipur and Naya Raipur (Urban Areas)*. Raipur, Chhattisgarh: State Forest Research and Training Institute, p. 80.

Dey, R.K., 2017. "Bird Diversity of Ratanpur Wetland Complex, Presentation Made at the Workshop on Conserving Wetland Wealth of Chhattisgarh". In Marothia, D.K., B. Gopal and S.K. Patil (eds.), *Report of the Workshop on Conserving Wetland Wealth of Chhattisgarh*. Naya Raipur: Chhattisgarh State Planning Commission.

Dolbear, F.T. Jr., 1967. "On the Theory of Optimum Externality". *American Economic Review*, 57: 90–103.

Dore, M.H.I., 1999. "The Economics of Well-Being: A Review of Post Welfarist Economics". In Dore, M.H.I. and T. Mount (eds.), *Global Environmental Economics Dore*. Oxford: Blackwell, pp. 21–38.

Dorfman, J., 1968. "Commons, John R". In Sills, D.L. (ed.), *International Encyclopedia of the Social Science*. New York: Macmillan and Free Press, vol. 3, pp. 22–24.

Downs, A., 1957. *An Economic Theory of Democracy*. New York: Harper and Row Publishers, Inc.

Drobak, J.K. and J.C. Nye, 1997. "Introduction". In Drobak, J.K. and J.C. Nye (eds.), *The Frontiers of the New Institutional Economics*. San Diego, CA: Academic Press.

Dugan, P.J., 1990. *Wetland Conservation: A Review of Current Issues and Action*. Gland: IUCN.

Dutt, N.V.B.S., 1964. "A Suggested Succession of the Purana formations of the Southern Part of Chhattisgarh, Madhya Pradesh". *Records of the Geological Survey of India*, 93: 143–148.

Edwards, V.M., 1996. *Managing the Commons: A Framework for the Analysis of Institutional Change and Its Application to the Management of the Multiple Use Commons of the New Forest*. Research Monograph, Dept. of Land and Construction Management, University of Portsmouth, p. 342.

Edwards, V.M. and N.A. Steins, 1996. "Developing an Analytical Framework for Multiple-Use Commons". In *Voices from the Commons, Sixth Annual Conference of the International Association for the Study of Common Property (IASCP)*. Berkeley: University of California, 5–8 June.

Edwards, V.M. and N.A. Steins, 1999. "A Framework for Analysing Contextual Factors in Common Pool Resource Research". *Journal of Environmental Policy and Planning*, 1: 205–221.

Ehrenfeld, J.G., 2000. "Evaluating Wetlands Within an Urban Context". *Urban Ecosystems*, 4: 69–85.

Elliott, J.E., 1978. "Institutionalism as an Approach to Political Economy". *Journal of Economic Issues*, 12: 91–114.

Ely, R.T., 1914. *Property and Contract*. New York: Macmillan Company.

Etiegni, C.A., K. Irvine, and M. Kooy, 2016. "Playing by Whose Rules? Community Norms and Fisheries Rules in Selected Beaches Within Lake Victoria (Kenya) Co-management". *Environment, Development and Sustainability*. doi:10.1007/s10668-016-9799-2.

Euisoon, S., M. Hufschmidt, Y.-S. Lee, J.E. Nickum, C. Umetsu and R. Gregory, 1997. Valuating the Economic Impacts of Urban Environmental Problems: Asian

Cities. Working Paper No.13., UNDP/UNCHS/The World Bank-UMP, 1818 H Street, NW Washington, DC 20433, USA, p. 123.

FAO, 1998. Wetland Classification and Characterization for Sustainable Agricultural Development. Harare, FAO Subregional Office for East and Southern Africa. www.fao.org.

Feeny, D., 1994. "Frameworks for Understanding Resource Management on the Commons". In Pomeroy, R.S. (ed.), *Community Management and Common Property of Coastal Fisheries in Asia and the Pacific: Concepts, Methods and Experience*. Manilla: ICLARM Conference Proceedings, ICLARM, vol. 45, 189 pp.

Finlayson, C.M., R. D'Cruz, N. Davidson (Co-Chair) Others, 2005. *Ecosystems and Human Well-Being: Wetlands and Water Synthesis*. Washington, DC: World Resources Institute.

Folke, C. and F. Berkes, 1995. "Mechanism that Link Property Rights to Ecological System". In Susan, H. and M. Munasinghe (eds.), *Property Rights and the Environment – Social and Ecological Issues*. Washington, DC: The Beijer International Institute of Ecological Economics and World Bank, pp. 121–138.

Foundation for Environmental Conservation Insights into Editorial: Time for a 'Sponge Cities' Mission in India. www.insightsonindia.com/2020/11/03/insights-into-editorial-time-for-a-sponge-cities-mission-in-india/.

Frenken, K. and I. Mahapatra (eds.), 2002. *Wetland Development and Management in SADC Countries*. Proceedings of a Sub-Regional Workshop, Harare, Zimbabwe, FAO Sub-Regional Office for East and Southern Africa (available at ftp://ftp.fao.org).

Furubotn, E. and S. Pejovich, 1972. "Property Rights and Economic Theory: A Survey of Recent Literature". *Journal of Economic Literature*, 10: 1137–1162.

Gadgil, M. and P. Iyer, 1989. "On the Diversification of Common-Property Resource Use by Indian Society". In Berkes, F. (ed.), *Common Property Resource: Ecology and Community-Based Sustainable Development*. London: Belhaven Press, pp. 240–255.

Galbraith, J.K., 1973. *The Economics and the Public Purpose*. Boston: Houghton Miftin Co.

Galbraith, L.M. and C.W. Burns 2007. "Linking Land-Use, Water Body Type and Water Quality in Southern New Zealand". *Landscape Ecology*, 22: 231–241.

Garg, J.K., 2017. "Geospatial Techniques for Wetland Inventory, Condition Assessment and Management. Presentation Made at the Workshop on Conserving Wetland Wealth of Chhattisgarh". In Marothia, D.K., B. Gopal and S.K. Patil (eds.), *Report of the Workshop on Conserving Wetland Wealth of Chhattisgarh*. Naya Raipur: Chhattisgarh State Planning Commission.

George, J.P., M.M. Ujjaini and K.C. Yadav, 2015. "Status of Wetlands and Wetland Birds in New Raipur, in Raipur, Chhattisgarh, India". *International Journal of Multidisciplinary Research and Development*, 2(9): 378–381.

Gevers, I., E.M.J. Koopmanschap, K. Irvine, C.M. Finlayson, and A.A. van Dam, 2018. "Capacity Development for Wetland Management". In Finlayson, C.M. et al. (eds.), *The Wetland Book*. https://doi.org/10.1007/978-90-481-9659-3_274.

Ghezae, N., 2009. *Natural Resource Tenure – A Crucial Aspect of Poverty Reduction and Human Rights*. Sida Studies, No. 23, Stockholm: Swedish International Development Cooperation Agency, p. 158.

Ghosh, D. and S. Sen, 1987. "Ecological History of Calcutta Wetland Conversion". *Environmental Conservation*, 14(3): 219–226.

Gibbs, C.J.N. and D.W. Bromley, 1989. "Institutional Arrangements for Management of Rural Resources: Common Property Regimes". In Berkes, F. (ed.), *Common Property Resource: Ecology and Community-Based Sustainable Development*. London: Belhaven Press, pp. 22–32.

GoCG, 2014. *Remote Sensing Data and Map of Ratanpur (District – Bilaspur) Water Bodies Developed by Chhattisgarh Council of Science and Technology*. Raipur: Department of Fisheries.

GoCG, 2018–19. *Economic Survey*, Directorate of Economics and Statistics.

GoCG, 2019. *Quest Chhattisgarh: Nature, Facts, Culture, History, Tourism Board*. Raipur, Chhattisgarh: GoCG.

GoCG, 2019–20. *Forest Management Information System (FMIS): 2019–20*. Atal Nagar, Nava Raipur, Chhattisgarh: Aranya Bhawan.

GoCG, 2019–20a. *Annual Technical and Administrative Report of Fisheries*. Raipur: State Department of Fisheries, Government of Chhattisgarh.

GoCG, 2019–20b. *New Fisheries Policy*. Raipur: State Department of Fisheries, Government of Chhattisgarh.

Goldberg, V., 1974. "Public Choice – Property Rights". *Journal of Economic Issues*, 8: 555–580.

Gonce, R.A., 1971. John R. "Commons' Legal Economic Theory". *Journal of Economic Issues*, 5: 80–95.

Gopal, B. (Compiler), 1995. *Handbook of Wetland Management*. New Delhi: World Wide Fund for Nature-India.

Gopal, B., 2002. "Management Issues of Aquatic Resources Including Wetlands". In Marothia, D.K. (ed.), *Institutionalizing Common Pool Resources*. New Delhi: Concept Publishing Company, pp. 657–671.

Gopal, B., 2003. "Wetlands, Agriculture and Water Resource Management: Need for an Integrated Approach". *International Journal of Ecology and Environmental Sciences*, 29: 47–54.

Gopal, B., 2013. "Future of Wetlands in Tropical and Subtropical Asia, Especially in the Face of Climate Change". *Aquatic Science*, 75: 39–61.

Gopal, B., 2015. *Wetland Conservation for Biodiversity and Ecosystem Services: Needs a Shift in Land and Water Resources Policies*. New Delhi: Policy Brief, National Institute of Ecology, September 2015, p. 16.

Gopal, B., 2018. "Science, Policy and Law of Wetland Conservation in India". *International Society of Limnology (SIL) News*, 72: 14–15.

Gopal, B. and M. Chauhan, 2001. "South Asian Wetlands and Their Biodiversity: The Role of Monsoons". In Gopal, B., W.J. Junk and J.A. Davis (eds.), *Biodiversity in Wetlands: Assessment, Functions and Conservation*. Leiden: Backhuys Publishers, vol. 2, pp. 257–275.

Gopal, B. and D.K. Marothia, 2015. *Integrating the Economics of Wetland Biodiversity and Ecosystem Services in Management of Water Resources of River Ken: A Study*. New Delhi: Economics of Ecosystems and Biodiversity (TEEB)–India Initiative, GIZ, and Ministry of Environment, Forest and Climate Change, Government of India, p. 77.

Gopal, B. and D.K. Marothia, 2016. "Seeking Viable Solutions to Water Security in Bundelkhand". *Economic and Political Weekly*, 51(44 & 45): 21–23.

Gopal, B. and D.K. Marothia, 2017. *Wetlands of Chhattisgarh, Status and Policy Perspective*. Chhattisgarh State Planning Commission, Yojana Bhawan, Naya Raipur, India, p. 12.

Gopal, B., A.P. Sharma, S. Sharma, S. Chhatterjee, D.K. Marothia, M.A. Hassan, and K.S. Rao, 2016. "Capacity Building in the Assessment of Biodiversity and Ecosystem Services for Conserving Wetlands for the Future-A Featured Article". *APN Science Bulletin* (issue 6, April), Asia Pacific NETWORK for Global Change Research, Japan, pp. 34–40.

Gopal, B.W.J., J.C.M. Finlayson and C. Breen, 2008. "Present State and Future of Tropical Wetlands". In Polunin, N.V.C. (ed.), *Aquatic Ecosystems: Trends and Global Prospects*. Cambridge: Cambridge University Press, pp. 141–156.

Government of India (GoI), Ministry of Environment and Forest, 2010. *Wetlands (Conservation and Management) Rules 2010*. Notification, New Delhi, 4 December 2010.

Government of India (GoI) and Ministry of Environment, Forest and Climate Change, 2017. *Guidelines for implementing Wetlands (Conservation and Management) Rules 2017*. Notification, New Delhi, 26 September, p. 51.

Gruchy, A.G., 1969. "Neoinstitutionalism and the Economics of Dissent". *Journal of Economic Issues*, 3: 3–17.

Gruchy, A.G., 1972. *Contemporary Economic Thought: The Contribution of Neoclassical Economics*. Clifton: Augustus M. Kelley.

GSI, 2005. *Geological and Mineral Map of Chhattisgarh (1:1000,000)*. Hyderabad: Geological Survey of India.

Guba, E.G., 1990. *Paradigm Dialog*. London: Sage,

Guba, E.G. and Y.S. Lincoln, 1994. "Competing Paradigms in Qualitative Research". In Denzin, N.K. and Y.S. Lincoln (eds.), *Handbook of Qualitative Research*. London: Sage, pp. 105–117.

Gunn, J., 2004. *Encyclopedia of Caves and Karst Science*. New York: Taylor and Francis, p. 1940.

Gupta, P., 1973. *Parchin Chhattisgarh (in Hindi)(Ancient Chhattisgarh)*. Allahabad: Leader Press.

Gupta, A.K., 1986. "Socio-ecology of Stress: Why Do Common Property Resource Management Projects Fail?" In *Proceedings of the Conference on Common Property Resource Management*. Washington, DC: National Research Council, National Academy Press, pp. 305–322.

Hanna, S.S., C. Folke and K-G. Mäler (eds.), 1996. *Rights to Nature*. Washington, DC: Island Press.

Harvey, D., 1996. *Justice, Nature and the Geography of Difference*. Oxford: Blackwell.

Herbert, E.R., P. Boon, A.J. Burgin, S.C. Neubauer, R.B. Franklin, M. Ardo´n, K.N. Hopfensperger, L.P.M. Lamers and P. Gell, 2015. "A Global Perspective on Wetland Salinization: Ecological Consequences of a Growing Threat to Freshwater Wetlands". *Ecosphere*, 6(10): 206. http://dx.doi.org/10.1890/ES14-00534.1.

Hettiarachchi, M., C. Mcalpine and T.H. Morrison, 2014. "Governing the Urban Wetlands: A Multiple Case-Study of Policy, Institutions and Reference Points". *Environmental Conservation*, 41(3): 276–289.

Hettiarachchi, M., T.H. Morrison and C. McAlpine, 2015. "Forty-three Years of Ramsar and Urban Wetlands". *Global Environmental Change*, 32: 57–66.

Holland, A., 2002. "Are Choices Trade-Offs?" In Bromley, D.W. and J. Paavola (eds.), *Economics, Ethics, and Environmental Policy; Contested Choices*. Oxford: Blackwell, pp. 17–34.

Huryn, A.D., C.W Burns, R. Allibone, B.H. Patrick and D. Scott, 2003. "Inland Waters and Wetlands". In Darby, J.T., R.E. Fordyce, A.F. Mark, P.K. Probert and C.R. Townsend (eds.), *The Natural History of Southern New Zealand*. Dunedin: University of Otego Press, pp. 237–265.

Hussain, S., 2020. "Planning Challenges and Implications of Sponge City Concept in Managing Urban Floods in Kerala". *International Journal of Research in Engineering and Science*, 8(12): 29–35.

Indian Space Research Organisation (ISRO), Space Applications Centre, 2010. *National Wetland Atlas: Chhattisgarh*. Ahmedabad: Space Applications Centre, p. 138.

Indrawan, M., M. Yabe, H. Nomura and R. Harrison, 2014. "Deconstructing Satoyama – The Socio-Ecological Landscape in Japan". *Ecological Engineering*, 64: 77–84.

International Union for Conservation of Nature (IUCN) (www.iucn.org).

International Union for the Conservation of Nature and Natural Resources (IUCN)/ United Nations Environment Programme (UNEP/World Wildlife Fund (WWF)), 1991. *Caring for the Earth: A Strategy for Sustainable Living*. Gland, Switzerland: IUCN/UNEP/WWF.

Irvine, K., 2018. "Editorial: Aquatic Conservation in the Age of the Sustainable Development Goals". *Aquatic Conservation: Marine and Freshwater Ecosystems*. https://doi.org/10.1002/aqc.3007.

Irvine, K., L. Castello, A. Junqueira and T. Moulton, 2016. "Linking Ecology with Social Development for Tropical Aquatic Conservation". *Aquatic Conservation: Marine and Freshwater Ecosystems*, 26: 917–941.

Isangkura, A., 1998. *Environmental Valuation: An Entrance Fee System for National Park in Thailand*. EEPSEA Research Report Series, Economy and Environment Programme for Southeast Asia, Singapore, p. 30.

Jagdalpuri, L., 1994. *Bastar: Itihas EvamSankriti (Hindi) (English – Bastar: History and Culture)*. Bhopal, India: Madhya Pradesh Hindi Granth Academy, p. 285.

Jentoft, S., 1989. "Fisheries Co-Management: Delegating Government Responsibility to Fishermen's Organizations". *Marine Policy*, 13: 137–154.

Jha, M., 1982a. *Complex Societies and Other Anthropological Essays*. New Delhi: Navrang, Ch 9.

Jha, M., 1982b. *Civilisation of Mithila and Mahakushal*. New Delhi: Concept Publishing Company, pp. 23–160.

Jha, M., 1982c. *India and Nepal: Sacred Centers and Anthropological Researches*. New Delhi: M.D. Pub. Pvt. Ltd., Ch 4, pp. 61–102.

Jha, M., 1997. *Anthropology of Ancient Hindu Kingdoms – A Study in Civilisation*. New Delhi: M.D. Pub. Pvt Ltd., pp. 115–116.

Johnsaon, G.L., 1962. "Interdisciplinary Approaches to Land Economics Research". In Ackerman, J., M. Clawson and M. Harris (eds.), *Land Economics Research*. Baltimore: Johns Hopkins University Press for Resources for the Future, pp. 197–209.

Jordan, A.J., 2001. "Environmental Policy (Protection and Regulation)". In Smelser, N. and P. Baltes (eds.), *International Encyclopedia of the Social and Behavioural Sciences* Oxford: Elsevier, vol. 7, pp. 4644–4651.

Joshi, D. and A. Ram Bhandari, 2016. "Shifting Paradigms in Wetland Governance: Shaping and Reshaping Conservation". *Journal of Forest and Livelihood*, 14(1), August.

Joshi, P.K. and A.K. Agnihotri, 1985. "An Assessment of the Adverse Effects of Canal Irrigation in India". *Indian Journal of Agricultural Economics*, 39(3): 528–536.

Kanel, D., 1974. "Property and Economic Power as Issues in Institutional Economics". *Journal of Economic Issues*, 8: 827–840.

Kapp, K.W., 1968. "In Defense of Institutional Economics". *Swedish Journal of Economics*, 70: 1–18.

Kashyap, S.P., 2004. *Ratanpur Ka Vaibhav* (Hindi) (*Splendor of Ratanpur*). Ratanpur (CG): Babu Reva Ram Sahitya Samiti Publication.

Katiha, P.K. and D.K. Marothia, 2006. "Valuation of Inland Open Water Fisheries Resources". In Vass, K.K., K. Mitra, V.R. Suresh, P.K. Katiha and N.P. Shrivastava (eds.), *River Fisheries in India: Issues and Current Status*. Barrackpore: Inland Fisheries Society of India, pp. 299–320.

Keil, R., 2003. "Urban Political Ecology". *Urban Geography*, 24(8): 723–738.

Kelso, M.M., 1977. "Natural Resource Economics: The Upsetting Discipline". *American Journal of Agricultural Economics*, 59: 814–823.

Khai, H.V. and Y. Mitsuyasu, 2014. "The Demand of Urban Residents for the Biodiversity Conservation in U Minh Thuong National Park, Vietnam". *Agricultural and Food Economics*, 2: 10.

Kielhorn, F., 1888. "Rajim Stone Inscription of Jagapala of the Kulachuri Year 896". *The Indian Antiquary*.

Kingsford, R.T., G. Bino, C.M. Finlayson, D. Falster, J.A. Fitzsimons, D.E. Gawlik, N.J. Murray, P. Grillas, R.C. Gardner, T.J. Regan, D.J. Roux and R.F. Thomas, 2021a. "Ramsar Wetlands of International Importance – Improving Conservation Outcomes". *Frontiers in Environmental Science* (9), March: 1–7. http://dx.doi.org/10.3389/fenvs.2021.643367.

Kingsford, R.T., C.A. McLoughlin, R. Brandle, G. Bino, B. Cockayne, D. Schmarr, T. Gotch, V. Norris and J. McCann, 2021b. "Adaptive Management of Malkumba-Coongie Lakes Ramsar Site in Arid Australia – A Free Flowing River and Wetland System". *Sustainability (Switzerland)*, 13: 3043–3043, http://dx.doi.org/10.3390/su13063043.

Kiser, L.L. and E. Ostrom, 1982. "The Three Worlds of Action: A Metatheoretical Synthesis of Institutional Approaches". In Ostrom, E. (ed.), *Strategies of Political Inquiry*. Beverly Hills, CA: Sage, pp. 179–222.

Kiser, L.L. and E. Ostrom, 1987. "*Institutional Economics, Economic Theory, Economic Behavior, Tragedy of the Commons, Action Research Workshop Institutional Analysis*". IAD framework paper presented at the Conference on Advances in Comparative Institutional Analysis, Dubrovnik, Yugoslavia, October 19–23.

Kneese, A.V., 1971. "Environmental Pollution: Economics and Policy". *American Economics Review*, 61: 153–166.

Krutilla, J.V., 1967. "Conservation Reconsidered". *American Economic Review*, 57: 776–786.

Kulke, H. and D. Rothermund, 1998. *A History of India*. Psychology Press. ISBN 978-0-415-15482-6.

Kumar, R. and E.J. James, 2012. "Inland Wetland Ecosystems". In Parikh, K.S., N.H. Ravindranath, I.K. Murthy, S. Mehra, R. Kumar, E.J. James, E. Vivekanandan and P. Mukhopadhyay (eds.), *The Economics of Ecosystems and Biodiversity – India: Initial Assessment and Scoping Report*. New Delhi: Working Document, MOEFCC, pp. 69–112.

Kumar, R., R.J. McInnes, M. Everard, R.C. Gardner, K.A.A. Kulindwa, H. Wittmer and M.D. Infante, 2017. "Integrating Multiple Wetland Values into Decision-Making". Ramsar Policy Brief No. 2. Gland, Switzerland: Ramsar Convention Secretariat.

Kumar, R., R.J. McInnes, C. Max Finlayson N. Davidson, D. Rissik, S. Paul, C. Lijuani, L. Yinru, S. Capon and S. Fennessy, 2020. "Wetland Ecological Character and Wise Use: Towards a New Framing". *Marine and Freshwater Research*. www.publish.csiro.au/journals/mfr.

Kumari, R., A. Sharma, S. Bhagta and R. Kumar, 2018. "River Bed Cultivation: A Kind of Vegetable Forcing for Remunerative Returns". *International Journal of Current Microbiology and Applied Sciences*, 7(4): 359–365.

Kumari, R., S.K. Shukla, K. Parmar, N. Bordoloi, A. Kumar and P. Saikia, 2020. "Wetlands Conservation and Restoration for Ecosystem Services and Halt Biodiversity Loss: An Indian Perspective (Chapter 6)". In Upadhyay, A.K., R. Singh and D.P. Singh (eds.), *Restoration of Wetland Ecosystem: A Trajectory Towards a Sustainable Environment*. Singapore: Springer Nature Pvt. Ltd., pp. 75–85.

Lal, H., 1916. *Descriptive List of Inscriptions in the Central Provinces and Berar*. New Delhi: Government Press. Nagpur Perspective, M.D. Pub. Pvt. Ltd., Ch 4, pp. 115–116.

Leibhafsky, H.H., 1976. "Commons and Clark on Law and Economics". *Journal of Economic Issues*, 10: 751–764.

Livingston, M.L., 1987. "Evaluating the Performance of Environmental Policy: Contributions of Neoclassical, Public Choice and Institutional Models". *Journal of Economic Issues*, 21(1).

Long, E.J., 1953. "Freedom and Security as Policy Objectives". *Journal of Farm Economics*, 35: 317–322.

Low, N. and B. Gleeson, 1998. *Justice, Society and Nature: An Exploration of Political Ecology*. London: Routledge.

Mahan, B.L., S. Polasky and R.M. Adams, 2000. "Valuing Urban Wetlands: A Property Price Approach". *Land Economics*, 76(1): 100–113.

Malhotra, K.C., 2002. "Sacred Groves as Common Property Resources: An Exploratory Study". In Marothia, D.K. (ed.), *Institutionalizing Common Pool Resources*. New Delhi: Concept Publishing Company, pp. 511–532.

Maltby, E. (ed.), 2009. *Functional Assessment of Wetlands: Towards Evaluation of Ecosystem Services*. Cambridge, UK and Boca Raton: Woodland Publishing, CRC Press.

Mandal, H., S. Mukherjee and A. Datta, 2002. *India – An Illustrated Atlas of Tribal World*. Kolkata: Anthropological Survey of India.

Marothia, D.K., 1981. *An Economic and Institutional Analysis of Soil Erosion on Agricultural Lands*. Ph.D. Dissertation, The University of Alberta, Edmonton, Canada.

Marothia, D.K., 1986. "Tank Irrigation in Chhattisgarh Region: Management and Policy Issues". *Indian Journal of Agricultural Economics*, 41(4): 538–539.

Marothia, D.K., 1988. "Afforestation of Bhata Soils: Socio-Economic and Management Aspects". *Indian Journal of Agricultural Economics*, 43: 305–309.

Marothia, D.K., 1989a. "Socio-Economic Rationales of Waste Land Afforestation". *Journal of Tropical Forestry*, 5(4): 343–350.

Marothia, D.K., 1989b. "Resource Ownership and Human Interaction: A Theoretical Exposition". *Annals of Arid Zone*, 28: 165–180.

Marothia, D.K., 1990. *Diagnostic Study of Bastar District to Prepare Comprehensive Agricultural Development Strategies using Natural Resource Base*. Planning Commission, Govt. of India Project, January–December 1990. Indira Gandhi Krishi Vishwavidyalaya, Raipur, p. 59.

Marothia, D.K., 1992a. *Village Ponds and Aquaculture Development: Issues of Inequality in Common Property Regimes*. Contributory paper, Third Conference of the International Association for the Study of Common Property, Washington, DC, U.S.A., September 18–20.

Marothia, D.K., 1992b. "Village Irrigation Tanks: Institutional Design for Sustainable Resource Use". *Agricultural Situation in India*, 47(6): 479–488.

Marothia, D.K., 1993a. "Rapporteur's Report on Tribal and Hill Economy". *Indian Journal of Agricultural Economics*, 47(3): 567–578.

Marothia, D.K., 1993b. "Property Regimes and Institutional Arrangements: Concepts and Their Relevance in Managing the Village Commons". *Indian Journal of Agricultural Economics*, 48(3): 557–565.

Marothia, D.K., 1994. *Decentralized Planning for Bastar District: An Agro-Climatic Regional Planning Approach (ACRP)*, A Project Report of Planning Commission, Government of India, p. 119.

Marothia, D.K., 1995a. *Urban Lake Pollution: Issues of Property Rights and Institutional Arrangements*. Invited Paper presented to Beijer Research Seminar organised by the Beijer Institute – The International Institute of Ecological Economics, The Royal Swedish Academy of Sciences, Stockholm, Sweden, held at Kota Kinabalu, Malaysia, 16–19 May 1995.

Marothia, D.K., 1995b. *Village Ponds and Aquaculture Development: Is Conflict Inevitable?* Invited paper presented to the National Workshop on Poultry, Fisheries and Food Processing Organized for the Honorable Members of Parliament by National Academy of Agricultural Research Management, Hyderabad, July 4–7, 1995.

Marothia, D.K., 1996a. "Cooperative Management of Minor Forest Products in Madhya Pradesh". In Singh, K. and V. Ballabh (eds.), *Cooperative Management of Natural Resources*. New Delhi: Sage Publications, pp. 92–108.

Marothia, D.K., 1996b. "Cooperative Management of Non-Timber Forest Products: Issues of Property Rights and Institutional Structures". In Rajgopalan, R. (ed.), *Rediscovering Cooperation Strategies for the Models of Tomorrow*. Anand: Institute of Rural Management, vol. 2, pp. 106–138.

Marothia, D.K., 1997a. "Agricultural Technology and Environmental Quality: An Institutional Perspective". Keynote paper, *Indian Journal of Agricultural Economics*, 52(3): 473–487.

Marothia, D.K., 1997b. "Institutional Arrangements in Common Fish Ponds". In Kerr, J.M., D.K. Marothia, K. Singh, C. Ramasamy and W.R. Bentley (eds.), *Natural Resource Economics: Theory and Application in India*. New Delhi: Oxford & IBH Pub. Co. pp. 463–476.

Marothia, D.K., 1997c. "Property Rights, Externalities and Pollution". In Agrawal, A. (ed.), *The Challenge of Balance*. New Delhi: Center for Science and Environment, pp. 290–298.

Marothia, D.K., 1997d. *Property Regimes in Medicinal and Aromatic Plan Resources: Some Policy Issues*. Keynote paper presented in National Conference on Health Care and Development of Herbal Medicines held at Indira Gandhi Agricultural University, Raipur, Chhattisgarh, 29–30 August 1997.

Marothia, D.K., 2001a. "Valuation of a Day-Use Recreation Site: Application of Alternative Estimation Techniques". *Indian Journal of Agricultural Economics*, 56(3): 313–324.

Marothia, D.K., 2001b. "Shared Management of Non-timber Forest Resources in Madhya Pradesh". *Indian Journal of Agroforestry*, 3(1): 1–10.

Marothia, D.K., 2002a. "Alternative Governance in Common Pool Resource". In Marothia, D.K. (ed.), *Institutionalizing Common Pool Resources*. New Delhi: Concept Publishing Company, pp. 701–716.

Marothia, D.K., 2002b. "Institutional Arrangements for Participatory Management: Initial Feedback from Central India". In Brennan, D. (ed.), *Water Policy Reform: Lessons from Asia and Australia*. Canberra: The Australian Centre for International Agricultural Research (ACIAR) Proceedings No.106, 2001, pp. 75–105.

Marothia, D.K. (ed.), 2002c. *Institutionalizing Common Pool Resources*. New Delhi: Concept Publishing Company.

Marothia, D.K., (Project Coordinator), 2002d. *Irrigation Tanks: Drought Coping Traditional Technology in Chhattisgarh*. IGKV-IRRI Project Report, Department of Agricultural and Natural Resource Economics, IGAU, Raipur, Chhattisgarh, India, p. 63.

Marothia, D.K., 2003a. "Domestic Trade in Aromatic Rice: Lessons from an Exploratory Study of Chhattisgarh". In Singh, R.K. and U.S. Singh (eds.), *A Treatise on the Scented Rices of India*. New Delhi: Kalyani Publishers, pp. 189–210.

Marothia, D.K., 2003b. "Enhancing Sustainable Management of Water Resource in Agriculture Sector: The Role of Institutions". Keynote Paper, *Indian Journal of Agricultural Economics*, 58(3): 406–426.

Marothia, D.K., 2003c. "Institutions for Common Pool Resources". In Suresh Pal, M., P.K. Joshi and R. Saxena (eds.), *Institutional Change in Indian Agriculture*. New Delhi: National Centre for Agricultural Economics and Policy Research Publication, ICAR, pp. 61–80.

Marothia, D.K., 2004a. *Impact Assessment of Technology Interventions and Crop Diversification in Tribal, Backward and Hilly Areas of India*. World Bank Funded Project under Jaivigyan National Science and Technology Mission on Household Food and Nutritional Security in Collaboration with ICAR, January 2000–December 2004.

Marothia, D.K., 2004b. *Minor Irrigation Water Storage Tanks – A Case Study of Bastar District, Chhattisgarh*. Research Report IWMI-TATA Water Policy Programme, Anand, Gujarat, India, January 3–4, p. 44.

Marothia, D.K., 2004c. "Institutional Structures for Governance of Freshwater Aquaculture: Lessons from Chhattisgarh". *Indian Journal of Agricultural Economics*, 59(3): 421–434.

Marothia, D.K., 2004d. "Restoration of Lake Ecosystem: An Environmental Economics Perspective". *International Journal of Ecology and Environmental Sciences*, 30(3): 197–207.

Marothia, D.K., 2004e. "Transforming Institutions for Sustainable Development of Natural Resources Programmes". *Agricultural Economics Research Review*, 17: 163–190.

Marothia, D.K., 2005a. "Institutional Reforms in Canal Irrigation System: Lessons from Chhattisgarh". *Economic and Political Weekly*, 40(28): 3074–3084.

Marothia, D.K., 2005b. "Common Property Resource: Managing the Unmanaged". In Sharma, B.R., J.S. Samra, C.A. Scott and S.P. Wani (eds.), *Watershed Management Challenges: Improving Productivity, Resources and Livelihoods*. Colombo: International Water Management Institute, pp. 157–172.

Marothia, D.K., 2005c. *Performance Evaluation of Culture Fisheries under Alternative Property Rights Regimes in Chhattisgarh*. Paper presented at IWMI-TATA Annual Partners' Research Workshop 2005 at IRMA, Anand, February 24–26.

Marothia, D.K., 2006a. "Alternative Institutional Structure for Sustainable Development of Common Pool Resources: Lessons from Chhattisgarh". In Parthasarathy, R. and S. Iyengar (eds.), *Development Paradigms and Challenges for Western and Central India*. New Delhi: Concept Publishing Company, pp. 393–412.

Marothia, D.K., 2006b. "Managing Culture Fisheries in Multi-use Water Bodies: A Property Rights Paradigm". In Vass, K.K., K. Mitra, V.R. Suresh, P.K. Katiha and N.P. Shrivastava (eds.), *River Fisheries in India: Issues and Current Status*. Barrackpore: Inland Fisheries Society of India, pp. 227–240.

Marothia, D.K., 2007. "Fish Farming in Small Multiuse Water Bodies: Water Sharing Mechanism and Its Implications". In Vass, K.K., N. Sarangi, K. Mitra, J.K. Jena, V.R. Suresh, N.P. Shrivastava and P.K. Katiha (eds.), *Water Management in Fisheries and Aquaculture*. Bhubaneswar, OR and Barrackpore: Inland Fisheries Society of India, W.B. and Association of Aquaculturists, pp. 151–159.

Marothia, D.K., 2009a. "Chhattisgarh Forest Policy at the Cross Road: An Institutional Interpretation". *International Journal of Ecology and Environmental Sciences*, 35(2–3): 219–230.

Marothia, D.K., 2009b. *Governance and Institutional Change in Traditional Commons: Lessons from Chhattisgarh, India*. Lead paper presented at the 4th World Congress on Conservation Agriculture, New Delhi, February 4–7, 2009. Proceeding of 4th World Congress on Conservation Agriculture, Indian Council of Agricultural Research, New Delhi, pp. 463–477.

Marothia, D.K., 2010a. "Decentralisation of Natural Resource Management in India: An Institutional Perspective". Presidential Address, *Indian Journal of Agricultural Economics*, January–March, 65(1): 1–34.

Marothia, D.K., 2010b. "Technological and Institutional Options for Common Property Resource Management in Rainfed Areas". *International Journal of Ecology and Environmental Sciences*, 36(1): 45–57.

Marothia, D.K., 2012a. "Performance of Culture Fisheries under Alternative Property Rights Regimes in Chhattisgarh". *International Journal of Ecology and Environmental Sciences*, 38(4): 163–207.

Marothia, D.K., 2012b. *Spread and Impacts of System of Rice Intensification: Drivers of Success and Failure in Chhattisgarh*. Invited paper presented in IWMI-TATA Policy Research Program, IRMA, Anand, November 28–30.

Marothia, D.K., 2014. *Urban and Peri-urban Wetlands of Chhattisgarh: The Dwindling Ecological Heritage*. Keynote Address delivered at the joint Workshop organized by INTACH and Pt Ravi Shankar Shukla University on Traditional Water Harvesting System and its importance in Urban Perspective, Also published in Central Chronicle, August 27, 2014.

Marothia, D.K., 2015. "Improving Water Sharing through Fostering Institutional Creativity: Lessons from Multi-Use Water Bodies Management". In Shrivastava, A.K., T.K. Datta, K.K. Vass, V.K. Gupta, and S. Ayyappans (eds.), *12th Agricultural Science Congress Proceedings*. Karnal: NDRI, February 3–6, 2015, pp. 304–311.

Marothia, D.K., 2016. "Institutional Reforms in Canal Irrigation System: Lessons from Chhattisgarh". In Shah, M. and P.S. Vijyashankar (eds.), *Water: Growing Understanding, Emerging Perspectives*. New Delhi: Orient Blackswan Pvt. Ltd. and *Economic and political Weekly*, 2016, pp. 75–101.

Marothia, D.K., 2017. *Governance for Water Security in 21st Century: Framing of Institutional Choices*, Third Dr. Rakesh Hooja Memorial Lecture delivered at

Indian Institution of Public Administration (IIPA), 22 November, Proceeding of the Lecture Published by IIPA, New Delhi, p. 51.

Marothia, D.K. (Chair), 2018. *Transforming Agriculture and Allied Sectors in Chhattisgarh: Report of the Task Force and Standing Working Groups.* Naya Raipur: Chhattisgarh State Planning Commission, Yojana Bhawan, pp. l–293.

Marothia, D.K., 2019. "Managing Wetland Ecosystems: A Polycentric Perspective". In Peshin, R. and Ashok K. Dhawan (eds.), *Natural Resource Management: Ecological Perspectives.* Switzerland, AG: Springer Nature, pp. 1–18.

Marothia, D.K., 2021. *Wetland-Agriculture Interactions: Sustaining Intergenerational Ecosystem Services.* Presidential Address delivered at the 29th Annual Conference of the Agricultural Economics Research Association (AERA), India, held at the Odisha University of Agriculture & Technology (OUAT) Bhubaneswar, 27 October 2021.

Marothia, D.K. and B. Gopal, 2014a. "Ganga: The Tragedy of Commons". *Central Chronicle*, Monday, May 26, 2014.

Marothia, D.K. and B. Gopal, 2014b. River Ecosystem: Managing the Unmanaged, *The Hitavada*, June 5, 2014.

Marothia, D.K., B. Gopal and S.K. Patil, 2017. *Conserving Wetland Wealth of Chhattisgarh, Report of the Workshop.* Naya Raipur: Chhattisgarh State Planning Commission, Yojana Bhawan, p. 24.

Marothia, D.K., A.P. Mukherjee and S.K. Sharma, 1987. "Canal Irrigation System: Socio-Economic and Management Issues". *Indian Journal of Agriculture Economics*, 42(3): 300.

Marothia, D.K. and D. Nandi, 1994. "Degraded Lands, Agro-forestry and Energy Needs: Issue and Development Strategies". In Singh, P., P.S. Pathak and M.M. Roy (eds.), *Agro-forestry System for Sustainable Land Use.* New Delhi, Oxford: IBH, pp. 250–262.

Marothia, D.K., S. Pandey and V.K. Choudhary, 2010. "Irrigation Tanks in Chhattisgarh: Traditional Technology for Sustaining Rainfed Agriculture". In Singh, S. and M.S. Rathore (eds.), *Rainfed Agriculture in India: Perspective and Challenges.* Jaipur: Rawat Pub., pp. 200–222.

Marothia, D.K. and W.E. Phillips, 1985. "Analysis of Natural Resource Problems Using Synthesis of Neoclassical and Institutional Economics". *Asian Journal of Economics and Social Studies*, 4: 263–281.

Marothia, D.K. and W.E. Phillips, 1990. "Economics of Soil Erosion Control from Farmers Perspective: A Peace River Survey". *International Journal of Ecology and Environmental Science*, 16(2): 85–96.

Marothia, D.K., R.K. Singh and Ajay Koshta, 2007. "Crop Diversification: Post Reforms Lesson from Chhattisgarh". *Agriculture Situation in India*, 64(6): 215–226.

Matthews, G.V.T., 1993. *The Ramsar Convention on Wetlands: Its History and Development.* Gland, Switzerland: Ramsar Convention Bureau.

Matthews, R.C.O., 1986. "The Economics of Institutions and the Sources of Growth". *Economic Journal*, 96: 903–918.

Maya, S., 2002. "Harvesting Water in Temple Tanks Through People's Participation: A Feasibility Study". *Nature Environment & Polution Technology*, 1(4): 375–378.

McClain, M.E. (ed.), 2002. *The Ecohydrology of South American Rivers and Wetlands.* Special Puplication #6 of the International Association of Hydrological Sciences (Access Book).

McGinnis, M. (ed.), 1999. *Polycentricity and Local Public Economies: Readings from the Workshop in Political Theory and Policy Analysis*. Ann Arbor: University of Michigan Press.

McNeill, D., 1999. "On Interdisciplinary Research: With Particular Reference to the Field of Environment and Development". *Higher Education Quarterly*, 53: 312–332.

Mehta, L., M. Leach, P. Newall, I. Scoones, K. Sivaramakrishnan and S-A. Way, 1999. *Exploring Understandings of Institutions and Uncertainty: New Directions in Natural Resource Management*. Sussex: Institute of Development Studies, University of Sussex, Brighton, DP 372.

Messouli, M., J.R. Holsinger and Y.R. Reddy, 2007. "Kotumsaridae, A New Family of Subterranean Amphipod Crustaceans from India with description of Kotumsaria Bastarensis, New Genus, New Species". *Zootaxa*, 1589: 33–46.

Millennium Ecosystem Assessment, 2005. *Ecosystems and Human Well-Being: Wetlands and Water: Synthesis*. Washington, DC: Water Resources Institute.

Mirashi, V.V., 1905. *Corpus Inscriptionium Indicarium Vol 4 Part 2 Inscriptions of the Kalchuri-Chedi Era*. New Delhi: Archaeological Survey of India, pp. 551–554.

Mirashi, V.V., 1955. *Corpus Inscriptionum Indicarum Vol. 4 Part 2*. New Delhi: Archaeological Survey of India.

Mirashi, V.V., 1957. "The Kalacuris". In Sharma, R.S. (ed.), *A Comprehensive History of India: AD 985–1206.4(Part-1)*. Indian History Congress/People's Publishing House. ISBN 978-81-7007-121-1.

Mishan, E.J., 1971. "The Postwar Literature on Externalities". *Journal of Economic Literature*, 9: 1–28.

Mishra, A., 1995. *Aaj Bhi Khare Hain Talab (in Hindi) (Ponds Are Relevant also Today)*. Nagpur: Swaraj Publication Group, p. 57.

Mishra, M., 2021. *Yield from the Floodplains: A Community of Migrant Seasonal Farmers, the 'Palaziye' Raise Summer Vegetables Annually on River Beds*. Gaon Connection-Your Connection with Rural India, 15th June 2021.

Mishra, R.N., 2015. *Chhattisgarh Ka Itihas (in Hindi) (History of Chhattisgarh)*. Raipur, Chhattisgarh: Shatakshi Publisher, pp. 93–99.

Misra, O.P., 2003. *Archaeological Excavations in Central India: Madhya Pradesh and Chhattisgarh*. New Delhi: Mittal Publications. ISBN 978-81-7099-874-7.

Mosse, D., 1997. "The Symbolic Making of a Common Property Resource: History, Ecology and Locality in a Tank-Irrigated Landscape in South India". *Development and Change*, 28: 467–504.

Mukherjee, R., 1942. *The Institutional Theory of Economics*. London: Macmillan and Co. Ltd.

Munasinghe, M., 1992. "Biodiversity Protection Policy: Environmental Valuation and Distribution Issues". *Ambio*, 21(3): 227–236.

Munasinghe, M., 1995. *Property Rights in a Social and Ecological Context: Case Studies and Design Applications*. Washington, DC: World Bank Publication, p. 215.

Murti, K.S., 1987. "Stratigraphy and Sedimentation in Chhattisgarh Basin". In Radhakrishna, B.P. (ed.), *Purana Basins of Peninsular India, Geological Survey of India Memoir*, 6: 239–260.

Myrdal, G., 1960. *Beyond the Welfare State*. New Haven: Yale University Press.

Myrdal, G., 1978. "Institutional Economics". *Journal of Economic Issues*, 12: 1–14.

Naidu, R., A.M.K. Bharos and S. Quader, 2019. *Birds of Chhattisgarh Report by Bird Count India 2017–2018*, Chhattisgarh Forest Department, GoCG, p. 29.

Natori, Y., 1999. *Information Sheet on Ramsar Wetlands: Yatsuhigata*. Gland, Switzerland: Ramsar Convention Bureau. www.wetlands.org/reports/ris/2JP007en.pdf (last accessed 22 April 2014).

Nayar, R., 2019. "Variation in Water Quality Index (WQI) by the Effects of Idol Immersion on the Water Quality Parameters of Dulhara Pond in Ratanpur in Chhattisgarh". *International Journal of Advanced Research*, 7(8): 1263–1270.

Nayar, R., 2020. "Impact on Ecological, Economical and Environmental Ecosystem by the Wetland Part of Religious Ponds of Ratanpur in Chhattisgarh, India. A Survey". *SSRG, International Journal of Nursing and Health Science*, 6(1): 52–59.

North, D.C., 1990. *Institutions, Institutional Change and Economic Performance*. Cambridge: Cambridge University Press.

Oakerson, R.J., 1986. "A Model for the Analysis of Common Property Problems". In *Proceedings of the Conference on Common Property Resource Management*. Washington, DC: National Academy of Sciences.

Oakerson, R.J., 1992. "Analyzing the Commons: A Framework". In Bromley, D.W. (ed.), *Making the Commons Work*. San Francisco: ICS Press, pp. 41–62.

Olson, M., 1971. *The Logic of Collective Action: Public Goods and the Theory of Groups*. Cambridge, MA: Harvard University Press.

O'Riordan, T. and A. Jordan, 1999. "Institutions, Climate Change and Cultural Theory: Towards a Common Analytical Framework". *Global Environmental Change*, 9: 81–94.

Ostrom, E., 1986. "An Agenda for the Study of Institutions". *Public Choice*, 48: 3–25.

Ostrom, E., 1988. "Institutional Arrangements and the Commons Dilemma". In Ostrom, V., D. Feeny and H. Picht (eds.), *Rethinking Institutional Analysis and Development*. San Francisco: ICS Press, pp. 101–139.

Ostrom, E., 1990a. *Governing the Commons: The Evolution of Institutions for Collective Action*. New York: Cambridge University Press.

Ostrom, E., 1990b. "Institutional Arrangements for Resolving the Commons Dilemma: Some Contending Approaches". In McCay, B.J. and J.M. Acheson (eds.), *The Question of the Commons: The Culture and Ecology of Communal Resources*, 2nd ed. Tucson: University of Arizona Press, pp. 250–265.

Ostrom, E., 1992a. *Crafting Institutions for Self-Governing Irrigation Systems*. San Francisco: ICS Press.

Ostrom, E., 1992b. "The Rudiments of a Theory of the Origins, Survival, and Performance of Common Property Institutions". In Bromley, D.W. (ed.), *Making the Commons Work*. San Francisco: ICS Press, pp. 293–318.

Ostrom, E., 1995. "Designing Complexity to Govern Complexity". In Hanna, S. and M. Munasinghe (eds.), *Property Rights and the Environment: Social and Ecological Issues*. Washington, DC: Beijer International Institute of Ecological Economics and the World Bank, pp. 33–45.

Ostrom, E., 1999a. "Coping with Tragedies of the Commons". *Annual Review of Political Science*, 2: 493–535.

Ostrom, E., 1999b. "Institutional Rational Choice: An Assessment of the Institutional Analysis and Development Framework". In Sabatier, P. (ed.), *Theory of Policy Process*. Boulder: West view Press, pp. 21–64.

Ostrom, E., 2005. *Understanding Institutional Diversity*. Princeton, NJ: Princeton University Press.

Ostrom, E., 2007. "Institutional Rational Choice: An Assessment of the Institutional Analysis and Development Framework". In Sabatier Paul, A. (ed.), *Theories of the Policy Process*. Boulder, CO: Westview Press, pp. 21–64.

Ostrom, E., 2010. "Beyond Markets and States: Polycentric Governance of Complex Economic Systems". *American Economic Review*, 100: 1–33.

Ostrom, E., J. Burger, C.B. Field, R.B. Norgaard and D. Policansky, 1999. "Revisiting the Commons: Local Lessons, Global Challenges". *Science*, 284: 278–282.

Ostrom, E. and S. Crawford, 1995. "A Grammar of Institutions". *American Annual Review of Political Science*, 89(3): 582–600.

Ostrom, E., T. Dietz, N. Dolsak, P.C. Stern, S. Stonich and E.U. Weber (eds.), 2002. *The Drama of the Commons*. Washington, DC: National Academy Press.

Ostrom, E. and R. Gardner, 1993. "Coping with Asymmetries in the Commons: Self-Governing Irrigation Systems Can Work". *Journal of Economic Literature*, 7(4): 93–112.

Ostrom, E., R. Gardner and J. Walker, 1994. *Rules, Games, and Common Pool Resources*. Ann Arbor: University of Michigan Press.

Ostrom, E. and L. Kiser, 1982. "The Three Worlds of Action: A Meta-Theoretical Synthesis of Institutional Approaches". In Ostrom, E. (ed.), *Strategies of Political Inquiry*. Beverly Hills: Sage Publications, pp. 34–35.

Ostrom, V., 1975. "Public Choice Theory: A New Approach to Institutional Economics". *American Journal of Agricultural Economics*, 57: 844–849.

Ostrom, V., 1976. "John R. Common's Foundations for Policy Analysis". *Journal of Economic Issues*, 10(4): 839–857.

Ostrom, V. and E. Ostrom, 1977. "Public Goods and Public Choices". In Savas, E.S. (ed.), *Alternatives for Delivering Public Services: Toward Improved Performance*. Boulder, CO: Westview Press, pp. 7–49. (Reprinted in McGinnis, M. (ed.), *Polycentricity and Local Public Economies: Readings from the Workshop in Political Theory and Policy Analysis*. Ann Arbor: University of Michigan Press, 1999, pp. 75–103.

Oviedo, G., S. Jeanrenaud and M. Otegui, 2005. *Protecting Sacred Natural Sites of Indigenous and Traditional Peoples: An IUCN Perspective*. Gland, Switzerland, Washington, DC: Earthscan.

Owens, S., 1997. "Negotiated Environments: Needs, Demands, and Values in the Age of Sustainability". *Environment and Planning A*, 29: 571–580.

Paavola, J., 2002. "Rethinking the Choice and Performance of Environmental Policies". In Bromley, D.W. and J. Paavola (eds.), *Economics, Ethics, and Environmental Policy: Contested Choices*. Oxford: Blackwell, pp. 87–102.

Palanisami, K. and R. Balasubramanian, 2002. "Management of Common Pool Irrigation Tanks of Southern India". In Marothia, D.K. (ed.), *Institutionalizing Common Pool Resources*. New Delhi: Concept Publishing Company.

Palanisami, K., R. Balasubramanian and A. Mohamed Ali, 1997. *Present Status and Future Strategies of Tank Irrigation in Tamil Nadu*. Coimbatore: Water Technology Centre, Tamil Nadu Agricultural University.

Palanisami, K. and K.W. Easter, 1984. "Irrigation Tanks of South: Management Strategies and Investment Alternatives". *Indian Journal of Agricultural Economics*, 39(2): 214–222.

Pandey, D.N., 2000. *Sacred Water and Sanctified Vegetation: Tanks and Trees in India*. Paper presented at the conference of the International Association for the Study of Common Property (IASCP), in the Panel "Constituting the Riparian Commons" (Bloomington, Indiana, USA, 31 May – 4 June 2000).

Panini, D., 1998. *The Ramsar Convention and the National Laws and Policies for Wetlands in India*. Gland, Switzerland: Ramsar Convention Secretariat.

Pant, N. and R.K. Verma (eds.), 2010. *Tanks in Eastern India: A Study in Exploration*. Lucknow: IWMI – TATA Water Policy Research Programme, Hyderabad and CDS, p. 55.

Parsons, K.H., 1941. "Social Conflicts and Agricultural Programms". *Journal of Farm Economics*, 23: 743–764.

Parsons, K.H., 1942. "John R. Commons' Point of View". *Journal of Land and Public Utility Economics*, 18: 245–266.

Parsons, K.H., 1974. "The Institutional Basis of an Agricultural Market Economy". *Journal of Economic Issues*: 737–757.

Parsons, W., 1995. *Public Policy*. Cheltenham, Glos: Edward Elgar.

Patel, J.G., 2018. *National Wetland Inventory and Assessment (NWIA)*. Ahmedabad: Space Applications Centre, February 2018.

Paul, M., M. Chanda and S.S. Gupta, 2011. "Strategy and Scenario for Wetland Conservation in India". *Chronicles of Young Scientists*, 2: 79–82.

Pearce, D.W. and R.K. Turner, 1990. *Economics of Natural Resources and the Environment*. The Baltimore: Johns Hopkins University Press, Ch 21, pp. 320–335.

Peters, P.E., 1986. *Concluding Statement in Proceedings of the Conference on Common Property Resource Management*. National Research Council. Washington, DC: National Academy Press, pp. 615–620.

Phillips, W.E., 1976. *The Conservation of the California Tule Elk: A Socio-economic Study of a Survival Problem*. Edmonton: University of Alberta Press.

Philips, W.E., D. DePape and L.R. Brook, 1975. *Water Induced Erosion on Privately Owned Agricultural Lands in the Peace River Region: An Institutional and Legislative Analysis*. Information Bulletin No. 2. Edmonton: Environment Conservation Authority, p. 77.

Phillips, W.E. and D.K. Marothia, 1982. "Integrated Area Development Approach to Rural Development: A Conceptual Framework". *International Journal of Agricultural Administration*, 9(3): 249-261.

Pinkerton, E. (ed.), 1989. *Cooperative Management of Local Fisheries: New Directions for Improved Management and Community Development*. Vancouver: University of British Columbia Press.

Pittock, J., M. Finlayson, A.H. Arthington, D. Roux, J.H. Matthews, H. Biggs, I. Harrison, E. Blom, R. Flitcroft, R. Froend, V. Hermoso, W. Junk, R. Kumar, S. Linke, J. Nel, da Cunha C. Nunes, A. Pattnaik, S. Pollard, W. Rast, M. Thieme, E. Turak, J. Turpie, L. van Niekerk, D. Willems and J. Viers, 2015. "Managing Freshwater, River, Wetland and Estuarine Protected Areas". In Worboys, G.L., M. Lockwood, A. Kothari, S. Feary and I. Pulsford (eds.), *Protected Area Governance and Management*. Canberra: ANU Press, pp. 569–608.

Porte, D.S. and S. Gupta, 2018. "Status and Diversity of Wetland Birds at Ancient Township Ratanpur, of Chhattisgarh, India". *Ambient Science*, 5(2): 12–17.

Pradhan, A. and S.K. Patil, 2017. "*Dalpat Sagar*. Fact Sheet-3". In Marothia, D.K., B. Gopal and S.K. Patil (eds.), *Conserving Wetland Wealth of Chhattisgarh, Report of the Workshop*. Naya Raipur: Chhattisgarh State Planning Commission, Yojana Bhawan, p. 24.

Pradhan, A., S.K. Patil, D.K. Marothia, T. Chandrakar, S.K. Nag and S.C. Mukherjee, 2019. "Effect of Catchment and Hydrology of Ancient Dalpatsagar Reservoir on Water Resources of Jagdalpur City, Bastar, India". *International Journal of Current Microbiology and Applied Sciences*, 8(1): 1797–1812.

Pradhan, R.K. and S.M. Agarwal, 1984. "Studies on Nemacheilus Evezardi Day from Kotumsar Cave 1-Chromatophores". *Geobios New Reports*, 3: 128–130.

Prasher, K., 2018. "The Case of India's Disappearing Wetlands". *The Times of India -Whether News*, (TWC India), 8 November 2018.

Radhakrishnan, M., T. Pathak, K. Irvine and A. Pathirana, 2017. "Scoping for the Operation of Agile Urban Adaptation for Secondary Cities of the Global South: Possibilities in Pune, India". *Water*, 9(12): 912–939.

Rahalkar, S. and A. Tiwari, 2014. "Role of Small Water Bodies in Sustaining Avian Diversity: A Case Study of Dulahara Pond Ratanpur (Distt. Bilaspur, CG)". *International Journal of Researches in Biosciences, Agriculture & Technology*, 2: 442–449.

Rahman, Md., Y. Jiang and K. Irvine, 2018. "Assessing Wetland Services for Improved Development Decision-Making: A Case Study of Mangroves in Coastal Bangladesh". *Wetlands Ecology and Management*, 26(4): 563–580.

Rahmani, A.R., A.M.K. Bharos, A. Manadvia, A. Vishwakarma and A. Bharos, 2018. "Threatened Birds of Chhattisgarh". In *Chhattisgarh Wildlife Society*. Raipur: Wildlife Society, p. 156.

Raipur Municipal Corporation, 2020. *Area of Lake Vivekananda*. Chhattisgarh: Municipal Corporation Office Raipur.

Raipur Smart City Limited (RSCL), 2016. *An Action Plan to Repair, Renovate, Restore and Rejuvenate Water Bodies Across the Raipur City During*. Raipur: Smart City Limited, p. 35.

Raju, K.V., G.K. Karanth, M.J. Bhende, D. Rajasekar and K.G. Gayathridevi, 2003. *Rejuvenating Tanks: A Socio-Ecological Approach*. Bangalore: Book for Change.

Ramakrishnan, M. and R. Vaidyanadhan, 2008. "Geology of India, Bangalore". *Geological Society of India*, 1 & 2: 552.

Ramjan, V.K. and A. Chhetri, 2018. "Production Technology of Cucurbits in Riverbeds". *Indian Farmer*, 5(4): 434–438.

Ramsar Convention, 2008. *Resolution X.27 Wetlands and Urbanization*. 10th Meeting of the Conference of the Parties to the Convention on Wetlands (Ramsar, Iran, 1971), "Healthy Wetlands, Healthy People", Changwon, Republic of Korea, 28 October to 4 November 2008.

Ramsar Convention, 2015. *Resolution XII.10 Wetland City Accreditation of the Ramsar Convention*. 12th Meeting of the Conference of the Parties to the Convention on Wetlands (Ramsar, Iran, 1971) Punta del Este, Uruguay, 1–9 June 2015.

Ramsar Convention's Fourth Strategic Plan, 2016–2024. *Scaling Up Wetland Conservation, Wise Use and Restoration to Achieve the Sustainable Development Goals*. Gland, Switzerland: Ramsar Convention Secretariat, July.

Ramsar Convention Resolution Vi, 1996. *Inclusion of Subterranean Karst Wetlands as a Wetland Type under the Ramsar Classification System*. Gland, Switzerland: Ramsar Convention Secretariat.

Ramsar Convention Secretariat, 2010a. *Designating Ramsar Sites: Strategic Framework and Guidelines for the Future Development of the List of Wetlands of International Importance. Ramsar Handbooks for the Wise Use of Wetlands*, 4th ed. Gland, Switzerland: Ramsar Convention Secretariat, vol. 17.

Ramsar Convention Secretariat, 2010b. *National Wetland Policies: Developing and implementing National Wetland Policies. Ramsar Hand Books for the Wise Use of Wetlands*, 4th ed. Gland, Switzerland: Ramsar Convention Secretariat, vol. 2, p. 60.

Ramsar Convention Secretariat, 2012. *Resolution XI.11, Principles for the Planning and Management of Urban and Peri-urban Wetlands*. 11th Meeting of the Conference of the Parties to the Convention on Wetlands (Ramsar, Iran, 1971), Wetlands: Home and Destination, Bucharest, Romania, 6–13 July 2012.

Ramsar Convention Secretariat, 2013. *The Ramsar Convention Manual: A Guide to the Convention on Wetlands (Ramsar, Iran, 1971)*, 6th ed. Gland, Switzerland: Ramsar Convention Secretariat.
Ramsar Convention Secretariat, 2016. *An Introduction to the Ramsar Convention on Wetlands*, 7th ed. (previously The Ramsar Convention Manual). Gland, Switzerland: Ramsar Convention Secretariat.
Ramsar Convention Secretariat, 2018. *Global Wetland Outlook: State of the World's Wetlands and their Services to People*. Gland, Switzerland: Ramsar Convention Secretariat.
Randall, A., 1971. "Market Solutions to Externality Problems: Theory and Practice". *American Journal of Agricultural Economics*, 54: 175–183.
Randall, A., 1974. "Coasian Externality Theory in a Policy Context". *Natural Resources Journal*, 14: 35–54.
Randall, A., 1975. "Property Rights and Social Microeconomic". *Natural Resources Journal*, 15: 729–748.
Randall, A., 1978a. "Property Institutions and Economic Behaviour". *Journal of Economic Issues*: 1–21.
Randall, A., 1978b. "Contemporary Issues in Natural Resource Economics: Discussion". *American Journal of Agricultural Economics*, 60: 290–292.
Rao, V.M. and M.G. Chandrakant, 1984. "Resource at the Margin, Tank-Irrigation Programme in Karnataka". *Economic and Political Weekly*, 19(26): A54.
Rawls, J., 1971. *A Theory of Justice*. Cambridge, MA: Harvard University Press.
Ray, R., M.D.S. Chandran and T.V. Ramachandra, 2014. "Biodiversity and Ecological Assessments of Indian Sacred Groves". *Journal of Forestry Research*, 25(1): 21–28.
Ray, R., and T.V. Ramchandra, 2010. "Small Sacred Grove in Local Landscape: Are they Really Worthy for Conservation?". *Current Science*, 98: 1078–1080.
Reddy, Y.R., 2006. "First Asian Report of the Genus Chilibathynella Noodt, 1963 (Bathynellacea, Syncarida), with the Description and Biogeographic Significance of a New Species from Kotumsar Cave India". *Zootaxa*, 1370: 23–37.
Reddy, Y.R. and D. Defaye, 2009. "Two New Parastenocarididae (Copepoda, Harpacticoida) from India: Parastenocaris Muvattupuzha N. Sp. from a River and P. Kotumsarensis N. Sp. from a Cave". *Zootaxa*, 2077: 31–55.
Ricker, W., 1962. *The Theory of Political Coalitions*. New Haven: Yale University Press.
Robinson, J., 1963. *Economic Philosophy*. Chicago: Aldine.
Robinson, J., 1971. "The Second Crisis of Economic Theory". *American Economic Review*, 53: 1–10.
Roy, A., 2009. "Why India cannot Plan its Cities: Informality, Insurgence and the Idiom of Urbanization". *Planning Theory*, 8: 76–87.
Roy, P.C., 1980. *The Coinage of the Kalchuris of Ratnapura: The Coinage of Northern India*. New Delhi: Abhinav Publications. ISBN 978-81-7017-122-5.
Ruddle, K., 1989. "Solving the Common Property Dilemma: Village Fisheries Rights in Japanese Coastal Waters". In Berkes, F. (ed.), *Common Property Resource: Ecology and Community-Based Sustainable Development*. London: Belhaven Press, pp. 168–184.
Ruge, C.F. and D.W. Bromley, 1979. *Property Rights and the First Economic Revolution: The Origins of Agriculture Reconsidered*. Working Paper No. 13. Madison, USA: Center for Resource Policy Studies, University of Wisconsin.
Sahu, M., 2017. *Survey of Ratanpur Birds during Mansoon Season*. Naya Raipur, Chhattisgarh: Preliminary Report submitted to Chhattisgarh State Planning Commission, p. 35.

Sahu, R. and S.S. Chandravanshi, 2017a. "A Case Study of Lotus Farming Community. Poster Presentation Made at the Workshop on Conserving Wetland Wealth of Chhattisgarh, Marothia". In Marothia, D.K., Brij Gopal and S.K. Patil (eds). *Proceeding of Workshop on Conserving Wetland Wealth of Chhattisgarh*. Naya Raipur: Chhattisgarh State Planning Commission, Yojana Bhawan, p. 24.

Sahu, R. and S.S. Chandravanshi, 2017b. "Makhana Cultivation Practices for Chhattisgarh. Poster Presentation Made at the Workshop on Conserving Wetland Wealth of Chhattisgarh". Marothia, D.K., B. Gopal and S.K. Patil (eds.), *Proceeding of Workshop on Conserving Wetland Wealth of Chhattisgarh*. Naya Raipur: Chhattisgarh State Planning Commission, Yojana Bhawan, p. 24.

Sahu, R. and B. Gopal, 2017. "*Lotus Fields of Dhamtari*. Fact Sheet-1". In Marothia, D.K., B. Gopal and S.K. Patil (eds.), *Conserving Wetland Wealth of Chhattisgarh, Report of the Workshop*. Naya Raipur: Chhattisgarh State Planning Commission, Yojana Bhawan, p. 24.

Sakai, S. and C. Umetsu (eds.), 2014. *Social-Ecological Systems in Transition*. Tokyo: Springer, vol. XI, p. 198.

Sakthivadivel, R., P. Gomathinayagam and T. Shah, 2004. "Rejuvenating Irrigation Tanks through Local Institutions". *Economic and Political Weekly*: 3521–3526.

Saleth, R.M., 1994. "Towards a New Water Institution: Economics, Law and Policy". *Economic and Political Weekly*, 29(39): A-147–A-155.

Saleth, R.M. and A. Dinar, 2004. *The Institutional Economics of Water: A Cross-Country Analysis of Institutions and Performance*. Cheltenham, UK: Edward Elgar.

Samaraweera, G.C. and D.K. Marothia, 2008. "User's Appraisal and Valuation of Changes in Renewable Natural Resources Status". *Indian Journal of Agricultural Economics*, 63(3): 415–425.

Sampath, M.D., 2001. *Epigraphs of Madhya Pradesh*. New Delhi: Archaeological Survey of India.

Samuels, W., 1972. "Welfare Economics, Power, and Property". In Wunderlich, F. and W.L. Gibson (eds.), *Perspective on Property*. University Park: Penn State University Press, pp. 20–32.

Savory, A., 1988. *Holistic Resource Management*. Washington, DC: Island Press.

Schlager, E., W. Blomquist and S.Y. Tang, 1994. "Mobile Flows, Storage, and Self-Organized Institutions for Governing Common-Pool Resources". *Land Economics*, 70: 294–317.

Schlager, E. and E. Ostrom 1992. "Property Rights Regimes and Natural Resources: A Conceptual Analysis". *Land Economics*, 68(3): 249–262.

Schmid, A.A., 1972. "Analytical Institutional Economics: Challenging Problems in the Economics of Resources for a New Environment". *American Journal of Agricultural Economics*, 54: 893–901.

Schmid, A.A., 1976. "The Economics of Property Rights: A Review Article". *Journal of Economic Issues*, 10: 159–168.

Schmid, A.A., 1978. *Property, Power and Public Choice: An Inquiry into Law and Economics*. New York: Praeger.

Schultz, T.W., 1968. "Institutions and Rising Economic Value of Man". *American Journal of Agricultural Economics*, 50: 1113–1122.

Sen, A., 1971. *Collective Choice and Social Welfare*. San Francisco: Sliver and Boyd.

Shaffer, J., 1969. "On Institutional Absolescence and Innovation". *American Journal of Agricultural Economics*, 5: 245–267.

Shah, T., 2003. "Who Should Manage Chandeli Tanks?" IWMI Tata Comment, No. 1, Anand.

Shah, T. and K.V. Raju, 1999. *Rajasthan Minor Irrigation Tank Rehabilitation Project: Socio-Ecological and Organisational Assessment*. New Delhi: Swedish International Development Agency.

Shah, T., R. Seenivasa, C.R. Shanmugam and M.P. Vasimalai, 2002. "Sustaining Tamil Nadu's Tanks: Field Notes on PRADAN's Work". In Marothia, D.K. (ed.), *Institutionalizing Common Pool Resources*. New Delhi: Concept Publishing Company.

Sharma, C.L., 2008. *Chhattisgarh ki Riyasate* (in Hindi) (*Princely States of Chhattisgarh*). Raipur: Grantha Academy.

Sharma, P., 2017. "*Wetland Complex of Ratanpur*. Fact Sheet-2". In Marothia, D.K., B. Gopal and S.K. Patil (eds.), *Conserving Wetland Wealth of Chhattisgarh, Report of the Workshop*. Naya Raipur: Chhattisgarh State Planning Commission, Yojana Bhawan, p. 24.

Shukla, B. and R. Shukla, 2018. "Urban Ponds and Lakes: Exploring Urban Ecology as a Part Natural Heritage". *Research Gate*, December: 1–16.

Singh, Deo, R.C., 1990. "Tribal Demography and Development". In Bose, A., U.P. Sinha and R.P. Tyagi (eds.), *Indian Association for the Study of Population*. New Delhi: B R Publishing Corporation.

Singh, K., 1994. *Managing Common Pool Resources: Principles and Case Studies*. New Delhi: Oxford University Press.

Singh, K., 1997. "Property Rights and Tenures in Natural Resources". In Kerr, J.M., D.K. Marothia, K. Singh, C. Ramasamy and W.R. Bentley (eds.), *Natural Resource Economics: Theory and Application in India*. New Delhi: Oxford & IBH Pub. Co., pp. 131–160.

Singh, K. and V. Ballabh (eds.), 1996. *Cooperative Management in Natural Resources*. New Delhi: Sage Publications.

Siy, R.Y. Jr., 1982. *Community Resource Management: Lesson from the Zanjera*. Quezon City, Philippines: University of the Philippines Press.

Smardon, R.C. (ed.), 1983. *The Future of Wetlands: Assessing Visual-Cultural Values*. Totowa, NJ: Allenheld-Osmun Press,

Smardon, R.C., 2006. "Heritage Values and Functions of Wetlands in Southern Mexico". *Landscape & Urban Plan*, 74(3–4): 296–312.

Smardon, R.C., 2009. *Sustaining the World's Wetlands: Setting Policy and Resolving Conflicts*. New York: Springer.

Smardon, R.C., 2015. "International Wetlands Policy and Management Issues". *National Wetlands Newsletter*, 37(3): 10–16.

Smith, V.K., 1996. *Estimating Economic Values of Nature*. Brookfield: Edward Elgar.

Söderbaum, P., 1986. "Economics, Ethics and Environmental Problems". *The Journal of Interdisciplinary Economics*, 1: 139–153.

Söderbaum, P., 1992. "Neoclassical and Institutional Approaches to Development and the Environment". *Ecological Economics*, 5(2): 127–144.

Solo, R., 1974. *The Political Authority and Market System*. Cincinnati: Southwestern.

Sugden, R., 1984. "Reciprocity: The Supply of Public Goods through Voluntary Contributions". *Economic Journal*, 94: 772–787.

Supreme Court of India, 2011. Bench of Justice Markandey Katju and Gyan Sudha Misra, Judgement of 28 January 2011, Civil Appellante Jurisdiction, Civil Appeal No. 1131/2011@SLP(C)NO. 3109/2011 (Arising out of Special leave petition (Civil) CC No. 19869 of 2010), Jagpal Singh & Ors vs State of Punjab & Ors on 28 January, 2011.

Swallow, B.M. and D.W. Bromley, 1994. "Co-management or No Management: The Prospects for Internal Governance of Common Property Regimes through Dynamic Contracts". *Oxford Agrarian Studies*, 22(1): 3–16.

Tang, S.Y., 1992. *Institutions and Collective Action: Self-Governance in Irrigation*. San Francisco: ICS Press.

Thakur, S.O., 2002. *Chhattisgarh ki Sabhyata (in Hindi) (Culture of Chhattisgarh)*. Raipur: Upkar Publications.

Timmons, J.F., 1972. "Public Land Use Policy: Needs, Objectives and Guidelines". *Journal of Soil and Water Conservation*, 27: 195–201.

Townsend, R.E., 1990. "Entry Restrictions in the Fishery: A Survey of the Evidence". *Land Economics*, 66: 359–378.

Townsend, R.E., 1995. "Fisheries Self-governance: Corporate or Cooperative Structures?" *Marine Policy*, 19: 39–45.

Townsend, R.E. and S.G. Pooley, 1995. "Distributed Governance in Fisheries". In Hanna, S. and M. Munasinghe (eds.), *Property Rights and the Environment – Social and Ecological Issues*. Washington, DC: The Beijer International Institute of Ecological Economics and the World Bank, pp. 33–40.

Tripathi, S. and N. Panigrahi, 2017. "Changing Sacred Status of Village Pond in the Context of People-Pond-Spirit Complex: An Empirical Study". *The Oriental Anthropologist*, 17(1): 149–161.

Tripathi, S. and S. Tripathi, 2012. *Chhattisgarh Vrihad Sandarbh (Hindi) (Chhattisgarh in Broad Context)*. Raipur: Upkar Publications.

Tugwell, R.G., 1922. "Human Nature in Economic Theory". *Journal of Political Economy*, 30: 317–345.

Tullock, G., 1964. *The Politics of Bureaucracy*. Washington: Public Affairs Press.

Uphoff, N., 1986. *Local Institutional Development: An Analytical Sourcebook with Cases*. West Hartford, CT: Kumarian Press.

Vaidyanathan, A., 1997. "Tanks and Tanks Irrigation". In Agrawal, A. (ed.), *The Challenge of the Balance: Environmental Economics in India*. New Delhi: Centre for Science and Environment, pp. 285–289.

Vaidyanathan, A., 1999. *Water Resource Management: Institutions and Irrigation Development in India*. Bombay: Oxford University Press.

Van, B.P.E. (ed.), 2011. *Karst Management*. New York: Springer Science Business Media B.V., Springer.

Veeman, T.S., 1978. "Water Policy and Water Institutions in Northern India: The Case of Groundwater Rights". *Natural Resources Journal*, 18: 569–587.

Verma, R.K., 2015. "Kalachuri Inscriptions: A Reflection of Dwindling Political Power (PDF)". *Veethika*, 1(3).

Vidyarthi, L.P., 1963. *The Maler: A Study in Nature-Man-Spirit Complex of a Hill Tribe*. Calcutta: Bookland Private Ltd.

Vidyarthi, L.P., M. Jha and B. Saraswati, 1979. *Sacred Complex of Kashi (The): A Microcosm of Indian Civilisation*. New Delhi: Concept Publishing Company.

Vishwakarma, A., A. Hemrom and K.C. Yadav, 2014. "Status of Terrestrial and Wetland Birds in Kawardha, Kabirdham District in Chhattisgarh, India". *International Journal of Scientific Research Publication*, 4(10): 1–7.

Von Oppen, M., 1985. *Tank Irrigation in South India: Adapting a Traditional Technology to Modern Socio-Economic Conditions*. Irrigation Symposium on Traditional Irrigation Schemes and Potential for their Improvement, Berlin, Federal Republic of Germany, pp. 109–118.

Von Oppen, M. and K.V. Subba Rao, 1980a. *Tank Irrigation in Semi-arid Tropical India. Part 1: Historical Development and Spatial Distribution*. Andhra Pradesh: ICRISAT Economics Programmes Progress Report 5, Patancheru.

Von Oppen, M. and K.V. Subba Rao, 1980b. *Tank Irrigation in Semi-arid Tropical India. Part II: Historical Development and Spatial Distribution*. Andhra Pradesh: ICRISAT Economics Programmes Progress Report 8, Patancheru.

Wallace, K.J., 2007. "Classification of Ecosystem Services: Problems and Solutions". *Biological Conservation*, (39): 235–246.

Walzer, M., 1983. *Spheres of Justice: A Defense of Pluralism and Equality*. New York: Basic Books.

Wanganeo, A., 2017. "*Intricacies of Natures Food & Fodder Bowl*. Presentation Made at the Workshop on Conserving Wetland Wealth of Chhattisgarh". In Marothia, D.K., B. Gopal and S.K. Patil (eds.), *Report of the Workshop on Conserving Wetland Wealth of Chhattisgarh*. Naya Raipur: Chhattisgarh State Planning Commission.

Ward, B., 1972. *What's Wrong with Economics?* New York: Basic Books.

Water, Land and Ecosystems (WLE) IWMI. WETWin, Enhancing Governance in Wetland Management, Factsheet-2.

Weimer, D.L., 1995. "Institutional Design: Overview". In Weimer, D.L. (ed.), *Institutional Design*. Boston: Kluwer Academic Publishers.

Wetlands International South Asia, 2018. *Wetlands for a Sustainable Urban Future, Sarovar*. Newsletter, December. New Delhi: WISA, vol. IV.

Wilber, C.K. and R.S. Harrison, 1978. "The Methodological Basis of Institutional Economic: Pattern Model, Story Telling and Holism". *Journal of Economic Issues*, 12: 61–89.

Wildfowl and Wetland Trust (WWT) Consulting, 2018. *Good Practices Handbook for Integrating Urban Development and Wetland Conservation*. Slimbridge, UK: WWT Consulting Limited, Wildfowl and Wetland Trust, p. 49.

Williamson, O.E., 1994. "The Institutions of Governance of Economic Development and Reform". In *Proceedings of the World Bank, Annual Conference on Development Economics*. Washington, DC: World Bank.

Wood, A. and G.E. van Halsema (eds.), 2008. *Scoping Agriculture-Wetland Interactions: Towards a Sustainable Multiple-Response Strategy*. FAO Water Report 33. Rome: FAO, Ramsar, WUR, IWMI, WA and WI, p. 177.

Yadav, K.C., 2017. "Romancing with Gidhwa Wetland: The Bird's Paradise. Presentation Made at the Workshop on Conserving Wetland Wealth of Chhattisgarh". In Marothia, D.K., B. Gopal and S.K. Patil (eds.), *Report of the Workshop on Conserving Wetland Wealth of Chhattisgarh*. Naya Raipur: Chhattisgarh State Planning Commission.

Young, O.R., 2002. "Institutional Interplay: The Environmental Consequences of Cross-Scale Interactions". In Ostrom, E., T. Dietz, N. Dolsak, P.C. Stern, S. Stonich and E.U. Weber (eds.), *The Drama of the Commons*. Washington, DC: National Academy Press, pp. 263–291.

Zingstra, H.L., 2009. "Wetlands and Agriculture – Relevance of Good Agricultural Practice and Wetland Management Guidelines for Harmonizing the Wise Use of Wetlands and Agriculture". *Research Gate*, January: 1–31.

Index

Note: Page numbers in **bold** indicate table and *italics* indicate figures. Page numbers followed by an "n" indicate notes

For Product Safety Concerns and Information please contact our EU representative GPSR@taylorandfrancis.com Taylor & Francis Verlag GmbH, Kaufingerstraße 24, 80331 München, Germany